Carbon Captured

American and Comparative Environmental Policy

Sheldon Kamieniecki and Michael E. Kraft, series editors

For a complete list of books in the series, please see the back of the book.

Carbon Captured

How Business and Labor Control Climate Politics

Matto Mildenberger

The MIT Press
Cambridge, Massachusetts
London, England

This book was set in ITC Stone Serif Std and ITC Stone Sans Std by Toppan Best-set Premedia Limited. Printed and bound in the United States of America.

Library of Congress Cataloging-in-Publication Data

Names: Mildenberger, Matto, author.
Title: Carbon captured : how business and labor control climate politics / Matto Mildenberger.
Description: Cambridge : MIT Press, 2020. | Includes bibliographical references and index.
Identifiers: LCCN 2019020113 | ISBN 9780262538251 (paperback)
Subjects: LCSH: Climatic changes--Political aspects. | Climatic changes--Political aspects--Australia. | Climatic changes--Political aspects--Norway. | Climatic changes--Political aspects--United States. | Climatic changes--Government policy. | Climatic changes--Government policy--Australia. | Climatic changes--Government policy--Norway. | Climatic changes--Government policy--United States.
Classification: LCC QC903 .M55 2020 | DDC 363.738/7456--dc23
LC record available at https://lccn.loc.gov/2019020113

10 9 8 7 6 5 4 3 2

Contents

Series Foreword

Climate change presents extraordinary challenges for our political systems. Forecasting the impacts of climate change is difficult. Policy action is rife with economic and political uncertainty. In many respects, climate change is the poster child for what some call the third generation of environmental policy problems. Its impacts are mostly, although not exclusively, long term and dispersed, and they are inevitably characterized by a high degree of uncertainty. Yet the costs of policy action tend to be short term and concrete enough for adversely affected interests, most notably the fossil fuel industry, to mobilize against mitigation efforts. In the case of the United States, efforts to control greenhouse gas (GHG) emissions are made even more difficult by an increasingly polarized national political system, even as considerable progress can be found at the state and local levels.

Climate change obviously requires meaningful global commitments and international cooperation. But domestic policy choices are just as politically difficult, if not more so. This is especially the case when newly emergent nationalist political forces in the United States and other nations inhibit governments' ability to find common ground and build the trust necessary for effective global solutions. That developed and developing nations face somewhat different kinds of costs and risks over time and have different capacities for action further compounds the problem's difficulty.

One would think that scientific consensus over the magnitude and likely impacts of climate change would improve prospects for policymaking. Yet, in the United States and many other nations, policy debates have not yielded the strong commitments that the 2015 Paris Agreement suggested were essential. Indeed, it is remarkable that, as of 2019, one of the two major political parties in the United States continues to question the existence of anthropogenic climate change. This party also seeks to curtail scientific research and public education necessary for the adoption of acceptable and effective climate solutions.

Given these conditions, Matto Mildenberger's book comes at an opportune time. He begins with the observation—and puzzle—that national policy responses to climate change have varied significantly, with some nations taking concrete and serious steps to reduce GHG emissions while other have not. He turns to comparative political analysis to address this puzzle, arguing that existing theories in the field are insufficient to explain this variation in climate policy timing and content. His empirical data come primarily from a detailed examination of climate policy in Australia, Norway, and the United States between the late 1980s and the adoption of the Paris Agreement in 2015. His contribution focuses on describing how "entrenched opponents" to climate reforms have captured the climate policymaking process. As he notes, the sober reality is that use of carbon-based fuels dominates the energy, transportation, and manufacturing sectors of national economies, making substantial change enormously difficult.

Mildenberger suggests that climate policy variation across countries can best be understood by observing how major economic and political actors align themselves during climate policy debates, and how that alignment interacts with domestic political institutions. He finds that the interests of workers and businesses in carbon-dependent economic sectors do not fall along usual ideological and partisan lines. Rather, he argues, climate change divides nations in new ways that reinforce policy inaction.

The opponents of climate action, Mildenberger maintains, benefit more than supporters from what he calls "double representation," which grants carbon polluters exceptional access to the policymaking process. He explains how domestic institutions can either augment or moderate this double representation of carbon polluters by structuring polluter access to climate policy design. Such access is molded by policymaking institutions, such as institutional interconnections between economic stakeholders and government decision makers. It is also molded by political organizations, for example through historical links between labor unions and political parties. The outcome is that no matter which party controls government, carbon polluters tend to have their interests met. Approved policies tend to shield these polluters from costs and emphasize modest incentives, subsidies, and voluntary action over tough regulation, technology mandates, and the imposition of carbon taxes. The author contends that the double representation of carbon polluters "is the single most important feature of climate policy conflict across advanced economies."

Readers will find that Mildenberger's richly detailed history and analysis of Australia, Norway, and the United States' climate policy actions, and shorter treatments of other countries' actions, offer new and significant insights into how developed nations have chosen to address the risks of climate change. The focus on double representation offers a new perspective on how and why political systems have failed to take the policy actions that nearly all observers believe necessary, and helps point toward future solutions to this political failure.

The book illustrates well our purpose in the MIT Press series in American and Comparative Environmental Policy. We encourage work that examines a broad range of environmental policy issues. We are particularly interested in volumes that incorporate interdisciplinary research and focus on the linkages between public policy and environmental problems and issues both within the United States and in cross-national settings. We welcome contributions that analyze the policy dimensions of relationships between humans and the environment from either a theoretical or empirical perspective.

At a time when environmental policies are increasingly seen as controversial and new and alternative approaches are being implemented widely, we especially encourage studies that assess policy successes and failures, evaluate new institutional arrangements and policy tools, and clarify new directions for environmental politics and policy. The books in this series are written for a wide audience that includes academics, policymakers, environmental scientists and professionals, business and labor leaders, environmental activists, and students concerned with environmental issues. We hope they contribute to public understanding of environmental problems, issues, and policies of concern today and also suggest promising actions for the future.

Sheldon Kamieniecki, University of California, Santa Cruz
Michael Kraft, University of Wisconsin-Green Bay

Co-editors, American and Comparative Environmental Policy Series

Acknowledgments

Fragrant little chips of history spewed from the saw cut, and accumulated on the snow before each kneeling sawyer. We sensed that these two piles of sawdust were something more than wood: that they were the integrated transect of a century; that our saw was biting its way, stroke by stroke, decade by decade, into the chronology of a lifetime, written in concentric annual rings of good oak.

Rest! cries the chief sawyer, and we pause for breath.

—Aldo Leopold, *The Sand County Almanac*

And now I too can pause for breath.

This project began at the Yale School of Forestry and Environmental Studies, a century after Aldo Leopold wandered through its halls. You still can't operate a cross-cut saw alone. And so it is that a great number of wonderful people have helped me hew this tree. To the scores of political actors, environmental advocates, labor leaders, and business officials who shared their time and experiences with me, thank you for welcoming me into your offices and homes. This work would not exist without your candor, trust, and insights. This work would also not exist without financial support from the Social Science and Humanities Research Council of Canada, the Yale Institute for Biospheric Studies, and the Yale Center for Business and the Environment.

Learning to read the tree rings of our political and policy world requires enormous care. I've had the great fortune of learning from some of the very best. Beth Savan gave me my first taste of social science and showed me how academic research can shed light on the problems of our time. Tad Homer-Dixon stretched my intellectual world over the years with his polymath interests and synthetic mind. He taught me how to take intellectual risks (and become a better writer!). One of my life's privileges was my work and friendship with Stephen Clarkson. Stephen offered me enormous responsibilities and, in meeting his expectations, I learned much about myself and the academic craft.

At Yale, I am grateful for Benjamin Cashore's boundless curiosity, warm friendship, and commitment to problem-oriented research. Jacob Hacker continuously pushed me to improve the empirical and theoretical rigor of this project. He exposed assumptions, asked difficult questions, and provided much-valued intellectual support. And Anthony Leiserowitz was an all-round exceptional mentor.

Here at UC Santa Barbara, I have found an unrivaled community of environmental colleagues, all of whom have shaped this project in large and small ways. Thanks to Eric R.A.N. Smith, Mark Buntaine, Sarah Anderson, Hahrie Han, Matt Potoski, Paasha Mahdavi, and David Pellow. Thanks also to the brilliant Kathy Harrison, Kate O'Neill, and Michaël Aklin who joined us at UCSB to workshop this manuscript. Thanks to Endre Tvinnerheim and Llewelyn Hughes, who provided helpful comments on the Norwegian and Japanese cases, respectively. Thanks to Michael Stone, Chase Foster, Alice Lépissier, and Geoff Henderson for their spirited critiques along the way. Lisa Camner McKay provided trenchant editorial comments, which improved my writing and arguments immeasurably. The index would not exist without the patience and care of Sydney Bartone. Alex Hertel-Fernandez read more chapter drafts than anyone else, yet somehow offered thoughtful advice across each iteration. And thanks, always, to my parents and brother for kindling in me a curiosity about life, and introducing me to the wonders of the natural world.

My real co-sawyer in this enterprise has been Leah Stokes. To have a partner with whom I can share the rhythms and joy and toil of this crazy enterprise has made all the difference. There is so much of her that enters into this work—her brilliant advice, her unconditional encouragement, her infectious energy. I would not have wanted to travel this path with anyone else.

Of course, no acknowledgments section would be complete without a few words about Delilah, who attempted on countless occasions to contribute to this text by running over my keyboard, purring. She bears sole responsibility for any typographic errors that remain! I forgive her, however, since her feral antics brought joy to the labors of writing.

Leopold wrote that the same tools "are requisite to good oak, and to good history." He could have just as easily added political science. So now, to follow his advice, comes the job of making wood.

1 The Puzzle of Climate Policy Action

On July 31, 2014, several thousand labor protesters descended on Pittsburgh. Workers from the United Mine Workers of America (UMWA) arrived in the largest numbers, joined by sisters and brothers from the International Brotherhood of Boilermakers, the International Brotherhood of Electrical Workers, the Utility Workers Union of America, and several railroad unions. These union protesters marched through the Steel City toward the William S. Moorhead Federal Building where, in an act of civil disobedience, they occupied the building's front steps. When the workers refused to leave, police arrested fourteen labor leaders, including UMWA's president, Cecil Roberts.[1] The occasion was an Environmental Protection Agency (EPA) hearing on the Obama administration's Clean Power Plan (CPP), a regulatory effort to reduce US carbon pollution. Industrial union leaders believed the proposal posed an existential threat to their jobs and communities.

Inside the Moorhead Federal Building, EPA administrators listened to testimony from the protesting workers' employers. Representing the coal industry, the American Coalition for Clean Coal Electricity (ACCCE) attacked the Clean Power Plan's legal basis and warned of double-digit electricity price increases.[2] Not all participants were united in their disapproval, however. Earlier, the agency had heard from the BlueGreen Alliance, a partnership of US labor unions and environmentalists, which lauded the EPA's proposal as a "significant step forward to tackle the impacts of climate change."[3] Numerous clean energy businesses also testified in strong support of the regulation. The Obama administration would push forward with the CPP, although both its efforts and the Trump administration's subsequent attempts to repeal the CPP became mired in legal conflict.

The United States was not the only country debating climate policy. On July 17, two weeks prior to the UMWA protest, the Australian Senate voted to repeal the Australian Carbon Pricing Mechanism (CPM). The CPM was a sweeping 2011 reform that imposed costs on Australian carbon polluters. It had been enacted by a previous Labor government in partnership with the Australian Green Party. In 2013, Labor lost its majority to the right-wing Liberal Party, led by climate skeptic Tony Abbott. To repeal the CPM, Abbott's Liberals needed Senate votes from Palmer United, a new political party founded by eccentric mining magnate Clive Palmer. Before entering politics, Palmer had refused to pay more than $8 million AUD (about $8.3 million USD at the time) in carbon taxes associated with his Queensland nickel mines, alleging the taxes were unconstitutional. In the Senate, Palmer's party helped kill the Australian carbon price entirely.

Around this time, climate policy conflict also flared in Norway. In October 2013, a center-right Norwegian government proposed modest cuts to a carbon polluter subsidy program. These cuts would have reduced government payments to domestic industries facing climate policy-linked electricity costs. Heavy industries protested the change, warning that factories would close. The country's main industrial unions threatened a nation-wide strike. Within weeks, the Norwegian government walked back its proposal.

These episodes of climate policy conflict are not exceptional. Carbon pollution is embedded in the electricity, transport, and manufacturing systems that fuel modern economies. Today, climate advocates seek to increase the costs associated with releasing carbon pollution into the atmosphere. Their attempts have sparked dramatic political conflict that divides political parties, labor unions, business stakeholders, and environmental advocates. The results of this conflict will shape billions of lives over the coming century. If enough countries enact ambitious climate reforms, global publics may avoid the worst projected impacts from climate change. However, if climate reforms continue to pose intractable political challenges, the economic and social costs will be catastrophic.

At the same time, even though climate change will harm all countries, national policy responses have varied. Some countries have undertaken substantial national action while others have done little. Surprisingly, we still lack comparative theories to explain this variation in climate policy timing and content. For all that we know about the dynamics of global climate negotiations—a topic that has been studied at length by international relations scholars—many climate researchers still "black box" national policymaking processes (Victor 2011).

This book joins an emerging literature to unpack this black box (Hughes and Urpelainen 2015; Raymond 2016; Karapin 2016; Rabe 2018; Lipscy 2019). Using in-depth analysis of climate reforms in Australia, Norway, and the United States between the late 1980s and the 2015 Paris Agreement, it offers a new theory to explain cross-national differences in climate policy timing and content. In doing so, it provides a new diagnosis for our catastrophic inability to reduce global carbon pollution, and offers guidance on how entrenched opposition to climate reforms might be disrupted.

My argument has two parts. First, I describe the distribution of climate policy preferences among major economic and political actors. Second, I show how this preference distribution interacts with domestic political institutions to explain variation in policy outcomes. Together, these analyses show how climate policymaking has been systematically captured by carbon-intensive business and labor actors.

I begin by showing how the climate threat's emergence revealed cross-cutting divisions within existing political and economic coalitions. Climate change divided workers in carbon-dependent industrial unions from workers in low-carbon sectors. It divided carbon-intensive businesses from businesses with small carbon footprints. And it split political actors on both the left and right with divergent ties to these divided labor and capital constituencies.

In principle, cross-cutting climate policy preferences could work to the advantage of either climate policy proponents or opponents. In practice, I find that opponents benefited more from this "double representation" by politicians on both the left and right, and by economic stakeholders aligned with both capital and labor. The dispersion of carbon polluters' political allies compounded status quo biases in public policymaking; it ensured that, no matter who controlled government, carbon polluters were accommodated in policy design. This book argues that the *double representation of carbon polluters* is the single most important feature of climate policy conflict across advanced economies.

I then describe how domestic institutions can either enhance or moderate this double representation by structuring polluter access to climate policy design. This access is shaped by policymaking institutions, such as institutional links between economic stakeholders and government policymakers. It is also shaped by political organizations, for instance through historical links between labor unions and political parties.

Climate policy enactment then proceeds along one of two stylized pathways. On the first path, double-represented polluters enjoy guaranteed access to climate policy design. Here, climate policy proposals that reach the political agenda are vetted by policy losers. Climate reforms shield producers from costs and prioritize policy carrots over sticks. On the second path, double-represented carbon polluters lack guaranteed access to climate policy design. Here, policy proponents sometimes propose reforms that impose significant costs on carbon-dependent economic actors. Threatened, policy losers mobilize conflict into the public domain to weaken the political incentives associated with policy enactment. Along this second pathway, climate reforms are more sensitive to consumer costs and can include more policy sticks.

In turn, climate policy content shapes policy timing. Policies that follow the first pathway rarely threaten the economic status quo. Once proposed, they often pass without controversy. By contrast, policies that follow the second pathway often fail in the face of opponent mobilization. Instead, it can take several attempts over multiple policymaking windows to pass reforms. As a result, countries that develop policies along the first pathway pass earlier climate reforms because, once proposed, these reforms are more likely to be enacted.

More broadly, this book advances a distributive-institutional account of climate politics. The scientific discovery of the climate threat revealed crosscutting climate policy preferences among political and economic actors. Domestic institutions conditioned the stability of these cross-cutting cleavages and shaped different economic stakeholders' influence on climate policy design. Policy conflict subsequently proceeded through one of two pathways. Understanding how national institutions and carbon polluters' double representation interacted to shape these pathways helps explain cross-national differences in climate policy timing and content. It sheds new light on why climate reform opponents remain entrenched in many advanced economies. And it helps us understand if—and when—advanced economies will confront the civilizational threat of climate change.

The remainder of this introductory chapter is structured as follows. First, I outline the puzzle of national climate policy action: why have some countries enacted climate reforms in the absence of an effective global climate agreement? Second, I offer a framework to study variation in climate policy outcomes. Third, I review existing explanations for this variation. Fourth, I preview my theory in greater detail and position its arguments within existing debates. Fifth, I outline the book's methodological approach. Finally, I outline the book's structure. In a chapter appendix, I provide details on the 101 interviews I conducted with senior politicians and policymakers in the process of researching this book.

The Puzzle of National Climate Policy Action

Atmospheric carbon pollution levels continue to increase. As of December 2018, atmospheric CO_2 concentrations had reached almost 410 parts per million (ppm), up from a preindustrial baseline of 280 ppm.[4] Scientists agree that atmospheric concentrations above 450 ppm will warm the planet more than 2°C on average, with serious consequences for economic prosperity, national security, and human welfare. Even at current pollution levels, shifts in global weather caused by climate change already threaten countries across the globe.

Because carbon pollution released into the atmosphere persists for over a thousand years, climate risk mitigation requires immediate reductions in annual carbon pollution flows. Yet, global efforts to negotiate a coordinated climate crisis response have repeatedly stalled (Barrett 2003; Victor 2011). The Kyoto Protocol, a 1997 global climate treaty, did not curb carbon pollution growth. The 2009 Copenhagen effort to negotiate a successor agreement collapsed. Some observers hope that the December 2015 Paris Agreement will succeed where previous efforts failed. Unlike the Kyoto Protocol, the Paris Agreement allows countries to propose their own carbon pollution targets; it does not specify minimum target ambition or prescribe how the international community will enforce national commitments. Yet, this approach's resilience is already being tested by the Trump administration's dramatic rejection of the Paris Agreement in June 2017.

In the absence of a binding global agreement, many scholars express skepticism that meaningful domestic reforms are possible. According to this received wisdom, it is irrational for any country to enact unilateral climate policies. Early actors would impose significant costs on their economies while receiving only a small fraction of the globally distributed benefits associated with their actions. Instead, every country should forgo costly domestic reforms to free-ride off the climate policy actions of others.

Yet, these predictions do not correspond to the empirical record of cross-national climate policymaking. The past two decades have witnessed substantial between-country and across-time variation in climate policy action, not convergent climate policy inaction. Even as the international community failed to negotiate a robust climate treaty, domestic climate reforms were actively debated at the highest political levels in every advanced economy. While these national policies remain collectively insufficient to mitigate dangerous, human-caused climate change, they have been something more than political greenwashing. Climate reforms have been the primary focus of multiple national elections, they have shifted the political fortunes of parties and elected leaders, and they have consumed the time and resources of bureaucrats and politicians during contentious, multiyear policy debates. In Norway, carbon taxes enacted in the early 1990s imposed substantial consumer costs; yet, efforts to extend costs to producers led to the collapse of a centrist government in the late 1990s. Three decades of legislative policymaking failed in the United States; yet, the Obama administration used regulatory levers to impose meaningful producer costs beginning in 2010. Meanwhile, in Australia three successive prime ministers and an opposition leader lost their jobs, in part, for their roles in enacting and then repealing a comprehensive climate reform. What accounts for this variation in climate policy timing and content?

Making Sense of Climate Policy Outcomes

Ideally, climate policies should be evaluated according to their ability to mitigate climate risks. Scholars could then identify the political conditions under which environmentally effective policies pass. However, carbon pollution levels are determined by diverse social, economic, and technological factors; as a result, it is not possible to reliably isolate the specific quantity of carbon pollution mitigated by particular climate reforms.

Greenhouse gas emissions pervade industrial economies as the by-product of transportation, energy, and manufacturing processes; thus, nearly every significant economic trend shifts carbon pollution patterns.

For example, global carbon pollution levels declined by 1.4 percent during the 2008–2009 financial crisis (Peters et al. 2012). In the 1990s, German carbon pollution plummeted as a result of post-reunification industrial restructuring in the former East Germany (Schleich et al. 2001).

This overdetermination of carbon pollution levels complicates analysis of climate policies' impacts (Christoff and Eckersley 2011). To date, ex-post evaluations of climate reforms remain ambiguous (Svendsen et al. 2001; Bruvoll and Larsen 2002; Hu et al. 2015). Some high-profile research does position environmental outcomes as its dependent variable; this work provides valuable information about the correlations between different institutional or political conditions and aggregate national pollution levels (King and Borchardt 1994; Neumayer 2003; Scruggs 2003; Bättig and Bernauer 2009; Jahn 2016). However, we cannot trace the causal pathways between political debates and policy outcomes when the relationship between policy outputs and policy outcomes remains unidentified.

By contrast, political scientists excel at studying the processes that shape climate policy outputs. Of course, even here we require conceptual precision. When virtually every economic policy has climatic implications, the concept of "climate policy" can itself become ambiguous (Harrison and Sundstrom 2010b; Dupuis and Biesbroek 2013). To describe variation in climate policy outputs, scholars must confront two challenges. First, given the diverse economic, environmental, and social policies that shape carbon pollution levels, they must define which policies are appropriately categorized as *climate* policies. Second, they require a conceptual framework to facilitate between-country and across-time climate policy comparisons.

Defining Climate Policy

In this book, I define as climate policies any efforts to deliberately reshape carbon pollution levels. While many economic policies or trends have unintended effects on carbon pollution, few analysts believe that "accidental" climate policymaking is sufficient to combat the climate crisis. For example, while the 2009 economic crisis curbed greenhouse gas emissions temporarily, carbon pollution rebounded by 5.9 percent in 2010 to negate crisis-induced pollution reductions (Peters et al. 2012). Instead, climate risk mitigation requires intentional efforts to increase the costs of carbon-intensive economic activity and/or reduce the costs of low-carbon economic activity. In the absence of such policy interventions, business-as-usual economic activity will not stabilize carbon pollution levels.

Climate policies can be deliberate without being publicly labeled as climate policies. Where climate policy has become politicized, officials may portray climate reforms as energy or fiscal interventions. For instance, regional US planners enacted climate adaptation policies in conservative areas by avoiding climate labels.[5] Similarly, federal bureaucrats during the George W. Bush administration disguised a gas tax increase as an opaque Renewable Fuel Standard (Breetz 2013). To classify climate policies in these cases, I look for evidence that policymakers intended to impose costs on carbon pollution, independent of how specific reforms were presented to the public.

Conceptualizing Climate Policy Variation

To study climate policy variation, we must conceptualize differences in policy outcomes across countries. However, simple policy metrics are often too blunt to support descriptive inferences about cross-national variation, let alone causal inferences about the over-time mechanisms that shape this variation. For example, just knowing the level of a country's carbon price does not tell us much about that country's policy regime. A $100 per ton carbon price imposed on an inelastic pollution source that comprises 5 percent of a country's emissions may be ineffective; a blanket $5 per ton price on 100 percent of a country's carbon pollution could meaningfully shift economic behavior.

Even simple policy shifts are routinely misinterpreted. For instance, Norway announced in 2012, to generous global press coverage, that it was doubling its carbon tax on the offshore oil sector. This tax increase would seem a clear-cut example of increased policy costs—but it was not. When Norway's domestic carbon pricing system was harmonized with the European Union Emissions Trading System (EU ETS) in 2008, offshore oil companies faced carbon liabilities under both domestic and EU-wide policies. Norwegian authorities lowered domestic tax levels to ensure that, combined, Norwegian and European costs matched preexisting liabilities. After the European carbon price collapsed, Norway's offshore sector suddenly faced net costs lower than they had *before* EU ETS implementation. In "doubling" Norway's carbon tax in 2012, Norway restored its historic policy regime, rather than ratcheting up costs on its offshore industry.

To more precisely describe climate policy outcomes, we must first differentiate policy timing from policy content. Policy timing is acutely important for climate change because climate risks accumulate through time in the absence of pollution mitigation. Second, we must disaggregate policy content into at least three linked dimensions: instrument choice, cost levels, and cost distributions.

Policy Timing

The timing of carbon emission policies varies across countries. While carbon taxes first emerged in Northern Europe, carbon taxes would later pass in the UK (2001), Ireland (2010), France (2013), Japan (2012), and Mexico (2014). The European Union introduced an emissions trading scheme in 2005, followed by New Zealand (2008), Australia (2011), and South Korea (2014). In parts of Scandinavia, legal authority to regulate greenhouse gas emissions using air pollution statutes was recognized in the 1990s but rarely used. Analogous authority was recognized by US courts in 2007 and was used for consequential rule-making by 2010.

Policy timing also requires attention to reform trajectories, not only the timing of initial enactment. Over time, policies may be repealed or modified. Australia passed an emissions trading scheme in 2011 but repealed this scheme in 2014. The Netherlands passed a carbon tax in 1995, but replaced this policy with the EU emissions trading system in 2005.

Policy Content

Policy content is more difficult to measure than policy timing. Policy content debates often center on *instrument choice*. Some countries enact carbon taxes while others enact emissions trading schemes. Some countries regulate carbon pollution through environmental performance standards while others rely on voluntary frameworks.

Both carbon taxes and emissions trading schemes are forms of carbon pricing. Carbon pricing policies raise the private costs of releasing carbon pollution to reflect the pollution's social costs. Carbon taxes directly increase carbon pollution's unit cost through a fixed price. An emissions trading system caps the economy-wide quantity of carbon pollution that can be released; the government then distributes a fixed number of pollution permits equal to this cap level. Individual polluters must acquire permits to cover their pollution, generating a carbon price through a pollution permit market. Economy-wide carbon pollution declines as the government reduces permit availability.

Other approaches to raising the costs of carbon pollution exist. Governments can directly set environmental performance standards—regulatory limits on pollution levels for particular economic activities. Governments can also change the relative costs of different technologies; for example, by subsidizing renewable energy or carbon capture and sequestration (CCS) technologies that keep carbon pollution from entering into the atmosphere in the first instance.

Analysts must consequently trace the complex blend of cost- and subsidy-based instruments that make up each country's climate policy regime. Cost-based policy "sticks" include carbon prices, technology mandates, penalties, and taxes. Subsidy-based policy "carrots" include subsidies, cost exemptions, and voluntary programs. Often, the balance between costs and subsidies is more politically consequential than a country's instrument choice (e.g., carbon taxation vs. emissions trading).

Beyond instrument choice, policy content also varies by cost or subsidization *levels*. Carbon taxes can be high or low; emissions trading schemes can induce high or low trading prices. As of April 2019, global carbon prices ranged from \$127 USD per ton of CO_2 on some Swedish pollution sources to less than \$0.01 USD per ton in the Ukraine. Similar variation characterizes climate subsidies. For example, global feed-in-tariff rates for distributed solar generation, a form of clean-energy subsidization, vary from several cents to over \$1 USD per kilowatt hour (kWh).[6]

Policies also vary by how they *distribute costs and benefits*. Even policies with identical carbon prices often include sectoral exemptions or compensation measures that shift the distribution of costs. Subsidy programs can also be directed toward specific economic actors while excluding others. For instance, among Northern European countries that implemented a carbon tax in the 1990s, the ratio of producers' costs to consumers' costs ranged from 1:1 in the Netherlands to over 1:9 in Denmark (Svendsen et al. 2001). In practice, this means that the actors who profit from releasing carbon pollution into our shared atmosphere do not always shoulder the associated climate policy costs.

The existence of this multidimensional climate policy variation precludes the development of simple indicators for climate policy content.[7] Figure 1.1 highlights variation in carbon pricing policies across OECD countries, visualizing the scope of between-country differences.[8] Variation in policy enactment is indicated by a country's presence or absence from

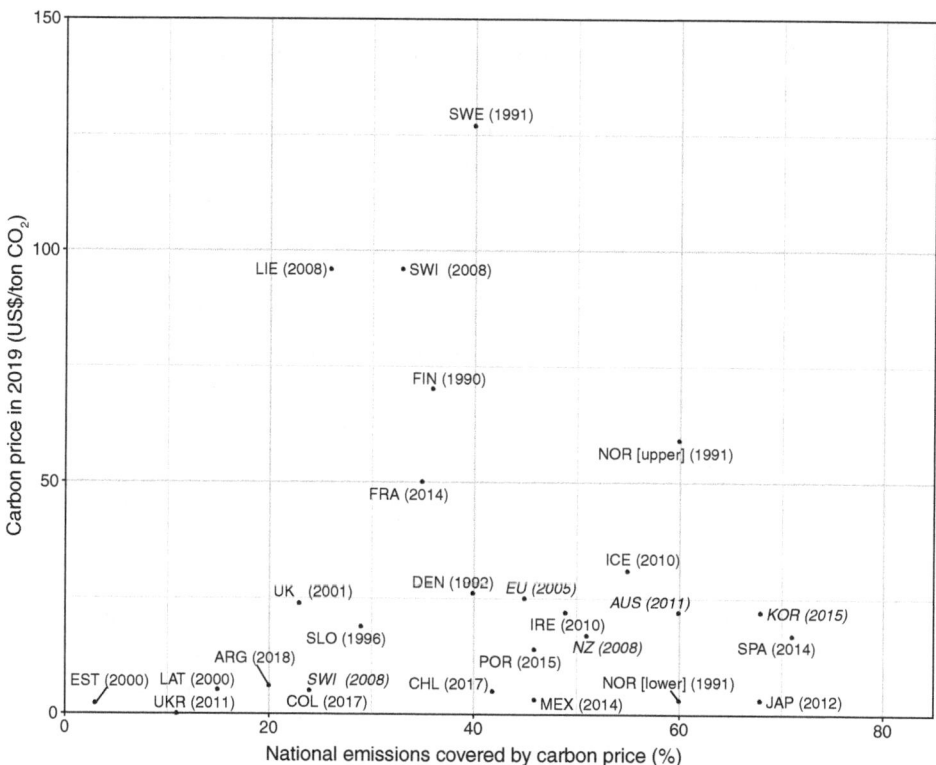

Figure 1.1

National carbon price levels as of April 2019, by percentage of economy covered by a carbon price. Countries with emissions trading instruments appear in italics; countries with carbon tax instruments are not italicized. Initial policy adoption year is noted in parentheses. The now-repealed Australian carbon price, discussed in chapter 6, is included for comparative purposes. The sectoral upper bound (max) and lower bound (min) of Norway's carbon tax, discussed in chapter 3, are presented separately. For other countries, only the highest tax rate is presented.

the figure. (Countries that have subnational carbon prices but not a single national price, such as Canada and the United States, are also excluded.) Variation in timing is provided in parentheses. Variation in instrument choice is indicated by font style. Cost variation is charted along the y-axis. Finally, the x-axis tracks sectoral coverage, a crude proxy for a policy's cost distribution. While this figure describes cross-national carbon prices, similar variation exists for subsidy-based and regulatory approaches to climate risk mitigation.

While climate policy content can be measured with attention to instrument choice, cost levels, and cost distributions, climate policy *ambition* is more difficult to specify. A more ambitious policy should, all else equal, be one that reduces more carbon pollution. Yet, as discussed earlier, carbon pollution levels are so overdetermined by diverse economic and social forces that *retrospective* causal identification of policy impacts remains difficult. Economists have offered evaluations of some policies, but these estimates are difficult to compare across countries and time. Nor can we reliably translate simple policy content metrics, such as figure 1.1's national carbon price, into units of carbon pollution reduced. Even identical carbon prices have different effects based on variation in sectoral cost exposure and sectoral differences in the elasticity of carbon-dependent activities.

Prospective models of policy impacts also face limits. Good economic models exist for some policies in some countries at some times; however, these models are not applied consistently across cases. Moreover, these models are sensitive to input assumptions, a feature exploited by interest groups seeking to shape policymaking. For example, estimated household costs associated with the 2009 Waxman–Markey climate bill in the United States ranged from $3,900 dollars in savings (environmentalist models) to $4,300 in costs (industry models).

Both retrospective and prospective measures of policy ambition also suffer from issues with incomplete data. Retrospective evaluation of policies that were not enacted is impossible. And prospective impact models for proposed (but unenacted) policies are rare, particularly in comparative perspective. However, measurements of both proposed and enacted policies would be necessary to properly explore the political drivers of policy ambition. Developing comparative metrics of policy ambition across countries and time, for both proposed and enacted policies, should be a future research priority.

Measures of *environmental* ambition can be contrasted with measures of *political* ambition. Many forms of carbon pollution are produced by concentrated interests with substantial political power (e.g., large industrial producers); other pollution results from the activities of diffuse actors who lack strong political representation. There is no simple relationship between the class of actors targeted by a policy and the scale of carbon pollution mitigated by that policy. Yet, how climate policies distribute costs across society still has long-term political ramifications (Jenkins 2014).

For instance, some policies may achieve minimal carbon pollution reductions while imposing disruptive costs on entrenched carbon-dependent interests. In doing so, these policies may shift political power from fossil fuel-dependent incumbents to clean energy entrants. In this sense, we can define *politically* ambitious policies as those that impose deeper costs on climate reform opponents, weakening their political influence over time. These *politically* ambitious policies may then facilitate *environmentally* ambitious policies during future climate policymaking rounds.

This book considers both environmental ambition with respect to changes in carbon pollution levels and political ambition with respect to changes in the distribution of political power. Both matter. Climate policymaking is not a one-shot game. It requires repeated rounds of climate reforms over decades. Consequently, we need to understand cross-national variation in the timing and content of climate policymaking efforts. We need to understand how climate reform efforts can reshape the political and economic power of entrenched opponents. And we must understand the conditions under which governments will act to protect our shared climate.

Theories of Climate Policy Action and Inaction

As figure 1.1 makes clear, substantial cross-national and over-time variation in national climate reforms exists. How can we explain this variation in policy timing and content? To date, three different intellectual traditions offer guidance: a first emphasizes the logic of collective action; a second highlights cognitive biases; a third invokes distributive conflict. Each tradition offers competing explanations for variation in climate policymaking outcomes, as summarized in table 1.1.

Collective Action

Standard accounts view climate change policymaking as a classic collective action problem. While there are significant collective benefits associated with climate mitigation, these benefits accrue to all countries regardless of whether or not they implement climate reforms. Every country subsequently faces the incentive to free-ride off others' mitigation actions. It is difficult to organize collective action when the costs are private but the benefits are not excludable.

Table 1.1

Explanatory drivers of climate policy outcomes

Factor	Examples
Collective action	
Free-riding incentives	Barrett 2003; Stern 2007
Issue linkages	Conconi and Perroni 2002; Kemfert 2004
Audience costs	Kroll and Shogren 2008
Individual psychology	
Cognitive biases	Feygina et al. 2010; Kahan 2015
Emotional responses	Leiserowitz 2006; Norgaard 2011
Distributive conflict	
Sectoral balance of power	Meckling 2011; Aklin and Urpelainen 2013
Capacity to self-organize	Svendsen et al. 2001; Harrison 2010a
Public mobilization	Harrison and Sundstrom 2010b; Hughes and Urpelainen 2015
Ideological conflict	Daugbjerg and Svendsen 2001; Lazyer 2012
Issue reframing	Raymond 2016
Policymaking institutions	Hatch 1995; Oshitani 2006; Finnegan 2019
Electoral institutions	Harrison 2010a; Hughes and Urpelainen 2015; Lipscy 2019; Finnegan 2019
Veto point density	Harrison and Sundstrom 2010b; Madden 2014
Bureaucratic power	Rabe 2004

In the absence of global institutions to facilitate climate cooperation, major economies will continue to unsustainably exploit the global atmosphere (Barrett 2003; Sandler 2004; Stern 2007). Consequently, collective action scholars suggest prescriptive institutional designs to mitigate free-riding incentives (Young 2002; Heitzig, Lessmann, and Zou 2011; Keohane and Victor 2011), from accountability mechanisms (Barrett and Stavins 2003; Newell 2008) to climate clubs (Victor 2011). They also emphasize how such factors as breakthrough technologies can shift incentives for climate cooperation (Barrett 2006; Hoel and De Zeeuw 2010; Urpelainen 2012).

Collective action accounts help explain climate policy inaction by outlining how free-riding concerns disincentivize domestic reforms. However, if individual countries shouldn't enact domestic reforms in the absence of strong global climate institutions, why then have some enacted costly, unilateral policies? These countries accept domestic economic costs without meaningfully reducing climate risks or fully capturing globally distributed mitigation benefits.

Collective action scholars explain these actions by theorizing linkages between global issue domains or by specifying cross-national differences in actor interests. Issue-linkage accounts emphasize how climate commitments can respond to international coalition building over unrelated issues; countries may undertake climate actions to support strategic relationships rooted in non-climate priorities (Kemfert 2004; Aklin 2016). Other scholars theorize differences between countries' preferences for global public goods, either as a result of differences in domestic audience costs (Kroll and Shogren 2008) or asymmetric costs and benefits (Barrett 2003; Wiener 2007). Yet, by invoking cross-national variation in domestic characteristics, these accounts mirror explanations emphasizing distributive conflict.

Individual Psychology

Other explanations link uneven climate policy outcomes to a series of psychological biases. These lead mass publics and political elites to misunderstand the climate threat and therefore misjudge the rational policy response (Feygina, Jost, and Goldsmith 2010; Weber and Stern 2011; Gifford 2011). Even experts struggle to evaluate complex climate risks characterized by interacting components and time delays (Sterman 2008). Individuals also engage in motivated reasoning, holding climate beliefs that serve their individual political or economic interests. For example, Kahan (2015) argues that climate beliefs reflect cultural identity, not climate science knowledge. Since climate impacts are complex, contradictory beliefs regarding threat severity can be validated through selective information retention (Dryzek 2013).

Beyond cognition, emotions also distort climate decision-making (Leiserowitz 2006). In a small coastal Norwegian town with high social capital, Norgaard (2011) documents widespread belief that human action is responsible for climate change, and strong support for climate change policy. Yet, she also finds motivated denial about the climate threat's severity, a paradox she links to the social regulation of negative climate-related emotions. Likewise, Milkoreit (2017) describes how cognitive and emotional considerations shape climate policy bargaining among political elites.

These psychological theories reveal important dynamics that shape climate policy conflict. However, cross-national differences in cognitive biases or affective engagement have not been identified that could explain cross-national variation in climate policymaking outcomes.

Distributive Conflict

Still other explanations emphasize the fundamentally distributive nature of climate policymaking. Climate policy involves a dramatic renegotiation of economic and social institutions. Consequently, climate reforms create new economic winners and losers. Material conflict between economic and political stakeholders are also accompanied by ideational and value conflict.

Distributive explanations for climate policy outcomes can be loosely grouped into accounts that emphasize the power of competing interest groups and accounts that emphasize the institutional terrain within which political interests engage.

Interest-Based Explanations

One group of distributive explanations focuses on the relative power of climate policy opponents and proponents, for instance the balance of power between "green" (low-carbon) and "brown" (carbon-intensive) economic actors (Harrison and Sundstrom 2010b; Hughes and Urpelainen 2015). Policy opponents face concentrated losses while policy benefits are diffusely distributed; since concentrated losers can organize more easily, they become overrepresented within climate policy debates (Svendsen et al. 2001; Harrison 2010a; Victor 2011). In this sense, policy opponents have greater voice because they are able to organize more effectively. In turn, the balance of power can shift when pro-reform interests develop their own lobbies. For example, where clean energy coalitions emerge, they counterbalance industrial lobbies to drive clean energy reforms (Aklin and Urpelainen 2013; Cheon and Urpelainen 2013). Similarly, "carbon coalitions" between environmental groups and pro-climate policy businesses can help drive emissions-trading reforms (Meckling 2011).

In these accounts, climate policy action is frustrated by entrenched policy opponents, not by free-riding concerns. For instance, US fossil fuel interests exploited regulatory, judicial, and ideational pathways to block US climate reforms over decades (Kamieniecki 2006; Layzer 2007; Karapin

2016). "Policy networks" that link carbon-intensive economic sectors with government officials also shape climate outcomes in many other countries (Kasa 2000; Daugbjerg and Svendsen 2001; Bailey et al. 2012). Correspondingly, differences in policy network membership may help explain climate policymaking variation (Daugbjerg and Svendsen 2001). Conversely, when the public mobilizes to support climate policy despite diffuse interests, this may increase the likelihood of climate policy enactment (Harrison and Sundstrom 2010b; Trumbull 2012; Hughes and Urpelainen 2015). Reforms can also be driven by empowered bureaucracies, particularly during moments where climate policy is less salient (Rabe 2004; Hughes and Urpelainen 2015).

Political conflict also unfolds over ideas. Layzer (2012) argues that conservative ideology is as important as economic interests in motivating anti-climate coalitions. Daugbjerg and Svendsen (2001) argue that social democratic parties passed early carbon taxes because economic interventionism was more consistent with left-leaning belief systems. Other research documents an ideological gradient in climate policy preferences at both elite and public levels (McCright and Dunlap 2011; Shipan and Lowry 2001). However, not all scholars find strong links between ideology and climate policy outcomes; Harrison and Sundstrom (2010a) note in passing the surprising absence of consistent left-right divides in global climate politics. By contrast, Raymond (2016) emphasizes a different form of ideational conflict. Studying regional US carbon pricing, he links policy enactment to "normative reframing" of the atmospheric commons as publicly owned. He argues that policies pass when advocates construct social understandings of polluter responsibility for public welfare.

Institution-Based Explanations

Policymaking institutions provide the routines, rules, and frameworks within which policy decisions are made. Institutions emerge from past episodes of distributive conflict even as they continue to structure the relative influence of present-day policy actors (Knight 1992). Majority policy support may not lead to enactment if political institutions still privilege the preferences of an institutionally advantaged minority. To explain climate policy variation, climate scholars place particular emphasis on policymaking institutions, electoral institutions, and veto point distributions.

Policymaking institutions are often classified according to their relative pluralism or corporatism (Schmitter 1974; Lijphart 1999). Pluralist institutions allow for competition between interest groups for policy influence, mediated by the policy interests of semi-autonomous state actors. Policymaking is adversarial, characterized by contestation between different interest coalitions. By contrast, corporatist institutions facilitate long-term policymaking by major economic stakeholders, often tripartite bargaining between labor, business, and the state.[9] In corporatist systems, business and labor groups are hierarchically organized into peak associations; these peak associations enjoy institutionalized access to government actors. Peak associations can also assume responsibility for regulating stakeholder compliance with government agreements (Smith 1993).

On balance, pluralist institutions are viewed as complicating climate reforms (Karapin 2016). By contrast, corporatist institutions are viewed as facilitating climate policymaking by nurturing trust between regulators and the regulated, focusing business attention on national rather than particularistic outcomes, and facilitating distributional loser compensation (Hatch 1995; Matthews 2001; Scruggs 2003; Christoff and Eckersley 2011; Finnegan 2019). In turn, early, even if weak, climate policy action can set the stage for incremental policy strengthening through time (Meckling et al. 2015; Finnegan 2019). Other scholars disagree. For example, Oshitani (2006) argues that corporatist institutions instead empowered climate policy losers in Germany and Japan.

Scholars also focus on the effect of national electoral systems on policy outcomes (Harrison 2010a; Christoff and Eckersley 2011). In proportional representation (PR) systems, Green parties are more likely to enjoy political representation than in first-past-the-post systems (Harrison and Sundstrom 2010b). In turn, Green Party presence increases climate reform probability (Neumayer 2003; Harrison 2010a; Hughes and Urpelainen 2015). Green parties can shape the agenda directly when they have legislative leverage and indirectly when they induce strategic repositioning by other parties to retain environmentally minded voters. Electoral institutions also matter when they shape elite prioritization of public goods provision (Lipscy 2019). Politicians in majoritarian systems cannot easily impose costs on diffuse consumers; however, politicians in non-majoritarian systems can maintain their electoral standing by directing policy revenues toward particularistic interests. Similarly, politicians in PR systems may be relatively insulated from consumer accountability; this could facilitate more ambitious cost imposition on consumer-facing sources of carbon pollution (Finnegan 2019).

A final literature highlights variation in policymaking veto points (Tsebelis 2002). The more veto points in a polity, the easier it is for opponents to block policy change. Systems with more veto players have weaker environmental performance (Jahn 2016), act later on climate change (Harrison and Sundstrom 2010b), produce fewer climate policy outputs (Madden 2014), and have more incremental policy content (Madden 2014). Similarly, Karapin (2016) emphasizes how US separation of powers persistently stymied US climate reforms.

Of course, institution-based explanations can interact with interest-based explanations in complex ways. Rabe (2018) charts the history of US and Canadian carbon pricing policies to profile their political contingency. He shows how climate policymaking involves the imposition of short-term costs in exchange for long-term benefits, a class of policy problems most policymaking and political institutions struggle with. Policies often fail because proponents oversell their reforms and give insufficient consideration to the difficult distributive conflicts that reforms trigger.

Despite the richness of distributive politics accounts, climate politics surveys often place more emphasis on collective action barriers and psychological biases (e.g. Giddens 2009; Hulme 2010). Recognizing this oversight, Victor (2011) criticizes the existing literature for "black boxing" national policymaking processes. Offering a theory of national climate regulation, he argues that countries enact policies that impose costs on poorly organized domestic actors while shielding organized actors. He emphasizes that the choice of policy instrument is contingent on the distribution of organized interests. This is a necessary insight for any theory of national climate policymaking. Yet, this account cannot fully explain why some countries *have* imposed costs on well-organized national actors. For instance, the Obama administration's Clean Power Plan threatened fossil fuel-dependent industries with substantial new regulatory burdens despite industry opposition. Instead, to fully explain cross-national and over-time variation in climate policy timing and content, we need a more complex theory of climate policy conflict.

How Business and Labor Control Climate Politics

This book explains variation in climate policy outcomes by highlighting the interaction between climate policy preference distributions and domestic political institutions. I develop my argument in two steps.

First, I highlight the distribution of climate policy preferences among political parties and economic interest groups. I argue that climate preferences cut across established economic and political coalitions. Both coal workers (through the UMWA) and coal businesses (through the ACCCE) mobilized against the Obama administration's Clean Power Plan. At the same time, other labor unions (through the BlueGreen Alliance) and businesses supported the reform. In Norway, both industrial unions and heavy industries banded together to protect the country's carbon pollution subsidies. And in Australia, both left- and right-leaning political coalitions navigated tensions between climate reformers and opponents during the late 2000s. These cross-cutting preferences are not structured by postmaterial value dimensions (cf. Inglehart 2015); instead, they are rooted in the distribution of economic benefits and costs. Climate policy conflict remains deeply material.

The presence of business opposition to climate policy is well understood. However, the presence of labor actors in opposition coalitions is often overlooked. This book draws out myriad examples. In Norway, peak labor groups mobilized to exempt onshore industries from the Norwegian carbon tax. In the United States, the AFL-CIO's Industrial Union Council lobbied for the anti-Kyoto Byrd–Hagel Resolution in the US Senate. And these coalitions often work in tandem with business. As Tim Phillips, the president of the Koch-funded Americans for Prosperity (AFP) lobbying network boasted: "We have a lot of private employee unions [who] know that the environmental agenda will kill their jobs. So we often ally with them to oppose the agenda coming from the climate change and global warming coalitions."[10]

In principle, cross-cutting preference distributions could benefit either proponents or opponents of climate policymaking. However, in most political systems it has privileged opponents. To enact climate reforms, advocates need to secure pro-reform coalitions at each step of the policy process. By contrast, opponents only need to control a single veto point. Because climate policy preferences cut across traditional political coalitions, opponents usually secure at least one such veto point, no matter which party controls the government. I describe this dynamic—where party factions on both sides of the ideological spectrum represent the interests of carbon-intensive constituencies within climate policy debates—as the *double representation of carbon polluters.*

However, institutional contexts are not identical across all countries. My argument also shows how domestic policymaking institutions and political organizations reinforce or moderate carbon polluters' double representation by structuring opponent access to climate policy design.

With respect to policymaking institutions, this book places particular emphasis on each country's corporatism or pluralism. Corporatist institutions, common in Europe, facilitate long-term collaborative policy bargaining by economic stakeholders; pluralist policymaking institutions, typical of the United States, Canada, and Australia, involve adversarial conflict between diverse-interest coalitions seeking to influence government decisions. In this book, I argue that corporatist carbon polluters enjoy more consistent access to government policymakers over time. By contrast, the influence of carbon-dependent economic actors is more variable in pluralist systems. When political allies control government, carbon polluters control policy design. However polluter interests may be marginalized when reform proponents set the political agenda.

With respect to political organizations, this book places particular emphasis on the links between political parties and carbon-dependent economic actors. In some countries, formal ties exist between economic stakeholders and political parties, as when the dominant left-leaning party is a direct offshoot of a country's labor movement. By contrast, in other countries, the labor movement is a constituency of left-leaning parties but lacks a formal relationship with party leadership. Where formal links between labor and left-leaning political coalitions exist, carbon-dependent unions have a more institutionalized voice in climate policy design.

Links between business associations and political parties are also consequential for climate policy conflict; however, I find less cross-national variation in this relationship. Because institutional ties between carbon-intensive businesses and political actors systematically stymie climate reforms across all countries, they do not easily explain cross-national differences in policy timing and content.

Pathways to Climate policies:

Depending on institutional context, climate policies can subsequently be enacted through one of two different causal pathways. A first pathway occurs when carbon polluters have direct influence on climate policy design. In these contexts, policy proposals shield carbon polluters from costs and tend to prioritize policy carrots over policy sticks. In the absence of a significant policy threat, carbon polluters have little incentive to mobilize policy conflict into the public domain (cf. Schattschneider 1960); climate policies that reach the political agenda pass with minimal controversy. An archetypal example of policymaking along this pathway was Norway's 1991 carbon tax. The policy was introduced with significant exemptions for Labor Party-allied onshore industries. Tax rates on the offshore oil industry did not threaten the industry's economic viability. Once proposed, Norway's carbon tax was enacted with minimal controversy.

A second pathway occurs when carbon polluters lack guaranteed influence on climate policy design. In these contexts, proponents may advance reforms that impose significant costs on carbon-dependent economic actors. Facing a policy threat, producers mobilize conflict into the public domain to weaken the electoral and legislative incentives associated with policy enactment. Often policy enactment fails in the face of this mobilization. For instance, in the United States, both the 1993 BTU energy tax and the 2009 American Clean Energy Security Act faltered after opponents expanded the scope of climate policy conflict. Yet, under some conditions, climate policies still move forward in the face of conflict. In the United States, the Obama administration persisted in its efforts to impose costs on carbon pollution using the Clean Air Act, despite intense opposition from some business and labor stakeholders.

Variation in policy timing and content stems from this interaction between carbon-polluter double representation and these institutionally mediated policy enactment pathways. When a country has a corporatist system and its political parties have strong links to carbon-dependent economic actors, the double representation of carbon polluters is strongly reinforced by domestic institutions. In these contexts, climate policies tend to be enacted along the first causal pathway. Proponents bring climate policies onto the political agenda, but they must design these policies in collaboration with the policies' economic losers. Consequently, proposals shield producers from costs and pose few threats to the economic status quo. In the absence of sustained opposition, policies are often enacted soon after their proposal. The result is early climate policy enactment.

By contrast, in pluralist systems and in countries where parties lack strong links to carbon-dependent economic actors, the double representation of carbon polluters is moderated. In these contexts, climate policies tend to be enacted along the second causal pathway, generating more variation in policy content. Sometimes, policy design shuts out climate advocates and weak policies result. Other times, advocates advance policies that pose real threats to carbon-dependent stakeholders. These policies may include more sticks and can involve substantial costs for carbon polluters. Policy opponents respond to this threat by mobilizing conflict into the public domain. Anticipating this conflict, proponents reduce consumer cost salience to make enactment easier.

Often, climate policies proposed along the second pathway fail: even when carbon polluters lack access to policy design, their interests are still represented by political allies across the ideological spectrum who maintain control over at least one veto point. However, on some occasions, more costly reforms can pass. Variability in proposal success also means that, in expectation, policy enactment occurs later along the second pathway; climate reformers may need multiple policymaking windows of opportunity before reform efforts succeed.

This argument departs from several of the explanations reviewed in table 1.1. Distributive accounts that focus on sectoral balance of power do not consider the cross-cutting nature of climate policy preferences. Accounts that emphasize ideological gradients ignore the distribution of climate opponents and proponents on both sides of the ideological spectrum. And both environmental advocates and Green parties play a more limited role in shaping climate policy timing and content than other theories presume. In chapter 2, I elaborate on these considerations and situate my theory more fully within existing debates.

My argument also has implications for global climate negotiations. To date, many climate proponents have focused on establishing international institutions to facilitate climate policy cooperation. These actors assume the most significant policy barrier is the absence of incentives to unilaterally reduce carbon pollution. Instead, I argue that climate policy inaction is not primarily rooted in free-riding concerns, but in patterns of institutionally entrenched reform opposition at the domestic level. The past two decades have witnessed serious efforts to enact major climate reforms in every advanced economy, even in the absence of a binding global climate treaty; yet, even powerful climate advocates have been routinely frustrated by the double representation of carbon polluters. I expand on this point—and its implications for climate policymaking—in chapter 8.

My argument also complicates the common assumption that institutions to facilitate collective action, such as democracy and corporatism, always promote better climate policy outcomes (Scruggs 2003; Bättig and Bernauer 2009). This is because institutions that promote collective action also facilitate the accommodation of distributional losers within policy design. By institutionalizing the voice of carbon-dependent actors, these institutions can reinforce the privileged influence of carbon-dependent economic actors. My conclusions thus echo a growing reevaluation of the welfare state (cf. Thelen 2014) to suggest there may be a class of public goods that even social welfare states are poorly equipped to deliver to citizens. Consensus-based policymaking, which facilitates the delivery of certain social protections, may simultaneously bias against non-incremental transformation of carbon-intensive economies.

Correspondingly, my findings also question whether carbon pricing is a politically viable approach to climate risk mitigation. By raising the salience and transparency of policy costs, carbon pricing may undermine advocates' ability to create durable reform coalitions. I suggest that climate policy debates have been too focused on economists' theories of "optimal" climate policymaking. Instead, substantially more attention must be paid to the political economy of climate reforms. Scholars and advocates must avoid automatic prioritization of the most "efficient" policy. Instead, they should focus on sequencing policy costs and benefits in ways that nurture the political coalitions necessary to pass ambitious reforms.

Finally, my argument also offers several surprising conclusions. It highlights how, contrary to popular accounts, climate policy conflict has been only loosely structured by traditional left-right ideological conflict. Instead, climate policymaking has generated tensions within both left- and right-leaning political coalitions. It also outlines why ambitious climate mitigation efforts come from actors not typically viewed as climate leaders. Many European countries are often portrayed as "green," but their domestic institutions may constrain deep decarbonization efforts. And, despite rhetoric by political actors on the global left, a wave of victories by left-leaning parties may usher in only weak levels of climate mitigation. Conversely, my theory of double representation outlines how unexpected actors could make important contributions to climate mitigation, including such perceived climate laggards as the United States and centrist political parties.

More broadly, this book contributes to a growing literature in comparative environmental politics. Important contributions have analyzed the relationship between cross-national institutional, cultural, and political differences and environmental outcomes (Enloe 1975; Lundqvist 1980; Vogel 1986; O'Neill 2000; Schreurs 2002; Scruggs 2003; Harrison and Sundstrom 2010b; Steinberg and VanDeveer 2012; Lipscy 2019). However, these texts have not yet cohered into a systematic account of cross-national climate politics. Conversely, the environment has not often been studied by scholars of comparative political economy, even though environmental harms compromise the well-being and security of communities across the planet. As a result, environmental policymaking has too rarely been placed in dialogue with core political science debates on the changing political economies of industrialized democracies (e.g., Esping-Andersen 1993; Hall and Soskice 2001). This intellectual isolation persists even though comparative social and environmental policy scholars study similar actors and institutions and share similar theoretical concerns. Such environmental goods as a stable climate or safe drinking water are foundational to public welfare; environmental conditions shape health, prosperity, and living standards in the industrialized and developing world alike (Heal 2000; Fitzpatrick and Cahill 2002).

Further, environmental policies have, over time, become increasingly central to the modern state's policy portfolio, a process that parallels welfare state development (Meadowcroft 2005, 2012). In fact, it is not difficult to echo Esping-Andersen (1993) and read three decades of environmental policymaking as debates over which environmental public goods are properly treated as social rights and the degree to which environmental goods and services should be decommodified (Duit 2016). Similarly, since the 1990s, environmental policy has been buffeted by the same liberalizing pressures that have structured social policy changes (Pierson 1994; Hajer 1995; Bernstein 2001; Newell and Paterson 2010; Layzer 2012; Thelen 2014). It should be no surprise that political scientists criticize the discipline's inattention to environmental issues, particularly climate change (Keohane 2015; Javeline

2014). This book helps build a bridge between these separate but interdependent conversations.

The Strategy of Evidence and Inference

To understand climate policy, we need to identify the causal mechanisms shaping climate policy conflict over time. At each step in a country's policymaking trajectory, diverse political pressures create demand for climate policy. Political actors can respond to these demands by supplying climate policy proposals. The details of these proposals create incentives for different economic and political stakeholders to mobilize. Shaped by this mobilization, climate policies pass or fail.

The object of this book's analysis is thus a complex, multidimensional policy process, sensitive to changes in actor and institutional configurations. This type of analysis resists quantification. Climate policy outputs cannot be reduced to unambiguous numeric measurements so long as comparative metrics do not exist for either proposed and enacted policies across countries and time. Further, there is a small number of advanced economies and strong climate policy interdependence among EU member states. This prevents the type of thoughtful cross-national statistical analysis that has been occasionally deployed in the welfare state literature (e.g. Scheve and Stasavage 2006; Iversen and Soskice 2006).

In the face of these constraints, qualitative forms of analysis are best positioned to rigorously unpack the dynamics of climate policy conflict. In particular, qualitative inquiry is uniquely positioned to analyze the substance of politics—how agents with power use it to shape the institutions and rules that structure political life (Pierson 2007). Accordingly, this book adopts a historical institutionalist approach, using process tracing to make rigorous causal inferences about the processes that shape climate reform trajectories over time. In an online Transparency Appendix (TRAX), I outline the methodological and inferential commitments of this approach for interested readers.[11] This appendix places particular emphasis on qualitative, small-n research as the basis for causal inference. Stated briefly, this book uses *within-case* analysis to understand the causal processes that shape climate policymaking. It then compares across cases to propose scope conditions under which these causal mechanisms matter. Finally, it probes scope condition generalizability using four out-of-sample shadow cases.

In my framework, causal mechanism identification occurs during within-case analysis. Consequently, it is important to choose cases that are substantively important for climate policymaking so that understanding these particular cases, themselves, sheds light on global climate risk mitigation. The book's three cases, Australia, Norway, and the United States, meet this substantive importance criterion. The United States was the world's largest carbon polluter until China surpassed it in 2007. Australia is the advanced economy with the most volatile climate policy regime. Norway is one of the advanced economies that took the earliest action to address climate change.

In contrast to within-case causal mechanism identification, which is immune to small-n selection biases, the elaboration of scope conditions involves cross-case comparisons and thus requires variation on a study's dependent variables. The choice of Australia, Norway, and the United States balances the substantive interest criterion with this dependent-variable variation criterion. Table 1.2 charts variation in climate policy outcomes across the book's three cases along the four climate-policy dimensions presented earlier. The empirical basis for these stylized facts will be elaborated in future chapters, but the table offers an initial overview of variation in policy timing and content. Note that some of the cells reflect relative attributes. For instance, almost every country tends to prioritize climate policy carrots over climate policy sticks; yet, this tendency is particularly pronounced in Norway and somewhat less pronounced in the United States.

For each case, I process-trace the trajectory of climate policymaking on the national political agenda from the emergence of climate change as a

Table 1.2

Climate policy outcomes in Australia, Norway, and the United States, through 2015

Dimension	Australia	Norway	United States
Policy timing	Late	Early	Late
Instrument choice	Mixed	Emphasis on carrots	Emphasis on sticks
Absolute cost levels	Low	Medium	Low
Relative cost distribution	Mixed	Emphasis on consumers	Emphasis on producers

political threat in the 1980s to 2015. To do so, I consulted as diverse a set of sources as possible. The history of climate policy is recent, and many relevant documents have not been released by, or even deposited in, archives. However, for some of the earliest episodes of policy conflict over carbon pricing, a limited US archival record can already be accessed, including records at the Clinton Presidential Library in Little Rock, Arkansas; the Ball State University Archive in Muncie, Indiana; and the Dirksen Congressional Research Center in Pekin, Illinois. I also extensively reviewed grey literatures, policy documents, and media reports in all three countries. Most critically, I conducted 101 interviews with politicians, bureaucrats, corporate officials, and social actors in all three countries during 2013 and 2014. The majority of these interviews were conducted during field research trips to Washington, DC; Oslo, Norway; and to Canberra, Sydney, and Melbourne, Australia. Interviews averaged just under one hour in length. The vast majority were recorded and then transcribed; for a few sensitive interviews, I took only handwritten notes, which I immediately transcribed. All interviews were conducted on the record but not for attribution. In most instances, interviewees were with high-ranking officials, often the most senior individual in an organization actively engaged in a particular policymaking debate. Interviews included former heads of states, party leaders, cabinet ministers, senior business executives, labor leaders, and environmental leaders. This high-level access to policymaking actors allowed me to reconstruct the fine details of political debates over climate policy in each country. I provide a key to these interviews as a chapter appendix, offering generalized descriptions of each interview subject without undermining any interviewee's confidentiality.

These diverse data allow me to reconstruct climate policymaking dynamics in each case. First, I evaluate the forces shaping the emergence of climate policy proposals onto national political agendas. Second, I evaluate the political coalitions that emerged in support of and in opposition to each proposal. Finally, I trace the causal processes through which proposals were enacted or failed. In each step, my analysis recognizes the interaction between the distribution of climate policy preferences and domestic political institutions. In the most general sense, I build from the observation that institutions condition the pathways through which economic stakeholders influence the policymaking process. In turn, I evaluate whether the presence of different causal pathways to policy enactment can help explain differences in climate policy outcomes.

To probe the generalizability of this book's theory, I next assess its explanatory power on four shadow cases. To select these cases, I focus on three institutional sources of variation that my theory draws attention to: a country's degree of corporatism or pluralism, the presence of strong, institutionalized linkages between labor and left-wing parties, and the presence of strong, institutionalized linkages between capital and right-wing parties. These conditions can be represented as a 2 × 2 × 2 cube (figure 1.2). However, as this book shows, close relationships exist between business interests and right-leaning political parties in every relevant case (and between business and left-leaning parties in most cases); the absence of variation on

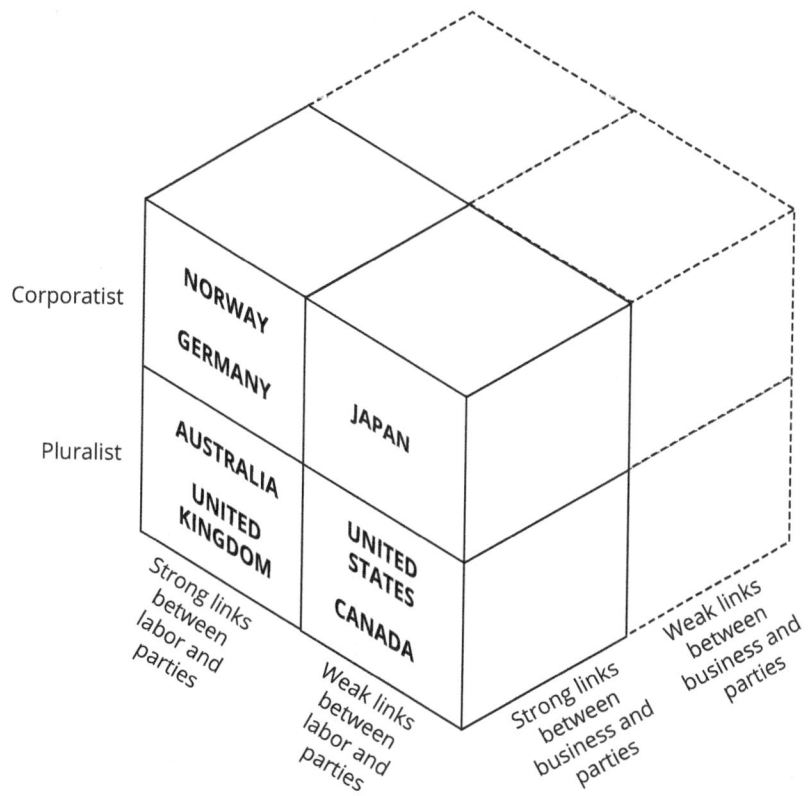

Figure 1.2
The book's primary and shadow cases, classified by sources of institutional variation that structure the double representation of carbon polluters

this dimension (see chapter 2) precludes predictive testing. Instead, I select shadow cases on only two dimensions: policymaking institutions and the presence of formal links between labor and the dominant left-leaning political coalition.

For each quadrant in figure 1.2, I choose the advanced economy with the largest annual carbon pollution level. Since these countries are the largest contributors to climate change, conditional on institutional arrangements, these are also the substantively most important cases we should want a theory of cross-national climate policymaking to explain. The corporatist shadow cases are Germany (with labor linkages) and Japan (without labor linkages). The pluralist shadow cases are the United Kingdom (with labor linkages) and Canada (without labor linkages).

Note that my primary cases were *not* selected based on this schematic. Rather, this schema is a function of variables uncovered by my primary casework. However, my three core cases can be retrospectively mapped onto this framework. Norway occupies the same cell as Germany, a corporatist country with strong links between labor and the dominant left-leaning political coalition. The United States occupies the same cell as Canada, a pluralist system with weaker political links to labor. Finally, Australia occupies the same cell as the United Kingdom, a predominantly pluralist system with strong labor links. None of the primary cases occupies the same cell as Japan.

All shadow casework was undertaken after the book's theory-building stage was complete to provide the most rigorous external test of the theory's propositions. Shadow case analyses relied primarily on secondary literatures.

Overview of the Book

The remainder of the book proceeds as follows. In chapter 2, I present the book's core theoretical claims. I contrast the trajectory of climate policymaking in Norway, Australia, and the United States to elaborate the logic of double representation. I consolidate evidence that points to the cross-cutting nature of climate policy preferences across different political and economic stakeholders, and I describe how the logic of double representation privileges the interests of carbon-dependent economic actors. I then show how cross-national variation in climate policymaking trajectories can be explained by the interaction of cross-cutting climate policy preferences and domestic institutions. I also contrast the explanatory power of my account against common alternative hypotheses. This chapter is presented before the empirical chapters that inform its analysis so that readers can make their own judgments about my theory's explanatory power when considering the empirical cases that follow. It nonetheless bears emphasizing that subsequent case chapters are not empirical tests of chapter 2; instead, they are the empirical and inferential basis of my theory.

Chapters 3 through 6 then trace in detail the trajectory of climate policymaking between 1990 and 2015 in Norway, the United States, and Australia. Together, these empirical chapters chart the complex trajectories of domestic climate reforms across three institutionally distinct countries. Norway acted early against the climate threat, but struggled to impose costs on many industrial interests. Carbon polluters blocked federal climate policy in the United States for almost two decades, but eventually the executive branch used regulations to impose significant costs on some carbon polluters during the Obama administration. Australia passed a major climate reform in 2011 that imposed a mix of consumer and producer costs; these reforms were repealed in 2014. Each of these chapters has been written so that it can also double as a self-standing narrative of climate policy debates in each country.

Chapter 3 considers the history of climate policymaking in Norway. Norway was the second country in the world to enact a carbon price, in 1991. However, efforts by climate policy proponents to increase tax rates throughout the 1990s were frustrated by a coalition of industrial unions and heavy industries that were allied with senior figures in both the Labor and Conservative parties. During the early 2000s, Norway mostly phased out its carbon tax and replaced it with an emissions trading scheme. From 2008, this scheme was harmonized with the European Union's system. Reluctant to impose significant costs on domestic producers, the Norwegian government instead subsidized carbon capture and sequestration technologies and invested in international carbon offsets.

Chapters 4 and 5 consider US climate policymaking. Chapter 4 traces US climate policy inaction through 2006. Over this period, federal efforts to enact climate reforms repeatedly stalled. The Clinton administration failed to increase the cost of carbon pollution as part of its ill-fated BTU tax proposal. While bipartisan climate policy proponents continued to press for domestic climate reforms throughout the Clinton and George W. Bush administrations, Democratic and Republican officials with ties to carbon polluting constituencies blocked all but voluntary programs.

However, beginning in the mid-2000s, the strategic climate policymaking context shifted, particularly after the landmark 2007 Supreme Court decision in *Massachusetts v. EPA*. That ruling created a viable executive branch strategy for climate policymaking in the absence of congressional approval. Chapter 5 traces the subsequent transformation of the United States from climate policy laggard to climate policy leader between 2007 and 2015. After a series of trial efforts, the House of Representatives passed a comprehensive climate reform bill in 2009; however, parallel legislative efforts stalled in the Senate. Instead, the Obama administration used the Clean Air Act to promulgate a series of sweeping carbon pollution regulations.

Chapter 6 charts climate policymaking in Australia. Early efforts to enact costly climate policies were blocked by a bipartisan coalition of industry-aligned Labor Party and Liberal Party members. By 2007, the political landscape had changed, and both Australian political parties went into the 2007 federal election proposing to establish an emissions trading scheme. The Labor government's subsequent Carbon Pollution Reduction Scheme (CPRS) died after the Liberal opposition reversed its policy support in 2009. However, after the 2011 federal election, a minority Labor government enacted an emissions trading scheme with support from Australia's Green Party. After returning to power in 2013, the Australian Liberals repealed this policy, replacing it with an abatement-purchasing approach to carbon pollution reduction.

Chapters 3 through 6 draw from inductive small-n analysis of climate policymaking across three advanced economies. Chapter 7 probes my theory's generalizability using a structured set of shadow cases: Germany, Japan, the United Kingdom, and Canada. I select these countries on the basis of their variation in policymaking institutions and left-leaning political party structure, two factors that my theory predicts should shape climate policymaking trajectories. Chapter 7 finds that this book's theory helps to explain otherwise puzzling features of the climate policymaking record in all four countries. Climate policy preferences in each country cut across preexisting political cleavages, including business organizations and labor movements. In turn, carbon-intensive economic interests in all four shadow cases enjoyed a double representation within national climate debates. In each country, political actors on both sides of the ideological spectrum with ties to carbon polluters blocked climate reforms.

Chapter 8 concludes with reflections on the book's policymaking implications. I marshal the book's evidence to argue that scholars must place more importance on the political economy of climate reforms. For instance, carbon pricing has been celebrated for its economic optimality; yet, the policy is politically suboptimal. By construction, putting an explicit price on carbon makes policy costs more salient than benefits, a feature that has been exploited by interest groups to undermine policymaking incentives. I also suggest that national distributive conflicts, not concerns over global free-riding, have been the binding constraint on global climate politics. I conclude by considering the conditions under which the double representation of carbon polluters can be disrupted.

More broadly, this book emphasizes how climate policy is fundamentally redistributive. It will require domestic political systems to redirect flows of economic resources in politically consequential ways. And global mitigation success will depend on climate reformers' ability to win dramatic distributive conflicts at home. By unpacking the logic of climate policy conflict, this book offers a contribution to their efforts.

A Note on Data Transparency

Both qualitative and quantitative scholars have increasingly focused on issues of data transparency. This book makes its sources as transparent as possible by providing a Transparency Appendix (TRAX) that follows American Political Science Association guidelines. This TRAX is available online at https://doi.org/10.5064/F6GYLSON, and also on the author's personal website. The TRAX offers more detail on the book's methodological approach and offers additional details on the book's sources. This includes expansions of many of this text's endnotes; an explanation precedes the Notes section at the end of this book. I also make my sources public, where possible, by depositing archival, media, and web materials at the Syracuse University Qualitative Data Repository (QDR), also available at https://doi.org/10.5064/F6GYLSON. My informant interviews were conducted under the promise of confidentiality, which prevents my including interview transcripts in this QDR. However, to increase transparency, I link specific inferences throughout this text to their interview source using the numerical codes presented in this chapter's appendix.

Chapter Appendix

Individuals were assigned numbers based on the order of interview transcription; references to informant interviews use these numbers. Individuals are identified on the basis of their most senior relevant position during relevant climate policymaking debates. For some individuals, this was not their position at the time of my interview.

- Interview 1, senior business lobby official, Washington, DC, October 16, 2013
- Interview 2, senior farm lobby official, Washington, DC, October 16, 2013
- Interview 3, US senior union official, email correspondence, October 22, 2013
- Interview 4, senior business lobby official, Oslo, November 7, 2013
- Interview 5, elected official, Oslo, November 7, 2013
- Interview 6, elected official, Oslo, November 6, 2013
- Interview 7, academic, Oslo, November 8, 2013
- Interview 8, two senior bureaucrats, Oslo, November 8, 2013
- Interview 9, senior union official, Washington, DC, October 18, 2013
- Interview 10, senior business lobby official, Oslo, November 11, 2013
- Interview 11, two senior union officials, Washington, DC, October 18, 2013
- Interview 12, senior environmental leader, Oslo, November 11, 2013
- Interview 13, senior policy advisor to political party, Oslo, November 12, 2013
- Interview 14, senior bureaucrat, Oslo, November 12, 2013
- Interview 15, two mid-level union policy advisors, Oslo, November 13, 2013
- Interview 16, senior religious official, Oslo, November 13, 2013
- Interview 17, mid-level policy advisor to political party, Oslo, November 15, 2013

- Interview 18, mid-level bureaucrat, Oslo, November 13, 2013
- Interview 19, academic, Oslo, November 13, 2013
- Interview 20, mid-level agency official, Oslo, November 15, 2013
- Interview 21, elected official, Oslo, November 15, 2013
- Interview 22, senior union official, Oslo, November 18, 2013
- Interview 23, mid-level business lobby official, Oslo, November 15, 2013
- Interview 24, energy consultant, Oslo, November 18, 2013
- Interview 25, two senior union officials, Oslo, November 18, 2013
- Interview 26, mid-level bureaucrat, Oslo, November 20, 2013
- Interview 27, mid-level bureaucrat, Oslo, November 11, 2013
- Interview 28, two senior environmental officials, Oslo, November 18, 2013
- Interview 29, two Norwegian farm lobby officials, Skype, November 27, 2013
- Interview 30, mid-level business lobby official, Oslo, November 19, 2013
- Interview 31, senior policy advisor to political party, Oslo, November 19, 2013
- Interview 32, senior policy advisor to political official, Oslo, November 20, 2013
- Interview 33, two mid-level bureaucrats, Oslo, November 20, 2013
- Interview 34, mid-level bureaucrat, Oslo, November 20, 2013
- Interview 35, senior union official, Oslo, November 20, 2013
- Interview 36, senior union official, Oslo, November 21, 2013
- Interview 37, senior business lobby official, Oslo, November 21, 2013
- Interview 38, senior policy advisor to political party, Oslo, November 21, 2013
- Interview 39, senior policy advisor to political party, Oslo, November 22, 2013
- Interview 40, senior union official, Oslo, November 22, 2013
- Interview 41, senior policy advisor to political party, Oslo, November 22, 2013
- Interview 42, Australian energy consultant, Skype, May 9, 2011

- Interview 43, senior political advisor to elected official, Sydney, February 6, 2014
- Interview 44, government advisor, Canberra, February 10, 2014
- Interview 45, senior policy advisor to political party, Canberra, February 10, 2014
- Interview 46, elected official, Canberra, February 11, 2014
- Interview 47, elected official, Canberra, February 11, 2014
- Interview 48, Australian elected official, phone, February 13, 2014
- Interview 49, senior agency official, Sydney, February 6, 2014
- Interview 50, senior bureaucrat, Sydney, February 7, 2014
- Interview 51, senior bureaucrat, Canberra, February 11, 2014
- Interview 52, mid-level bureaucrat, Canberra, February 11, 2014
- Interview 53, elected official, Canberra, February 18, 2014
- Interview 54, senior business lobby official, Melbourne, February 19, 2014
- Interview 55, two senior bureaucrats, Canberra, February 18, 2014
- Interview 56, senior bureaucrat, Canberra, February 13, 2014
- Interview 57, senior Australian bureaucrat, Skype, February 13, 2014
- Interview 58, senior union official, Melbourne, February 24, 2014
- Interview 59, elected official, Canberra, February 13, 2014
- Interview 60, senior environmental official, Canberra, February 18, 2014
- Interview 61, energy consultant, Melbourne, February 25, 2014
- Interview 62, mid-level environmental official, Melbourne, February 20, 2014
- Interview 63, mid-level bureaucrat, Melbourne, February 26, 2014
- Interview 64, farm lobby official, Melbourne, February 21, 2014
- Interview 65, senior bureaucrat, Melbourne, February 21, 2014
- Interview 66, senior business lobby official, Melbourne, February 21, 2014
- Interview 67, senior business lobby official, Sydney, February 27, 2014
- Interview 68, senior policy advocate, Canberra, February 13, 2014
- Interview 69, senior bureaucrat, Sydney, February 27, 2014

- Interview 70, senior environmental official, Sydney, February 26, 2014
- Interview 71, senior environmental official, Sydney, February 28, 2014
- Interview 72, senior union official, Sydney, March 3, 2014
- Interview 73, senior union official, Sydney, March 3, 2014
- Interview 74, senior policy advocate, Sydney, March 12, 2014
- Interview 75, senior policy advocate, Melbourne, March 11, 2014
- Interview 76, senior environmental official, Skype, March 5, 2014
- Interview 77, senior business leader, Melbourne, March 7, 2014
- Interview 78, senior environmental official, Sydney, February 27, 2014
- Interview 79, mid-level bureaucrat, Oslo, November 21, 2013
- Interview 80, senior bureaucrat, Skype, March 26, 2014
- Interview 81, economists, Skype, May 9, 2014
- Interview 82, Australian elected official, phone, May 20, 2014
- Interview 83, senior policy advisor to elected official, phone, May 20, 2014
- Interview 84, senior union official, Washington, DC, August 13, 2014
- Interview 85, policy advocate, Cambridge, MA, August 26, 2014
- Interview 86, senior bureaucrat, Washington, DC, September 9, 2014
- Interview 87, senior policy advisor to elected official, Washington, DC, September 11, 2014
- Interview 88, senior policy advisor to elected official, Washington, DC, September 12, 2014
- Interview 89, senior bureaucrat, Washington, DC, September 12, 2014
- Interview 90, mid-level environmental official, Washington, DC, September 15, 2014
- Interview 91, mid-level environmental official, Washington, DC, September 15, 2014
- Interview 92, senior US policy advisor to elected official, phone, October 2, 2014
- Interview 93, US environmental leader, phone, October 10, 2014
- Interview 94, senior policy advisor to elected official, phone, October 23, 2014
- Interview 95, elected official, Washington, DC, September 17, 2014

- Interview 96, senior policy advisor to elected official, Washington, DC, September 16, 2014
- Interview 97, mid-level US bureaucrat, phone, November 7, 2014
- Interview 98, elected official, Cambridge, MA, January 28, 2015
- Interview 99, senior policy advisor to elected official, Washington, DC, September 16, 2014
- Interview 100, elected official, New Haven, CT, April 14, 2015
- Interview 101, senior US political staffer, phone, March 28, 2018

2   The Logic of Double Representation

New issues join the political agenda in diverse ways. Some issues emerge as the result of multigenerational efforts to organize around new political demands, for instance the abolition of slavery (Sinha 2016). Other issues gain prominence as the composition of the voting electorate changes: in many countries, social spending increased as women's enfranchisement expanded (Aidt and Dallal 2008; Iversen and Rosenbluth 2010). Still other issues develop quietly outside the spotlight before catapulting to political relevance as the result of political, social, or economic crises.

The climate threat disrupted politics along a different path when a new scientific discovery exposed differences in the material interests of otherwise aligned political actors. For the greater part of the twentieth century, the greenhouse gas emissions associated with economic activity did not constrain economic development. Odorless, invisible, and posing few known risks, carbon pollution levels increased across advanced economies.[1] This process of "carbon lock-in" entrenched carbon-intensive technologies at the heart of the modern economy, from systems of transport to energy production to manufacturing (Unruh 2000). In Norway, carbon-intensive manufacturing facilities were central to postwar economic growth; offshore oil and gas later emerged as critical national industries. In Australia, coal and gas extraction and combustion were central to modern economic development. In the United States, oil, gas, and coal deposits laid the foundation for regional economic development. In all three countries, household transportation and energy consumption grew to depend on fossil fuel combustion.

The process that made carbon-intensive technologies central to economic growth in industrialized economies created many economic actors whose profitability depended on their ability to release carbon pollution cost-free into the global atmosphere. These actors included the businesses whose capital was invested in carbon-intensive economic activities, as well as the workers whose jobs depended on these businesses. Of course, not all industries produced carbon emissions; other capital was invested in less carbon-intensive economic activity and other workers developed skills aligned with these lower-carbon sectors.

Beginning in the 1980s, climate science advances pushed climate change onto the political agenda. The growing realization that carbon pollution threatened human welfare also threatened the economic actors who profited from releasing carbon pollution into the atmosphere. The scientific discovery of climate change thus exposed latent differences in the material interests of preexisting political coalitions. Abruptly, coalitions of capital and labor with otherwise similar policy preferences found themselves divided by climate policy.

This exogenous reveal of latent policy preferences sets climate politics apart from other contemporary political challenges. In many issue domains, political coalitions adapt incrementally to new policy demands. As issue salience increases, small initial differences in coalition membership can amplify over time to create dramatic political splits (Stimson 2015). In these instances, preferences tend to develop endogenously as a function of existing coalition membership. By contrast, the distribution of climate policy preferences was structured around economic and political decisions that predated the climate threat's scientific discovery.

This chapter advances a theory of climate policymaking that emphasizes the interaction between these cross-cutting climate policy preferences and domestic political institutions. First, I describe similarities in climate policy preference distributions among interest groups and political parties across countries. I describe how, because climate policy preferences cut across existing political cleavages, policy opponents became embedded within both left- and right-leaning political coalitions. I describe this dynamic as the *double representation of carbon polluters*, since political actors on both sides of the ideological spectrum came to represent the interests of carbon-dependent economic actors within climate debates. The double representation of carbon polluters helps account for climate policy inaction across time and countries because it compounds status quo biases in public policymaking. With opponents represented in most major parties, it is difficult for proponents to change current policy, no matter which party controls the government. I argue that double representation is a central feature of all climate policy conflict across advanced economies.

At the same time, this book's empirical chapters reveal substantial cross-national and over-time variation in policy enactment records. To understand this variation, we need to understand the interaction between the distribution of climate policy preferences—that is, the double representation of carbon polluters—and each country's domestic policymaking institutions. The second half of this chapter outlines how differences in climate policymaking trajectories reflect institutional differences in carbon polluter influence on policy design. Domestic political institutions can either reinforce or moderate polluter representation by structuring polluters' access to policy design venues. This access is shaped by policymaking institutions, for instance through systematic links between economic stakeholders and policymakers; and through political organizations, for instance through historically contingent links between labor or business associations and political parties. I show how the degree of polluter access to the policy process shapes the pathway through which opponents attempt to block climate reforms, which in turn influences policy content and timing.[2]

The Cross-Cutting Distribution of Climate Policy Preferences

The first step in explaining climate policy conflict involves understanding how climate policy preferences cut across preexisting economic and political coalitions. Consider a simple model in which national political actors at time t are sorted into two camps on the basis of their preference for an economic policy bundle. Assume that you can place left-leaning parties and labor unions at one end of this continuum and right-leaning parties and business organizations at the other end of this continuum.

Now suppose that, at time $t + 1$, an exogenous shock exposes latent heterogeneity in actor preferences. This shock introduces a second, cross-cutting political dimension. Members of both left- and right-leaning political coalitions now find themselves with divided preferences over the new issue, even as they maintain homogenous preferences with respect to the old issue. Subsequent political and ideological debates may transform this new issue cleavage. Political actors may sort by party around the new issue. They may construct political and social understandings of the issue dimension in an effort to expand their coalition size. Or the new cleavage may prove durable if political institutions entrench the presence of cross-cutting voices within existing political organizations. Analyzing the politics of a cross-cutting issue's emergence thus requires attention to both the initial distribution of policy preferences and the pathways through which institutions stabilize or reshape these preference distributions through time.

Now consider the initial distribution of climate policy preferences. Across advanced economies, the climate threat's emergence divided preexisting coalitions of labor and capital, creating new cleavages within economic coalitions that otherwise enjoyed shared material interests. This distribution of interests is presented schematically in figure 2.1. Distributional losers from climate policy included both workers whose jobs depended on carbon-intensive industries as well as business owners whose capital was invested in these carbon-intensive sectors. Analogously, economic winners

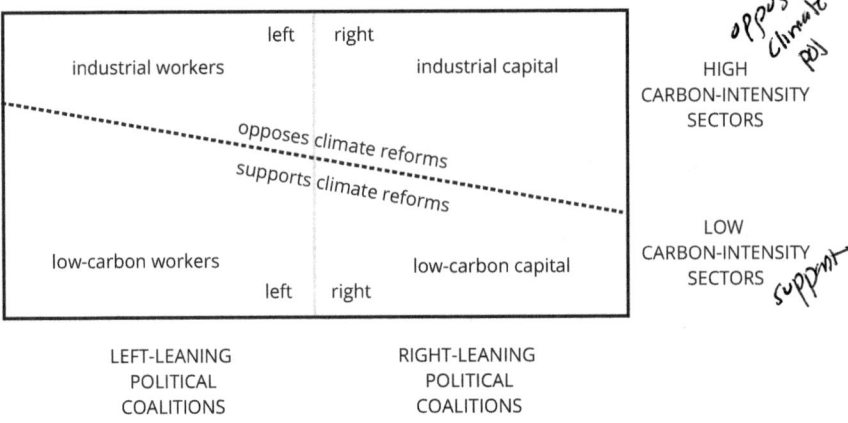

Figure 2.1
Climate change as a cross-cutting issue. Economic winners include both businesses and workers in low-carbon economic sectors, who therefore support climate reforms. Economic losers from climate policy include both workers whose jobs depend on carbon-intensive industries and carbon-intensive business owners, who therefore oppose climate reforms. Cross-cutting preferences can still be subject to an ideological gradient, with a larger support coalition on the left despite the presence of supporters and opponents on both sides of the ideological spectrum.

included both businesses and workers in low-carbon economic sectors. Chapters 3 through 6 provide extensive evidence of these cross-cutting cleavages in climate policymaking debates in Norway, Australia, and the United States. Coal and oil industries, and their economic dependents, mobilized against climate reforms in all three countries. At the same time, economic winners such as clean energy, climate-vulnerable industries such as insurance, and low-carbon knowledge and service sectors all championed reforms. Responding to the same economic incentives as business owners, labor unions also intervened on both sides of climate reform debates, splitting public sector "white collar" unions from industrial "blue-collar" unions.

With both labor and business communities divided in their climate policy preferences, political parties representing these economic constituencies also struggled to manage within-party climate policy tensions. Before the climate threat's emergence, carbon dependence was not a salient source of political conflict. However, as chapters 3 through 6 emphasize, climate change subsequently created splits among both left- and right-leaning political coalitions. In the United States, successive presidential administrations, from Reagan to George W. Bush, experienced sharp internal divisions over climate reforms. While a partisan gradient always characterized climate preferences in the US Congress, a vocal Republican minority embraced climate policymaking. At the same time, a vocal Democratic minority aligned with workers and industries in carbon-intensive regions to undermine policy action. In Australia, early Labor cabinets split over the climate issue, a cleavage that later defined factional climate policy conflict within both the Labor and Liberal parties. In Norway, both Labor and Conservative parties refused to impose costs on domestic industrial actors; the costliest policy proposals were advanced by small parties from across the ideological spectrum that lacked strong ties to carbon-intensive constituencies.

In this book, I argue that this cross-cutting distribution of climate policy preferences is the single most distinguishing feature of climate policy conflict across advanced economies. Importantly, this argument is distinct from the literature on postmaterial values (cf. Inglehart 2015). Postmaterialist scholars also argue that environmental debates generate cross-cutting political preferences. However, they argue these preferences reflect nonmaterial values and beliefs. By contrast, the cleavages highlighted in this book are deeply material. They emerged because the climate threat fractured within-coalition material interests, not from the introduction of new value dimensions at particular levels of economic development.

Climate policy scholars have not always recognized this feature of climate politics, often assuming that climate policy preferences correlate with left-right ideology. For instance, Daugbjerg and Svendsen (2001) describe carbon taxation as fundamentally redistributive, arguing that preferences for carbon pricing naturally follow from ideological posture. Other scholars highlight links between conservative activists and climate skepticism (Oreskes and Conway 2010; McCright and Dunlap 2011; Brulle 2014), as well as low climate policy support among global conservatives (Tranter 2011; Clements 2012).

Yet, these accounts face two weaknesses. First, in most countries for most of the recent past, there have been significant factions of climate supporters and climate opponents among both left- and right-leaning political coalitions. While climate support is structured along an ideological gradient, with stronger climate policy support on the left, a vocal minority within both left-leaning and right-leaning coalitions hold countervailing preferences. To accurately characterize climate policy conflict, we must recognize these within-coalition cleavages.

Second, this perspective is biased towards recent US political dynamics. When President Clinton quietly floated a carbon tax in 1993, a bipartisan coalition mobilized in opposition, one defined by regional material interests rather than partisan affiliation. Likewise, emissions trading was initially a Republican-sponsored idea, opposed by such climate policy advocates as Vice President Al Gore.

Over the past decade, US partisan sorting has created stronger alignment between climate policy preferences and party ideology.[3] But this alignment should not be viewed as the necessary result of ideological coherence between anti-climate and conservative beliefs. Instead, the relationship between ideology and environmentalism varies by country (Nawrotzki 2012; Jahn 2016; McCright, Dunlap, and Marquart-Pyatt 2016; Harring and Sohlberg 2017). As I argue, extreme partisan sorting of US climate policy preferences simply reflects the absence of institutional features that, in other countries, stabilized cross-cutting climate policy preferences over time.

The Double Representation of Carbon Polluters

In principle, cross-cutting climate policy preferences could work to the advantage of either climate policy proponents or opponents. For instance, policy advocates might find it easier to assemble reform coalitions if they can rely on support from pro-climate factions within their own political party *and* opposition parties. Such cross-cutting policy preferences might also enhance the durability of climate reforms by making it difficult to assemble policy repeal coalitions. Yet, I find little empirical evidence for either possibility. Pro-climate coalitions on both the right and left in Australia, Norway, and the United States have persistently struggled to enact costly climate reforms, even when reform coalitions had bipartisan or multipartisan membership. At best, cross-cutting preferences helped proponents force climate policies onto national political agendas. Mostly, however, they systematically privileged policy opponents by reinforcing status quo biases in public policymaking. Cross-cutting policy preferences have thus allowed carbon-dependent business and labor to capture the climate policymaking process.

In most political systems, proponents need a pro-reform coalition at each step in the political decision making process, including political coalitions across each government branch. By contrast, policy opponents need only a single veto point to block ambitious climate reforms (Baumgartner et al. 2009). Since the initial status quo in all countries was climate policy absence, policy proponents inherently faced higher organizational demands than opponents.

Cross-cutting policy preferences compounded status quo biases because, in their presence, climate reform opponents usually had access to at least one veto point. Carbon-dependent economic interests were not limited to a narrow set of political actors; instead, they became organically embedded within most major political coalitions and many important interest groups. Consequently, the political allies of policy losers were positioned to undermine costly reforms, independent of which political party controlled government. I describe this dynamic as the double representation of carbon polluters in climate politics. Carbon polluters enjoy double representation

Dauble rep.

because cross-cutting climate policy preferences disperse reform opponents across economic interest groups and political parties.

Chapters 3 through 6 offer myriad examples of double representation in action. Vice President Gore privately promoted carbon taxation as part of the Clinton administration's 1993 deficit reduction package. However, the policy was viewed as politically infeasible because his copartisan and lifelong coal champion, West Virginia Senator Robert Byrd, held a de facto veto as chair of the Senate Committee on Appropriations. During the late 2000s, US cap and trade proponents had to accommodate the concerns of senior Democratic politicians with carbon-intensive constituencies. In Australia, the Liberal Party narrowly ousted pro-climate leader Malcolm Turnbull in 2009 after he negotiated with a Labor government over a federal carbon pricing system. In Norway, industrial Labor leaders systematically undermined recommendations by a 1995 Green Tax Commission to protect onshore industry from climate policy costs. Concurrently, peak unions and business associations collaborated with Labor and Conservative legislators to *carbonize* the Norwegian electricity system by constructing new gas-fired power plants.

At the same time, my focus on "double" representation is not intended to reduce climate politics to a binary conflict between only two opposing groups. As this book's empirical case studies emphasize, representation comes in myriad forms. At some moments, carbon-intensive interests directly control policymaking. Sometimes they are able to influence a political party's policymaking agenda without controlling the policymaking process. Sometimes they may indirectly shape political debate by mobilizing conflict into the public sphere. Moreover, interest groups don't stay in their ideological lanes. Union actors can influence right-leaning debates; business actors interact with left-leaning politicians who are sensitive to industrial jobs in their constituencies. And, of course, the mechanisms of interest group representation are also shaped by institutional context. Yet, through all of these processes, the concept of double representation draws attention to the ways in which carbon polluters enjoy influence across the political spectrum and diverse stakeholder coalitions.

Connections between carbon-dependent economic interests and political actors also exist because of the fluid interchange of personnel between political organizations on all sides of the ideological divide and carbon-intensive economic groups. The leader of the Australian Labor Party until

2019, Bill Shorten, was head of the industrial Australian Workers Union from 2001 through 2007. In Norway, Labor politician Stein Lier Hansen was the State Secretary for the Environment from 2000 to 2001; in 2004, Hansen became head of Norsk Industri, which represents Norwegian carbon-intensive process industries. In the United States, the chair of the Council of Environmental Quality during the George W. Bush administration, Phillip Cooney, was previously an American Petroleum Institute lobbyist.

The influence of climate reform opponents is further enhanced by the path-dependent importance of many carbon-intensive economic actors to national and regional economic development trajectories (Kasa 2000). Consequently, carbon-intensive economic interest groups often hold a more powerful position than their current economic importance might suggest. Consider the case of Norway, where climate reforms divided pro-climate public sector unions from industrial unions concerned with protecting carbon-dependent jobs.[4] Despite a small ecosystem of peak labor organizations, the peak labor union LO (*Landsorganisasjonen i Norge*) enjoyed the strongest institutional ties to the Labor Party and senior government leaders (Allern 2010). LO consistently sided with its industrial membership on domestic climate policy, opposing costly policies for Norwegian industry.[5] As one industrial union official pointed out: "even if we are not the largest [member union], industry is always perceived as very important by LO."[6] By contrast, the peak labor association UNIO consistently supported costly climate reforms but only dates to 2001 in its current organizational form. The association lacked LO's privileged access to the Labor Party and instead relied on LO to represent its interests within Labor Party debates. The result was a marginalization of UNIO's pro-climate positions within party discussions. I find, across all three book cases, that the growing economic importance of low-carbon actors has not automatically translated into increased political influence.

In short, double representation is a basic feature of climate policy conflict across advanced economies. Double representation emerged because climate change exposed cross-cutting cleavages within left- and right-leaning political and economic coalitions. In turn, to understand how double representation shaped climate policy timing and content, we must examine how these cross-cutting cleavages were reinforced or moderated by institutions over time.

Synchronization in Climate Proposal Timing

The timing of climate policy enactment varies considerably. However, there has still been substantial empirical convergence around *when* climate policy proposals were introduced across countries. This is the result of global events that focused national policymakers in many countries on the climate threat. For instance, the 1988 Toronto conference on global warming catalyzed debates over carbon pollution reduction targets within the Australian and Norwegian cabinets and the US executive branch. Between 1990 and 1994, all three countries considered carbon taxation, although the idea gained traction in Norway alone. Later, Norway, the United States, and Australia all debated emissions trading during the 2000s. In short, all three countries responded to the climate threat with loosely similar instruments at similar times. The features of these policy instruments varied substantially, including cost levels and cost distributions. Still, divergence in policy enactment timing has tended to reflect variation in enactment success, not variation in policy proposal timing.

This cross-national convergence in proposal timing also depends on cross-cutting climate preference distributions. When climate change enters public debate as a result of UN climate conferences, Intergovernmental Panel on Climate Change (IPCC) climate science assessment reports, and the like, national climate advocates in many countries can use the policymaking windows offered by these events to advance reform proposals. Because proponents are distributed across the political system, proposal timing is rarely constrained by which party holds power during a global focusing event. This leads to the apparent synchronization of climate policy proposals. However, advocates' ability to propose policies—and thus generate cross-national convergence in proposal timing—does not imply convergence in enactment timing; the agenda-setting capacity of proponents does not disrupt double-represented opponents' persistent ability to block costly reforms.

Other research corroborates my finding that the timing of climate reform proposals are synchronized across countries as a result of global focusing events. Time series studies of global media coverage show common spikes in climate change attention (Boykoff 2011). Global climate opinions also follow convergent trends (Scruggs and Benegal 2012), which can lead to similarities in the content of policy proposals across countries. Transnational relationships between leading climate opponents and proponents reinforce these patterns. Senior Australian political officials repeatedly invoked Al Gore's film, *An Inconvenient Truth*, to explain domestic climate change salience. Prominent American climate activists James Hansen and Bill McKibben were both awarded the Norwegian Sofieprisen for sustainable development. And both Republican Senator Chuck Hagel and Democratic Congressman John Dingell attended an Australian climate skeptics conference before Kyoto (Hamilton 2007). Through these networks, climate policy ideas have been continuously exchanged across advanced economies.

Of course, idiosyncratic events—from elections to local weather conditions—also shaped climate policymaking in all three countries. Extreme US weather in 1988 increased the media's attention to congressional testimony on the climate threat. A major Australian drought focused political attention on climate change during the 2007 election cycle. Yet, I still find that cross-national variation in climate policy enactment is largely a function of enactment rates. Consequently, this book gives particular consideration to differences in the causal pathways through which reforms are enacted after they are proposed.

Pathways to Policy Enactment

My empirical analysis finds that climate policies are enacted through one of two causal pathways, and that these pathways help explain variation in climate reforms. In one pathway, carbon-dependent economic actors enjoy substantial influence over climate policy content, reinforcing their double representation. A second pathway occurs when carbon polluters lack direct influence on climate policy design; under some circumstances, this moderates double representation. I illustrate the core features of each pathway in figure 2.2.

Along the first pathway, carbon-intensive economic interests directly influence climate policy design. Even if cross-cutting policy preferences generate opportunities for climate policy proposals to reach the political agenda, advocates must design these policies in collaboration with policy losers. Consequently, policy proposals introduced along the first pathway shield carbon-dependent economic actors from concentrated costs. In the absence of a significant policy threat, carbon-dependent economic actors lack incentive to mobilize policy conflict into the public domain (cf. Schattschneider 1960). Thus, these climate policies are often enacted with

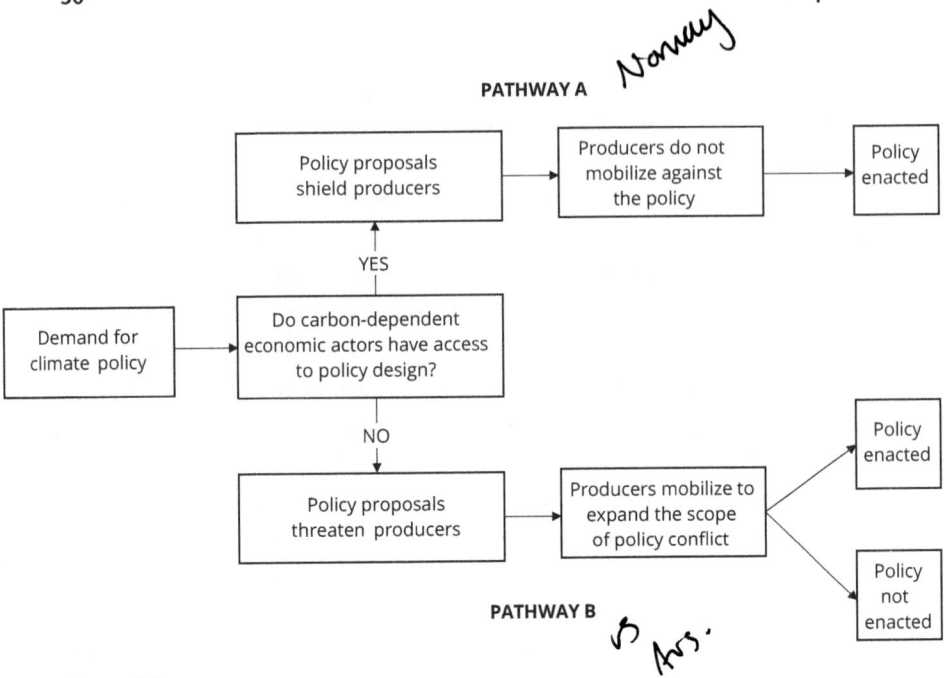

Figure 2.2
Two pathways to national climate policy enactment

minimal political controversy. An archetypal example of climate policy-making along this pathway is Norway's 1990 carbon tax enactment. The tax was introduced with significant exemptions for Labor-allied onshore industries, and so did not face substantial opposition.

Along the second pathway, carbon-intensive economic interests lack direct institutionalized access to the policy process. As a result, policy proposals that are introduced can include concentrated costs on carbon-dependent economic actors, which threatens the economic viability of major carbon polluters. They respond with efforts to mobilize climate policy conflict into the public domain, either through direct legislative lobbying or indirect efforts to shape public preferences. This weakens the electoral or legislative incentives associated with policy enactment. Often, climate reforms that follow this pathway fail, in part because of this mobilized opposition. For instance, in the United States, both the BTU tax proposal in 1993 and emissions trading legislation in 2009 failed after climate policy opponents expanded the scope of policy conflict. However, under some conditions, climate policies are still enacted despite opponent mobilization. For example, the Obama administration persisted in its efforts to impose costs on carbon pollution through the Clean Air Act, despite intense mobilization from labor and business groups who faced serious economic costs from the administration's efforts.

This second pathway is consistent with research that emphasizes the importance of consumer cost salience for climate reforms (Raymond 2016; Rabe 2018). It is also consistent with conditional accounts of business influence that emphasize efforts to shift public opinion through business-funded think tanks and messaging campaigns (Smith 2000; Oreskes and Conway 2010; Layzer 2012). In many settings, public campaigns against climate science may signal the tenuousness of interest group influence rather than special interest dominance of the climate policymaking process.

Determinants of the Policy Pathway

The pivot between these two pathways is the relative influence of carbon-dependent actors on climate policy design. Where carbon polluters have access to policy design, the first pathway tends to prevail. Where their access is limited, the second prevails. To understand variation in climate policy enactment, we must consequently understand variation in this interest group access. This book's empirical cases highlight two institutional factors that shape carbon polluter access to policy design: one, a country's relative pluralism or corporatism; and two, the presence of organizational links between political parties and carbon-dependent economic stakeholders.

First, national policymaking systems structure carbon polluter access to policy design. In corporatist systems, carbon-dependent unions and businesses enjoy guaranteed access to government decision makers. For example, Norwegian policy reforms often begin with expert commissions that include representatives from peak labor and business associations. Policy proposals are then developed in close consultation with economic stakeholders.[7]

By contrast, in pluralist policymaking settings, carbon polluter access to policy design is contingent on having political allies in government since there are no institutional venues that guarantee particular interest groups access to decision-making. When allies control the political agenda, policy proposals will closely reflect carbon polluter interests. For example, fossil fuel interests strongly guided executive branch climate policy during the George W. Bush administration. However, this access can reverse when climate reform proponents seize control of the policymaking agenda. In pluralist settings, carbon-dependent economic actors can occasionally find themselves shut out of climate policy design.

Second, formal institutional links between political parties and economic interest groups shape interest group access to policy design. This variation is more dramatic on the left. In Norway and Australia, the dominant left-leaning political coalition emerged as an outgrowth of national labor movements in the early 1900s; union officials continue to enjoy privileged positions within party structures. In Norway, union officials sit on the Central Executive Committee (*Sentralstyret*) of the Labor Party and they consult with government officials in weekly meetings of the party–union Cooperation Committee (*Samarbeidskomiteen*).[8] These meetings create a privileged venue for Norwegian industries to influence policy design (Kasa 2000).

Australia has lower union density than Norway; yet, because the Australian Labor Party emerged as the labor movement's political arm, unions retain significant influence through personal relationships between party leaders and union officials.[9] Many party officials have personal union ties. By contrast, the US Democratic Party has never been a "labor" party although it is associated more with workers' interests than the Republican Party is. As a result, the US labor movement lacks the institutionalized influence within the Democratic Party that labor movements in Norway and Australia enjoy.

Less cross-case variation exists in links between carbon-intensive economic actors and right-leaning political coalitions. In Norway, peak business associations hold traditionally strong ties to the Norwegian Conservative Party (*Høyre*). In Australia, groups such as the Australian Chamber of Commerce and Industry are viewed as political extensions of the Liberal Party.[10] In the United States, state-level Republicans filed legal briefs directly authored by the energy industry in a coordinated effort to weaken Obama-era federal energy and climate regulation.[11] This constant business influence exists despite variation in the formal institutional links enjoyed by business actors, reflective of the structural power of business across advanced economies. Because businesses provide employment to the public, election-seeking leaders are sensitive to business needs on both the left and right (cf. Lindblom 1982).

The ties between carbon polluters and right-leaning political coalitions are consequential to climate policymaking in all three cases, and they help explain the failure to impose costs on many carbon-intensive businesses. However, the absence of meaningful variation in the influence of business on party politics means this factor is less useful for explaining between-country differences in reform timing and content. In other words, links between carbon-intensive capital and political actors help explain climate policy inaction, but they are less central to explanations for cross-national differences in climate policy action.

Table 2.1 summarizes differences in national political institutions across these dimensions for the book's three cases. (These three dimensions also structure shadow case selection in chapter 7.) This variation leads to different constraints on carbon-intensive stakeholders' double representation. Of the three countries, carbon polluter access to policymaking is the most stable in Norway, reinforced by corporatist institutions and close ties among business, labor, and political parties. In Australia, the pluralist policymaking system sometimes disrupts carbon polluter access to policy design; however, institutional ties between political parties and carbon-intensive interest groups continue to guarantee polluters a voice, no matter if the left or right controls government. In the United States, a pluralist policymaking system creates space for climate policy proponents to propose costly policies for carbon polluters. While close ties remain between the business community and political parties, weaker ties between the US labor movement and the Democratic Party sometimes creates space to disrupt carbon polluters' double representation.

Institutional variation does not deterministically shape policy outcomes; thus, pluralist and corporatist systems don't impose a fixed level of carbon polluter access to policy design. Instead, institutional variation shapes the probability that particular carbon-dependent economic interests will

Table 2.1

Institutional variation across cases

Dimension	Australia	Norway	United States
Policymaking institution	Pluralism	Corporatism	Pluralism
Ties between labor and left-leaning coalitions	Strong	Strong	Weak
Ties between business and right-leaning coalitions	Strong	Strong	Strong

enjoy policy design access. For example, Norwegian climate policy proponents have occasionally pushed reforms against major carbon producers' preferences, as was the case during the late 1990s when the Bondevik government attempted to expand the sectoral coverage of the country's carbon tax. However, these efforts systematically failed.

Institutional context does dictate the *variability* of carbon polluter access. In corporatist systems, the access of carbon-intensive economic actors is stable over time because the system is designed to ensure major economic stakeholders always have a seat at the table. By contrast, in pluralist systems, the access of carbon-intensive economic interests waxes and wanes. When allies are in power, the sector may enjoy deeply privileged access to policy design, as was the case for the oil industry during the George W. Bush administration (and most carbon polluters during the Trump administration). This access can be more dramatic than in corporatist systems because polluter influence is not moderated by pressure to compromise with other economic or environmental stakeholders. Conversely, when climate policy advocates are in power, carbon polluters may sometimes find themselves with minimal influence.

This variability has consequences for patterns of interest group mobilization around climate change. In corporatist systems, carbon polluters' constant access to policy design moderates future policy threats and reduces incentives for carbon polluters to preemptively mobilize policy conflict into the public domain. In pluralist systems, carbon polluters' direct access varies from high to low depending on government composition. This instability ensures that *future* climate policy threatens polluters even during periods when carbon-intensive actors control the political agenda. This creates incentives for carbon polluters to mobilize conflict into the public domain to reduce electoral incentives to pass climate reforms. Consequently, climate change becomes more highly politicized.

Explaining Variation in Climate Policy Timing and Content

Contrasting climate policymaking records in Australia, Norway, and the United States, I argue that national institutions shape carbon polluter access to climate policy design. Cross-cutting preferences still shape distributive climate conflict in all three countries; what differs is the degree to which institutions stabilize the expression of these preferences over time.

Comparing the institutional contexts in Norway, Australia, and the United States suggests a series of propositions that outline how the interaction between climate policy preference distributions and national institutions shape policy timing and content. These propositions describe my theory's scope conditions: when and under what circumstances should we expect that different causal pathways to climate policy enactment realize. Collectively, they cover each dimension introduced in table 1.2: policy timing, instrument choice, absolute cost levels, and relative cost distribution.

> Proposition 1—Policy Timing: Climate policy is enacted earlier in systems where the double representation of carbon polluters is strongly reinforced by political institutions.

When the double representation of carbon polluters is strongly reinforced by domestic institutions through time, carbon polluters have more reliable access to climate policy design. Policy proposals do not impose threatening costs on producers, and policy enactment proceeds through the first pathway. Importantly, once introduced, policies are likely to be enacted without significant political conflict. By contrast, where carbon polluters lack guaranteed policy design access, more threatening policies are sometimes introduced onto the political agenda. Carbon-dependent economic interests mobilize against these policies to undermine legislative and electoral incentives for policy enactment. Facing this opposition, many policy proposals fail. Proponents must then wait for a new policymaking window to emerge, often years later. In this way, the politicization of climate policy and the mobilization of policy opponents delay enactment timing along the second pathway.

This proposition implies that the content of policies enacted earlier differs from the content of policies enacted later. Proposals that are enacted early are much more likely to have followed the first pathway, meaning they were written in collaboration with carbon-intensive actors and thus did not impose threatening costs on carbon polluters. This was the case in Norway, where early policy enactment depended on domestic policies that shielded politically connected carbon polluters from costs. Conversely, policy inaction in other countries reflected stronger ongoing policy threats (even if unrealized) for carbon-dependent economic actors. These policies could sometimes be designed with weaker input from carbon polluters. Thus, when climate policies were eventually enacted in pluralist cases (United States, Australia), they sometimes included higher costs for carbon polluters.

Proposition 2—Policy Content: Double representation of polluter interests skews domestic climate reforms toward carrots and away from sticks.

While double representation privileges the policymaking influence of distributional losers, climate policy supporters are also distributed across political coalitions. Consequently, climate policy demands by proponents must be reconciled with co-partisan allies who are aligned with carbon polluters. Technological subsidies and other types of policy carrots can often bridge these within-coalition tensions. In Norway, left-leaning politicians pushed carbon capture and sequestration (CCS) technologies to bridge some constituencies' desire to extract more offshore gas with other constituencies' desire to mitigate climate change. Coal-allied Democrats pushed similar technology-centric policies during US emissions trading debates. And the Australian Clean Energy Futures Package included expansive investments in industrial infrastructure to soften the impact of its carbon price. Facing cross-cutting preferences, coalitions often moderate internal tensions with a focus on carrots over sticks.

While all countries prioritize carrots over sticks, the tendency is less pronounced where double representation is more weakly institutionalized; in these settings, proponents sometimes exclude carbon polluters from policy design. When this happens, they attempt to impose stronger costs on carbon polluters. Australian climate reforms included policies that aimed to close the country's most inefficient coal facilities. The US Clean Power Plan proposed state-level electricity standards that would have disadvantaged coal by limiting the carbon-intensity of energy production. By contrast, the Norwegian government has had the legal authority to regulate carbon pollution since the mid-1990s but has instead preferred incremental or voluntary agreements with major onshore polluters.

Proposition 3—Policy Content: Where the double representation of carbon polluters is strongly reinforced, climate policies privilege producers; where it is weaker, policies are designed to reduce consumer costs. Absolute cost levels tend to be higher in producer-shielding systems, but are applied selectively to avoid serious threats to existing carbon polluters.

In the context of double representation, climate policymaking involves accommodating distributional losers, particularly industrial producers. In some cases, producers are accommodated through weak policies that impose few costs on any actor, for instance when policy measures are voluntary.

In cases where policy costs are higher, these costs are often imposed on diffuse, unorganized consumers. For example, the Norwegian carbon tax imposed relatively high costs on household energy and transportation pollution.

When the influence of policy losers is disrupted, climate policies can impose higher producer cost burdens. This disruption is more likely in systems with weakly reinforced carbon-polluter double representation. Further, when policy advocates enact climate policy despite mobilized opposition by producers, policies can include more *consumer* compensation. Producers contest policy by increasing consumer cost salience, so proponents work to preempt or weaken this salience through policy design. This producer versus consumer orientation of climate policy across different institutional contexts echoes other policy domains (cf. Prasad 2013).

By contrast, the relationship between double representation and policy cost levels is complex. The book's highest cost levels (measured in terms of price per ton CO_2) are present in corporatist Norway, where the double representation of carbon polluters is reinforced by domestic institutions. Yet, these costs levels have been applied selectively. Only some industries (e.g., offshore oil companies) and consumers face serious burdens; decisions about which industries are covered by particular policies were calibrated to avoid non-incremental disruptions to the status quo. By contrast, while overall cost levels are lower in Australia and the United States, costs have been more evenly distributed and, occasionally, posed stronger threats to entrenched incumbents.

What claims can be made about political ambition on the basis of these propositions? When higher costs are imposed strategically on sectors that can easily accommodate them, it seems misleading to classify these policies as ambitious. As discussed in chapter 1, this means that measurements of absolute costs—for example, the national carbon price—offer uneven proxies for policy ambition. Instead, we must also consider the degree to which policies reshape the distribution of political power and set the stage for more aggressive future pollution reductions. Generally, in settings with strongly institutionalized double representation, carbon-dependent economic actors are protected from consequential costs. Irrespective of short-term reform impacts, these actors maintain their profitability and policymaking influence. They remain powerful actors with a stake in carbon pollution, and can continue to constrain future efforts to ratchet up environmental ambition. By contrast, as the institutionalization of double representation weakens, more threatening policies can theoretically reach the policymaking agenda even though, to date, few such costly proposals have been enacted. Such costly policies might weaken polluter profitability and reshape the distribution of power for future reform debates. Chapter 8 offers a broader reflection on my theory's implications for political ambition and discusses prospects for effective climate risk mitigation in a world where carbon polluters continue to capture climate policymaking through their double representation.

Theoretical Implications

This book sides with arguments that emphasize the importance of policymaking institutions to climate policy outcomes (e.g., Hatch 1995; Oshitani 2006), but its conclusion as to which institutions best support effective climate reforms departs from previous work. For some scholars, institutions such as democracy and corporatism facilitate ambitious environmental outcomes because they promote collective action (Scruggs 2003; Bättig and Bernauer 2009). My research does not cleanly support this relationship. As I argue, institutions promoting collective action also tend to accommodate distributional losers; they thus reinforce the privileged influence of carbon-dependent economic actors on climate policymaking.

In part, this difference may reflect the issue domain. Previous literature on the environment and institutions has focused on a broad set of environmental issues (Scruggs 2003; Jahn 2016). Climate change may interact with policymaking institutions in a different fashion to these other environmental challenges. Corporatism involves consensus-making around the economic status quo (Schnaiberg 1980). But climate reforms demand an uprooting of this status quo. Correspondingly, we cannot assume that the welfare state is a climate-friendly state (Meadowcroft 2005). Instead, the very conditions that allow some policymaking institutions to guarantee certain social rights may complicate efforts to guarantee climate safety.

Instead, scholars must consider the interaction of climate policy preferences and institutions over time. Climate policymaking is not a one-shot game. It demands repeated efforts by proponents over decades to ratchet up policy ambition. Some scholars thus view cost imposition on consumers and generous concessions to industry as the basis for long-term decarbonization, pointing to aggregate levels of policy stringency in corporatist countries (Finnegan 2019). In this view, corporatist systems have an enhanced capacity to compensate distributional losers, which sets the stage for long-term decarbonization. High short-term costs on consumers will ultimately facilitate subsequent producer costs. However, if short-term climate policies do not undermine carbon polluter influence, the dynamics described in this book caution against assumptions that incremental approaches will support the rapid decarbonization necessary to manage the climate crisis. If climate reforms demand political confrontation with entrenched opponents, institutions that support non-incremental cost imposition may ultimately perform better.

The double representation of carbon polluters also speaks to broader comparative political economy debates about long-term policy planning. Climate change is just one instance of a policy challenge that creates intertemporal trade-offs; it forces politicians to choose between imposing costs in the short term (via policy) versus in the long term (via unmitigated climate impacts). Policy scholars too rarely consider the politics of intertemporal distributive conflict, focusing instead on cross-sectional disputes at a single point in time (Jacobs 2016). In contrast to such other policy domains as tax and social policy, this book finds little empirical evidence that policymakers in *any* institutional context have prioritized future public needs over short-term political considerations. Instead, the double representation of carbon polluters parallels work on the role of entrenched interest groups in focusing policymaking on short-term costs at the expense of long-term public welfare (Jacobs 2011).

The book also raises fundamental questions about carbon pricing as a climate policymaking strategy. To date, climate policy design has been dominated by economic concerns over efficiency and least-cost pollution reduction. Yet, efforts to pass "optimal" carbon pricing policies have not meaningfully reduced climate risks. In chapter 8, I argue that scholars should focus their attention on a policy's impact on political power distributions, not its theoretical economic properties. Our strategic efforts must be grounded in a sophisticated understanding of the political economy of climate reforms. In this way, because carbon pricing raises the relative salience of costs, particularly consumer costs, it is politically suboptimal. Climate policy necessarily involves massive redistribution of wealth and economic opportunity. In contrast to carbon pricing, policies that foreground benefits and make consumer costs more opaque have a greater chance of providing the public with meaningful climate protection.

Finally, this book adds nuance to existing explanations for cross-national climate policy variation, as reviewed in chapter 1. While echoing these existing theories' emphasis on distributive conflict, my theory of double representation extends and refines existing accounts in several important ways.

First, a focus on the power balance between green and brown industries correctly highlights distributive conflict (e.g., Meckling 2011; Aklin and Urpelainen 2013); however, it ignores the cross-cutting distribution of climate policy preferences. As this book makes clear, simple differences in either the type or quantity of a country's carbon-intensive interests cannot easily account for policy timing or content. For example, the 2011 adoption of an Australian carbon pricing regime did not coincide with either short- or long-term shifts in the economy's carbon-intensity or carbon polluter lobbying. Emissions grew in all economic sectors, except agriculture, between 1990 and 2011. This included a 61 percent increase in energy emissions and a 35 percent increase in industrial process emissions.[12] During this period, Australian capital continued to invest in coal extraction and other fossil fuel developments. In short, changes in the prevalence or economic importance of climate policy losers did not precede policy enactment. Instead, our theories must examine the interaction of different preference coalitions with political institutions.

At the same time, my account reinforces the importance of within-industry divisions (Meckling, Sterner, and Wagner 2017; Rabe 2018). Some of these divisions result from cross-cutting differences in material interests. Yet, political expression, even in the shadow of double representation, occurs within a strategic context. As political actors came to anticipate reform enactment, some endorsed limited actions to forestall even costlier future reforms. For instance, some carbon polluters joined clean energy companies in lobbying for US emissions trading during the late 2000s. This strategic accommodation (cf. Hacker and Pierson 2002) made climate policy supporters out of some business opponents. Industrial unions also engaged in strategic accommodation by participating in reform coalitions.

Second, some scholars point to environmentalists' importance in shaping policy outcomes (Svendsen et al. 2001; Harrison 2010a). This book supports the obvious claim that environmental advocates are important actors in shaping the climate policymaking agenda. However, the role of mainstream environmental pressure groups is less prominent than one might expect. Carbon pricing debates in the early 1990s were driven by political leaders with environmental preferences; environmentalist involvement tended to follow, not lead, these political actors. For instance, green groups supported the Clinton administration's BTU tax proposal, but only after Gore and other administration officials committed themselves to the policy. In Australia, climate reforms were not an integrated focus of the environmental lobby until the mid-2000s. In Norway, mainstream environmental groups shied away from radical efforts to impose costs on domestic polluters throughout the 1990s and 2000s.

Weak environmentalist influence on climate policy debates partly reflects strategic choices by the environmental movement. Often, environmental NGOs engage in climate policymaking through joint coalitions with workers or capital. These efforts include the US Climate Action Partnership (USCAP) and the Australian Southern Cross Climate Coalition (SCCC). However, these coalitions don't just include economic actors interested in pushing costly climate policies. They also include carbon-dependent unions or businesses whose support for climate policymaking is induced by a desire to forestall even costlier reforms. Consequently, these coalitions mask competing policy preferences. All coalition participants want to shape climate reforms; however, participants bring different economic interests to the table. In many cases, environmentalist lobbying efforts within these coalitions have been constrained by carbon-intensive actors' economic concerns. These coalitions have also ignored policy issues where cross-cutting preferences blocked consensus, as with coal policies for the SCCC or Keystone XL pipeline support for the BlueGreen Alliance. These coalitions thus moderated the intensity and content of environmentalist demands.

Third, my arguments probe the limits of ideological explanations for climate policymaking (e.g., Daugbjerg and Svendsen 2001; Layzer 2012). Previous work suggests an ideological climate preference gradient in many countries that corresponds to the intensity of a country's climate reform policy; yet, this perspective obscures substantial conflict within both right- and left-leaning coalitions over climate reforms. Even in countries where the ideological gradient over climate policy has become steeper, this has been the result of climate policy conflict as much as it has been the conflict's source.

As a result, shifts in climate policies are not always linked to government ideology. Instead, climate policy preferences remain ideologically cross-cutting. This includes significant factional support for climate policymaking within governments that have "brown" political reputations. While the Australian Liberals repealed the country's carbon price in July 2014, they were only one vote away from keeping a pro-climate leader in 2009. Moreover, the costliest climate policies do not always come from left-leaning political actors. In countries like Norway, centrist parties have proposed the most disruptive reforms, opposed by a joint coalition of labor and conservative parties.

Ideological gradients in climate policy support are most pronounced in the United States, though extreme climate-related polarization is relatively recent. I suggest that this polarization is consistent with the causal mechanisms outlined in this text. As in other advanced economies, the climate threat initially exposed cross-cutting cleavages in the United States; however, these cleavages eroded in the absence of institutions that reinforced them through time. US policy opponents lacked guaranteed access to climate policy design and so have always faced an elevated policy threat. Their persistent efforts to undermine climate reforms provided an incentive for many Republicans to oppose climate policy. In turn, even carbon-intensive labor groups began to align with Republicans, at least on climate issues. By contrast, tighter institutional links between labor movements and left-leaning politicians in Australia stabilized ties between carbon-dependent constituencies and the Labor Party despite similar opponent mobilization.

Fourth, this book downplays the importance of explanations that highlight the relationship between proportional representation systems and Green parties (e.g., Harrison 2010a; Hughes and Urpelainen 2015). Green-oriented parties have enjoyed significant representation in Australia and Norway as a result of electoral systems that give more voice to smaller parties; however, this political voice never disrupted carbon-polluter double representation. Prominent red–green (that is, Labor Party–Green Party) political coalitions emerged in both Norway and Australia during the 2000s, as detailed in chapters 3 and 6, but had only limited effects on climate policy outcomes. In Australia, the Labor Party initially refused to negotiate with Green Party senators over its Carbon Pollution Reduction Scheme

(CPRS), preferring to abandon the policy rather than compromise with the Greens. In 2010, however, a minority Labor government found itself dependent on Green Party support. This pushed emissions trading back onto the policy agenda and facilitated the eventual adoption of national climate reforms in 2011. Still, while Green representation shaped climate policy timing, it had less impact on climate policy content. Climate policymaking during Norway's red–green coalition between 2005 and 2013 played out similarly. Attempts by the green-leaning Socialist Left Party to ratchet up cost levels were frustrated by persistent opposition from Labor's carbon-dependent constituencies.

If the main barrier to climate policy enactment is an absence of climate advocates, then the presence of Green parties would reshape distributive climate policy conflict. However, advocates are already distributed across both left- and right-leaning political coalitions as a result of cross-cutting climate policy preferences. Instead, the major barrier to policy enactment is opponents' dispersal across and within existing political coalitions. The presence of red–green coalitions does not change this persistent presence of policy opponents within the "red" wing of red–green coalitions. Because they do not address this primary policy barrier, Green parties do not disrupt the logic of double representation.

Fifth, the book supports the importance of veto point density to policy outcomes (e.g., Harrison and Sundstrom 2010b; Madden 2014). Veto points matter because the logic of double representation disperses climate policy opponents throughout the political system whether left- or right-leaning political coalitions control government. However, this book suggests an expansive read of what constitutes a consequential veto point. For example, Madden (2014) argues that, as the number of veto players increases, countries will emphasize more incremental and diffuse cost policies. By this argument, the United States, as one of the most veto-dense advanced economies, should have the most incremental policies. However, I find that incremental policymaking is most prevalent in Norway. Climate policymaking veto points cannot be reduced to counting the number of policymaking process steps. Other important vetoes stem from advocates' and opponents' presence within existing political coalitions. The presence of these interests can be more or less durable depending on whether domestic institutions reinforce the double representation of carbon polluters. Where there are structural ties between Labor parties and industrial unions, industrial unions can de facto veto costly climate reform. However, the United States lacks such an "industrial union" veto within the Democratic Party. This allowed the Obama administration to pursue comparatively non-incremental climate reforms.

Finally, this book's theory agrees that bureaucratic power matters for policymaking (cf. Rabe 2004), but not as a simple function of issue salience. Climate reforms tend to involve such a substantial renegotiation of a country's economic institutions that distributional losers will invariably mobilize conflict into the public domain. These efforts raise issue salience. It is difficult to imagine a scenario similar to early US state-level climate reforms where climate debates can now occur under the radar. Instead, one of the crucial ways that bureaucratic power matters for climate policy-making debates relates to the transfer of policymaking authority between ministries. For example, efforts to enact emissions trading in Australia involved a transfer of bureaucratic authority from the environment ministry to core economic ministries, most notably the Department of Treasury. In Norway, the Ministry of Finance looms large over carbon pricing policies. In many ways, the United States stands unique among advanced economies because its environmental bureaucrats are also institutionally empowered to manage climate policymaking through the Environmental Protection Agency.

Of course, these theoretical conclusions draw primarily from detailed examination of three cases: Norway, the United States, and Australia. Do they generalize to other advanced economies? Chapter 7 takes up this question through an analysis of four shadow cases. I find that the empirical records of climate policymaking in Germany, Canada, Japan, and the United Kingdom are all consistent with the logic of double representation. Moreover, my theory helps to explain a number of otherwise puzzling features of climate policymaking in each country. However, before probing the book's generalizability, we must first examine its empirical basis. Chapters 3 through 6 ground this chapter's theoretical claims in detailed climate policy histories of Norway, the United States, and Australia.

3 Climate Policy Cooperation in Norway

The social democracies of Northern Europe have reputations as environmental policy leaders (Andersen and Liefferink 1999; Kelemen and Vogel 2010). Carbon pricing policies emerged in Nordic countries in the early 1990s, with all five of Esping-Andersen (1993)'s social democratic welfare states enacting a carbon tax by 1994. These policymaking efforts preceded the rest of Europe by almost ten years and came nearly two decades before many non-European countries enacted carbon pricing reforms. More recently, Northern European countries have embraced far-reaching carbon pollution reduction goals. Denmark plans to phase out coal usage by 2030. Norway has committed to national carbon neutrality by 2030. Sweden intends to produce no net carbon pollution by 2050.

Why were Northern European democracies early climate policy actors? Norway and its neighbors are no different from other advanced economies on many relevant dimensions. Fossil fuels play a central role in Norway's economic prosperity. The Norwegian finance minister has, recently, come from a political party that denies climate changes. Fifteen percent of the Norwegian public are unconcerned about the environment: only Austria, Denmark, and Sweden have lower concern among advanced economies; and 15 percent of Norwegians don't believe humans are causing climate change, second only to Australia's 17 percent (Tranter and Booth 2015).

Yet, Norway was the second country in the world to introduce a carbon tax, in 1991. This policy remained largely unchanged through the 1990s, despite several efforts to increase its sectoral coverage. Beginning in the early 2000s, Norway began to phase out its carbon tax system and replaced it with domestic emissions trading. During the late 2000s, Norway also

Figure 3.1
Milestones in Norwegian carbon pricing, 1990–2015

made significant domestic investments in carbon capture and sequestration (CCS) technologies, low-carbon transportation, and emissions reduction initiatives in tropical forests. Figure 3.1 charts milestones in this history of Norwegian climate policymaking between 1990 and 2015.

In this chapter, I trace the trajectory of Norwegian climate policymaking through 2015. I argue that Norwegian policy leadership on carbon pollution reduction did not reflect a unique distribution of climate policy preferences. In Norway, just as in the United States and Australia, the climate threat's emergence split stakeholder groups and political parties by exposing latent differences in the material interests of otherwise aligned economic actors. Furthermore, Norway's corporatist institutions ensured that national economic stakeholders, including carbon-intensive businesses, enjoyed sustained access to Norwegian decision makers. Thus, carbon polluters directly participated in policy design. In this way, Norwegian institutions facilitated early policy timing by proposing policies that contained only modest climate policy costs.

Over time, efforts by Norwegian climate advocates to impose higher costs on domestic carbon pollution were systematically blocked by political officials on both the left and right in coordination with union and business allies. The result has been a domestic policymaking trajectory that has achieved significant incremental reforms, but has never imposed threatening costs on entrenched Norwegian carbon polluters. Instead, climate policies have systematically accommodated double-represented carbon polluters through a focus on subsidy-based policies and consumer-facing costs while protecting carbon-intensive jobs and capital.

The Emergence of Norwegian Carbon-Intensive Sectors

Norway industrialized later than its European neighbours. Consequently, carbon-polluting economic activity was not central to the Norwegian economy until the early twentieth century, when foreign capital began developing energy-intensive fertilizer and metal processing industries (Hodne 1975; Moen 2009). These developments exploited cheap hydropower from waterfalls scattered across Norway's fjord-carved coastline. After, Norway passed "concession" laws in 1909 and 1917 that effectively nationalized hydropower development, low-cost energy came to be viewed as a national public good (Hodne 1975; Wicken 2009).

Beginning in the late 1940s, the Norwegian Labor Party and its union allies intensified efforts to expand hydropowered industrial employment without regard for cost or electricity prices (Midttun 1988; Kasa and Underthun 2010). Within a decade, this "onshore" industrial sector expanded dramatically. By the late 1970s, 40 percent of Norwegian exports, 10 percent of industrial employment, and 12 percent of GNP were linked to process-based industrial sectors (Midttun 1988). In many countries, energy-intensive industries release carbon pollution directly through manufacturing *and* indirectly through consumption of carbon-intensive electricity. In Norway, process industries rely on clean hydropower and do not generate indirect carbon pollution. However, they remain significant direct sources because metal processing, fertilizer manufacturing, and cement production all release carbon pollution as a by-product.

The second half of the twentieth century saw Norway develop its second major industrial sector: offshore oil and gas extraction. Since the early 1970s, exploration in the North Sea and the Norwegian Sea have led to significant oil discoveries. By 2015, Norway was the world's fourteenth largest oil producer, extracting over two million barrels per day. It was also the world's third largest gas producer, extracting almost 400 billion cubic feet annually. As of 2015, Norway's oil and gas industry generated 40 percent of national exports and 15 percent of the country's GDP.[1] The Norwegian government maintains a significant direct stake in oil and gas production through two-thirds ownership of the state-owned Equinor (formerly Statoil). Norway also maintains a significant indirect stake in oil production through marginal tax rates of up to 78 percent on oil and gas extraction profits. These revenues fund an enormous sovereign wealth fund, valued at over $850 billion USD in 2014, that covers budget deficits. Under international greenhouse gas accounting rules, Norway is not responsible for carbon pollution associated with combustion of its exported oil or gas. However, fossil fuel production also creates domestic carbon pollution during the extraction process.

Norway's political parties were founded to represent the country's important economic interests. The Labor Party (*Arbeiderpartiet*) was founded in 1887 with a dual mandate to represent labor's political interests and coordinate labor constituencies in the workplace. These roles were shared by LO (*Landsorganisasjonen i Norge*), the country's major peak labor organization that was established in 1899 (Allern 2010). The Labor Party and LO remained institutionally interconnected; by convention, senior LO leaders are members of the Labor Party's national executive committee and, usually, the Labor Party Congress's Election Committee (Allern 2010). Further, regular meetings of a Cooperation Committee (*Samarbeidskomiteen*) bring Labor prime ministers and senior party officials together with union leaders to discuss the party's agenda. Today, LO remains the country's largest peak labor organization, representing about nine hundred thousand workers. Peak labor organizations, in turn, comprise multiple member unions in specific economic sectors.

On the business side, the Norwegian Employer's Confederation (*Norsk Arbeidsgiverforening* or NAF) was founded as a strategic response to LO in 1900, followed by the Federation of Norwegian Industries (*Norges Industriforbund*) in 1919. These associations merged to form the Confederation of Norwegian Enterprise (*Næringslivets Hovedorganisasjon* or NHO) in 1998, which remains Norway's dominant peak business association. Norway's Conservative (*Høyre*) Party lacks institutional links to peak business associations analogous to Labor's links with LO; however, the Conservatives grew increasingly close to NAF and its successor organizations over the twentieth century (Allern 2010).

The development of carbon-intensive industries enjoyed support from political parties across the ideological spectrum. Critically, coastal industrial expansion was spearheaded by Labor-led governments with support from opposition Conservatives (Kasa 2000). Thus, factions of the Norwegian left that are deeply committed to industrial development are still described as "*Kraftsosialists*" (electric power socialists). Kraftsosialists believe Norway should develop its economy by exploiting energy resources; they tend to relegate environmental issues as secondary concerns.[2] This industrial expansion, in turn, embedded carbon-intensive economic interests across the Norwegian political landscape.

Today, most Norwegian industrial unions are affiliated with LO, including the United Federation of Trade Unions (*Fellesforbundet*), which represents workers from onshore Norwegian industries, and Industri Energi, which represents offshore oil and gas workers. LO also includes a number of major nonindustrial unions, including the Norwegian Union of Municipal and General Employees (*Fagforbundet*). By contrast, many other public sector and professional workers have chosen to affiliate with competing peak labor associations.[3]

Among business organizations, industrial sectors are almost exclusively affiliated with the NHO, notably Norwegian Industry (*Norsk Industri*) and the Norwegian Oil and Gas Association (*Norsk Olje & Gass*). By contrast, a growing number of small and medium-sized service sector businesses affiliate with a competing peak business association, the Enterprise Federation of Norway (*Virke*), founded in 1990.[4]

Table 3.1 summarizes Norway's economic stakeholder landscape. While both carbon-intensive industrial actors and low-carbon service sectors are represented by large peak business and labor associations, carbon-intensive interests remain embedded within the oldest and most powerful Norwegian peak associations. As I now show, these carbon polluters were strongly positioned to shape corporatist climate reforms in the presence of crosscutting climate policy cleavages.

Table 3.1

The landscape of large Norwegian economic stakeholders

Peak business associations	Peak labor associations
NHO (Industry) 23,000 companies 540,000 jobs dates back to 1900	**LO (Industry + Service)** 600,000 members dates back to 1899
Virke (Service) 17,000 companies 220,000 jobs dates back to 1990	**UNIO (Service)** 320,000 members dates back to 2001

The Emergence of Climate Change on the Policymaking Agenda

The earliest Norwegian climate policy debates were not structured by ideological divides. As climate science exposed the climate threat, the Norwegian left and right initially responded with similar levels of concern. During
Labor Prime Minister Gro Harlem Brundtland's first term from 1986 to
1989, Norway set its first emissions reduction commitments: stabilize carbon pollution at 1989 levels by the year 2000. Brundtland enjoyed a green
reputation, having led the United Nations World Commission on Environment and Development; that commission's final report, published in 1987
while Brundtland was prime minister, widely influenced global thinking on
sustainable development and became eponymous with her name.

The Norwegian Conservatives also embraced environmental reforms
during this time, partly to avoid losing environmentally minded voters
impressed by Brundtland's green profile.[5] At the Conservative's pre-election
conference in 1989, the party's youth branch (*Unge Høyre*) successfully
championed an environmental platform. Right-leaning opposition parties
then criticized Brundtland's emissions reduction target as insufficiently
ambitious. The 1989 stabilization target was ultimately a Conservative-
advanced compromise that surpassed Labor's initial proposal (Hovden and
Lindseth 2002).

After the elections in 1989, the new Conservative coalition government
continued to promote environmentally friendly policies. Prime Minister
Jan Syse established Norway's Environmental Tax Committee (*Miljøavgifts-
utvalget* or ETC)[6] and gave it a broad mandate to investigate environmental
taxation's potential in Norway, including as a climate policy tool. As a tech-
nocratic expert commission, common within Norway's corporatist system,
the ETC's membership was dominated by economists from government
ministries and universities (Moe 2010). The committee reviewed theoretical literatures on environmental taxation and ran macroeconomic models
of the Norwegian economy under different tax regimes.

Four months after the ETC's establishment, Prime Minister Syse requested
its confidential input during the 1991 budget planning process (Moe 2010).
In an April 1990 interim report, the ETC recommended Norway establish a
carbon tax on all fossil fuels, with full exemptions for shipping and North
Sea oil extraction and partial exemptions for inland shipping, fisheries,
and process industries. Exemptions were conceived as a temporary tool to

protect specific sectors from competitiveness concerns until other countries enacted comparable policies (Moe 2010). The scope of these potential reforms concerned both peak business and labor groups. Seeking to
preempt potential costs, LO and NHO ran a public campaign throughout
1990 to oppose carbon taxation in the metal processing sector. At the same
time, carbon-dependent economic actors worked internally with political
allies to block government consideration of ETC-endorsed sectoral taxes
(Kasa 2000).[7]

In the face of this opposition, Syse's 1991 carbon tax proposal departed
from ETC recommendations and was limited to gas and mineral oils. However, before the policy could be enacted, the Syse government fell on an
unrelated political conflict related to Norway's proposed EU membership.
A new government in late 1990, again with Brundtland as prime minister, kept Syse's proposed carbon tax in its new budget. As with the Syse
proposal, Brundtland's carbon tax had narrower sectoral coverage than
the ETC recommendation. It continued to exempt many carbon pollution
sources, including gas use in refineries, fishing fleets and—bowing to pressure from industrial unions—land-based industrial processes. Carbon tax
levels were differentiated across sectors, ranging from $20 to $50 US per
ton CO_2, depending on fuel type and end use (e.g., household versus industrial consumption).[8] Generally, consumers enjoyed fewer exemptions than
industrial actors, who were shielded from virtually all costs.

In contrast to the Syse proposal, Brundtland expanded the tax's sectoral coverage to include offshore oil sector emissions. This expansion was
opposed by the Conservatives (Kasa 2000).[9] However, it is not clear how
consequential the policy's costs were to industry; offshore industries saw
the tax as capturing more oil-production profits without changing the sector's growth trajectory.[10] Since 1990, only a handful of investment decisions can be directly traced to the carbon tax, and these pale in comparison
to substantial growth in offshore carbon pollution during this period.[11] In
other words, it is not clear that the offshore carbon tax can be understood
as having imposed substantial sectoral costs.

In this way, Norway's carbon tax passed easily and with little conflict
precisely because it shielded politically connected industrial producers
from costs. ETC recommendations that threatened carbon-intensive interests embedded within the Syse and Brundtland coalitions were dropped
during policy design. Likewise, a 1992 tax reform proposal to increase the

carbon tax rate on mineral oils was withdrawn after industrial protest; instead, industrial carbon tax exemptions were *expanded* in December 1993 (Reitan 1998).[12] Eventually, for every unit of cost imposed on producers, the tax system imposed between two and three units of cost on consumers (Svendsen et al. 2001). Norway's carbon tax thus shielded producer interests, which were well represented in Norway's corporatist political institutions, and imposed most of its costs on consumers. As a Center Party (*Centerpartiet*) politician criticized, "Gro Harlem Brundtland and the powerful business leaders have pushed all the green taxes over to ordinary citizens ... this is a blue tax, Gro."[13]

Climate Policy Conflict during the 1990s

Throughout the 1990s, Norwegian climate policymaking continued to be driven by both Labor and Conservative party efforts to represent the interests of carbon-intensive LO (labor) and NHO (capital) interests in policy design debates. Green interests outside the Labor and Conservative parties occasionally managed to push consequential climate reforms but saw their proposals systematically blocked by polluter-allied politicians. While pro-climate voices emerged within both the Labor and Conservative parties, they could not overcome the policymaking vetoes of co-partisans aligned with the country's carbon-intensive unions and businesses. This double representation of carbon polluters played out during debates over both carbon tax reforms and gas-fired power plant construction.

Carbon Tax Reform Efforts

A major effort to expand the carbon tax's sectoral coverage begin in late 1994 when the Socialist Left party (*Sosialistisk Venstreparti*) forced Brundtland's minority Labor government to establish a Green Tax Commission (*Grønn skattekommisjon* or GTC) as a condition of their support for the government's budget. The GTC was tasked with investigating changes to the tax system to promote environmental policies. The Socialist Left had increased its commitment to environmental policies over the preceding years.[14] This commission exposed splits in Labor Party preferences, with some members' material interests colliding with others' environmental policy preferences (Lund 2009). The GTC's mandate reflected these tensions, but ultimately stressed that Norwegian energy production and export should be safeguarded by any reform proposals.[15] To help bridge these internal tensions in the Labor Party, the GTC was also tasked with investigating job expansion through environmental policymaking, as raising employment was also a government priority at the time.[16]

Like many corporatist expert commissions, the GTC incorporated economic stakeholders to consult on policy design. It included representatives from business, labor, and environmental organizations, including LO, NHO, aluminum giant Norsk Hydro, and environmental NGO Friends of the Earth (*Naturvernforbundet*). The GTC's mission was complicated by divergent policy preferences among its membership. An alliance between LO, NHO, Norsk Hydro, and Ministry of Trade and Industry bureaucrats pushed policies to shield industrial producers from climate mitigation costs (Lund 2009).[17] These opponents took this position despite cross-cutting organizational preferences: within both LO and NHO, energy-intensive industries battled other economic sectors to control the peak organization's GTC interventions. And, in each case, industrial sectors successfully dominated the content of LO and NHO's GTC proposals (Kasa 2000).

By contrast, a coalition of environmentalists, Ministry of Finance, and Ministry of Environment officials advocated sweeping carbon tax reforms to impose costs on all carbon polluters. Reform proponents had diverse motives. Environmental groups wanted ambitious climate risk mitigation. Finance officials advocated efficient policies that would not discriminate between different pollution sources; they believed the market should decide where emissions reductions were undertaken. Other bureaucratic agencies were motivated by distributional concerns: for instance, a Ministry of Transport representative worried that industrial tax exemptions would increase the transport sector's relative emissions-reduction burden (Kasa 2000).

The GTC chair, Labor Party insider Bernt Lund, worked to broker a compromise between the alliance of industrial labor and capital on the one hand, and environmental policy proponents on the other.[18] While the commission could agree on technical design issues, conflict over sectoral coverage proved more difficult to resolve. By mid-1996, a consensus emerged among many GTC members about the importance of implementing a carbon tax in Norway that was based on the emissions intensity of different economic activities, with minimal sectoral exemptions. While some GTC members favored high prices, negotiations landed on a compromise proposal of 50 NOK per ton CO_2 ($7.13 USD) to recognize industrial actors' competitiveness concerns (Lund 2009).[19] This proposal would apply equally to almost all industries and Norway's carbon polluters would thus face some costs.

Norwegian producer interests within the GTC viewed this emerging package as an economic threat. LO, NHO, and Norsk Hydro representatives, as well as their bureaucratic allies in the Ministry of Trade and Industry, rejected the proposal. Industry worried the tax would curtail onshore process industries.[20] While carbon polluters enjoyed access to policy design through GTC membership, the industry faced, for the first time, a real policy threat. To moderate this threat, carbon-dependent interests expanded the scope of policy conflict, mobilizing publicly and preemptively against the GTC proposal. Several weeks before the release of the commission's report, the NHO leaked to the media that an undifferentiated carbon tax had consensus support among the commission's bureaucrats, economists, and environmentalists (Reitan 1998; Kasa 2000). Carbon-intensive labor and business groups made dramatic public statements to oppose potential reforms, even holding a press conference to decry the still-unpublished proposal (Reitan 1998).

Close institutional links between these carbon-dependent actors and government officials on both sides of the ideological divide quickly neutralized the policy threat. First, governing Labor Party officials marginalized the GTC within the Norwegian policymaking process. Later, Labor and Conservative actors jointly mobilized with LO and NHO allies to block a GTC-inspired tax reform package that threatened costs for both offshore and onshore carbon polluters.

In the short term, senior Labor Party officials protected carbon-dependent Norwegian workers by watering down the GTC's impact. Both Prime Minister Brundtland as well as Minister of Trade and Industry Jens Stoltenberg resisted the idea of a universal carbon tax, even if the tax rate was set as low as 25 NOK per ton CO_2, which was half the rate recommended by the GTC (Lund 2009). Brundtland and Stoltenberg also weakened the symbolic consensus achieved by nonindustry GTC members by instructing commission members who worked for government ministries to withhold endorsement of any carbon tax in the final report.[21] By Norwegian standards, this was a dramatic intervention in the policy design process, catching pro-climate Labor members off guard (Lund 2009).[22] Government bureaucrats were appointed to the GTC as independent experts, not as government representatives. When appointed to an expert commission as private citizens, bureaucrats traditionally made recommendations as they saw fit.[23] Yet, bureaucrats on the GTC were forced by senior government members to adopt an agnostic position on the desirability of a universal carbon tax, despite their private support for the proposal.[24] The Ministry of Environment representative was particularly aggrieved by this political muzzling, but his attempts to write a dissent were blocked by Minister of Trade and Industry Stoltenberg (Lund 2009). The overall effect was to undermine public perceptions that most GTC members were in favor of universal carbon taxation.[25] While a majority on the commission still endorsed reform, the lack of support from ministry bureaucrats and opponent-driven media coverage diluted the weight of the GTC's eventual report.

Over the subsequent months, Brundtland and her Labor successor as prime minister, Thorbjørn Jagland, ignored the GTC report entirely.[26] While climate policy proponents may have shaped commission proposals, they had not reshaped the cross-cutting representation of carbon-intensive capital and labor across the Norwegian left and right. Instead, in a quick retreat from the GTC proposals, Jagland initiated conversations with Norwegian industry about voluntary emissions reductions in lieu of formal carbon pricing.

The GTC proposal resurfaced in 1997 after Norway elected a new centrist government led by Christian Democrat (*Kristelig Folkeparti*) leader Kjell Magne Bondevik. The Christian Democrats were joined by the Liberal (*Venstre*) and Center (*Centerpartiet*) parties in a center-right government. Both the Christian Democrats and Liberals had long-standing programmatic commitments to environmental policy. Senior Christian Democrats describe Christian values of stewardship as motivation for their party's pro-environmental positioning.[27] A coalition of three minor centrist parties, Bondevik's cabinet lacked the institutionalized ties to carbon polluters that had characterized previous Labor and Conservative cabinets throughout the 1980s and 1990s; even the new minister of trade and industry had only weak ties to industrial labor and capital. Many business organizations suddenly found themselves looking in from the outside as the Ministry of Finance developed a new carbon tax proposal in April 1998 (Kasa 2000). Positioned as a formal response to the GTC, the Bondevik government proposed a 100 NOK ($13.51 USD) per ton CO_2 carbon tax with no sectoral exemptions.[28] The Bondevik proposal promised to tax all economic activity on its true carbon pollution intensity; all Norwegian actors would have to internalize the negative externalities associated with their carbon pollution. Moreover, it was explicitly conceived as a form of carbon-polluter cost imposition.[29] Still, the package included transition assistance for some sectors, including the onshore industry, through 2010.[30]

Despite this compensation, a coalition of Labor and Conservative policymakers joined with peak labor and business associations to oppose the policy, including direct lobbying of other Conservative and Labor party members (Kasa 2000).[31] However, business opposition was not universal. Instead, the episode revealed latent, cross-cutting differences in sectoral economic interests. The service-oriented Federation of Norwegian Enterprises (*Handels-og Servicenæringens Hovedorganisasjon* or HSH) broke from the NHO to voice support for the Bondevik proposal. HSH released a policy brief calling for universal green taxes at a well-covered press conference, deeply antagonizing NHO.[32] As a senior business official recalled, "There was a sort of reaction we were stepping on their toes when we argued for universal taxes because then we argued for their sectors. Everyone should be arguing for themselves, and we were arguing for the total."[33] LO joined NHO in criticizing HSH for overstepping its place in the stakeholder landscape.[34] HSH's perceived transgression for speaking out of turn underscores the degree to which industrial labor and capital enjoyed a structural position of power within Norwegian energy and climate policymaking during this period.

While pro-climate forces were able to put carbon tax reforms on the agenda, they did not have a broad enough coalition to enact them. The Bondevik government did not enjoy a majority in parliament, thus Labor and Conservative opposition members together were able to block the tax proposal. Ultimately, Bondevik achieved only a minor tax code amendment to extend the carbon tax's coverage to limited forms of air and sea transport at 100 NOK per ton of CO_2. The Bondevik government was also forced to concede to opposition demands to transition *away* from carbon taxes altogether to focus instead on emissions trading. Industrial actors believed that they would be able to successfully lobby for generous pollution allowances under an emissions trading system, which would lower producer costs. Such a policy could transfer the distributive burden of Norway's domestic policy onto others.

Responding to this parliamentary charge, the Bondevik government appointed yet another expert commission, the Quota Commission

(*Kvoteutvalget*).[35] The Quota Commission reported back to parliament in December 1999, advocating an ambitious emissions trading scheme to begin in 2008 that would cover 90 percent of the Norwegian economy.[36] However, the commission was split on its preferred approach to allowance allocation.[37] In most ways, the Quota Commission replayed the GTC debates in the context of a different policy instrument. However, the Bondevik government could not organize a political response to the report, losing the confidence of parliament in early 2000.

Conflict over Gas-Fired Power Plants

At the same time that double-represented carbon polluters frustrated carbon tax reforms, they mobilized to *increase* the economy's carbon-intensity though the construction of new gas-fired power plants. Norway was unique among industrialized economies for having a virtually carbon-free electricity sector, with just under 97 percent of domestic electricity production generated by hydropower, which produces no carbon pollution. However, as the country discovered large offshore gas reserves, the development of gas-fired electricity became a policymaking focus during the early 1990s. In 1994, Statoil (the state-owned oil and gas producer), Norsk Hydro (one of the world's largest aluminum producers), and Statkraft (the state-owned electricity company) established Naturkraft, a joint venture to build two gas-fired power plants on the country's west coast at Kårstø and Kollsnes. These plants would *carbonize* Norway's electricity sector relative to the country's hydropower baseline. For instance, the Kårstø plant would produce 1.2 million tons of carbon pollution annually, almost 3 percent of Norway's total emissions in 2000.[38]

In 1995, the Labor-led parliament approved tax breaks for gas-fired power plant construction (Hovden and Lindseth 2002) and, in 1996, it issued a construction license to Naturkraft. However, the Norwegian environmental community protested this decision, with ENGO Nature and Youth (*Natur og Ungdom*) organizing civil disobedience at the proposed construction sites while other groups circulated a country-wide petition to revoke the license, collecting 100,000 signatures.[39] Senior bishops in the Church of Norway called for church members to pray against the plants' construction.[40] By May 1997, public support for the plants had reversed, with a plurality opposed to the developments.[41]

The gas plant controversy underscored the absence of a clear ideological divide in Norwegian climate debates, as proponents and opponents of the gas plant construction existed on both the left and right of the ideological spectrum. For example, on the left, the Labor government's gas plant decision divided the labor movement. The Union of Municipal Employees (*Kommuneforbund*), which was affiliated with LO, asked its 230,000 members to engage in civil disobedience, defying LO's leadership.[42] However, opposition from service and public sector unions did little to influence the contours of an industrial union-dominated discussion.[43] In a 1997 keynote speech to the LO Congress, Labor Prime Minister Jagland criticized environmentalist antagonism to the plants.[44]

Prime Minister Jagland attempted to bridge political tensions within his own party by suggesting the plants be built immediately, but that their climate impacts could be mitigated through CCS at an unspecified future point.[45] Under Labor's direction, the Ministry of Petroleum and Energy then denied appeals from diverse environmental groups over Naturkraft's license to operate the plants, rejecting CO_2 emissions concerns.[46]

As part of its license, the Ministry of Petroleum and Energy's decision asserted for the first time that carbon dioxide should be considered as a criteria pollutant under the country's existing air pollution legislation.[47] This punted consideration of the gas plants' climate impacts to the Norwegian Pollution Control Authority (*Statens forurensningstilsyn* or SFT), the agency responsible for these permits. SFT immediately denied its ability to delay plant construction on a carbon pollution basis once a Ministry of Petroleum and Energy license existed.[48] As the agency pointed out, Norwegian law requires discharge permits for all air pollution but these permits are rarely refused for licensed industrial activity.[49]

The Jagland government lost the 1997 election before making a final gas plant determination. Likely, Jagland postponed the decision to avoid antagonizing either the industrial or environmental factions of his party in advance of the fall election.[50] Post-election, Bondevik reversed Labor's course. While the Labor government had promised the future implementation of CCS technology, Bondevik insisted that carbon pollution permits require CCS technology immediately. Bondevik thus invoked the latent regulatory authority the Jagland government had been reluctant to use. The eventual SFT permit, issued in 1999, required that 90 percent of the plant's emissions be reduced as an operational requirement.[51]

The Bondevik government's use of latent regulatory power to require CCS marked the apex of reformers' efforts to impose direct costs on Norwegian carbon polluters. As with Bondevik's carbon tax proposal, this effort emerged from a center-right minority coalition with fewer ties to peak labor and business associations. Yet again, this reform coalition lacked the political power to enact their costly policies. Bondevik's decision to force CCS technology on gas-fired power plants precipitated his government's fall in March 2000 when Labor and Conservative politicians banded together to void the SFT permit and demand the immediate construction of plants without CCS. The centrist Bondevik government thus fell over its refusal to *carbonize* the Norwegian electricity system, the first democratic government in the world to lose power as a result of domestic climate policy conflict. The Bondevik coalition was replaced by a minority Labor government under former Minister of Trade and Industry Jens Stoltenberg. Stoltenberg quickly issued Naturkraft an unconditional permit to discharge carbon pollution and proceed with plant construction.

These brief efforts by the short-lived Bondevik government to expand sectoral coverage of Norway's carbon tax and mitigate carbon pollution in the country's electricity system highlight the persistent structural power of carbon polluters within Norway's policymaking system. When the Christian Democrats led a coalition government, carbon polluters faced a period of substantial policy threat, as their insider access to decision makers somewhat weakened. However, pro-climate voices still found their preferences blocked by the strong veto of double-represented carbon polluters in both the Labor and Conservative parties. They were able to exploit their structural position to deflect efforts to impose costs on carbon-intensive economic activity.

Climate Policy Conflict during the 2000s

Norwegian climate policymaking throughout the 2000s continued to be constrained by carbon polluter representation on both the left and right. During this period, strong pro-climate factions emerged within successive Norwegian governments. However, carbon polluters systematically used their political access to block costly reforms. Because Norwegian policymaking processes institutionalized their influence, economic losers' preferences became the binding constraint on climate reform content. Only policies acceptable to embedded carbon-intensive interests were enacted. These included a pivot from domestic carbon taxes to an EU-harmonized cap and trade scheme, a heavier focus on international rather than domestic carbon pollution reductions, and an expansion of subsidy-based climate policies. Because these reforms posed minimal economic threat to domestic carbon polluters, they enjoyed broad support across government, industry, and labor.

The Pivot to Emissions Trading

When the Labor Party's Stoltenberg took office in 2000, his government embraced conversion of Norwegian carbon taxes to emissions trading. In a June 2001 White Paper, Labor declared that Norway should meet its climate commitments using a domestic emissions trading scheme, accompanied by voluntary carbon-pollution-reduction agreements with select industries. His plan also endorsed construction of new gas-fired power plants (Hovden and Lindseth 2002). In short, the plan was relatively unambitious, catered to the needs of Labor's industrial base, and prioritized sectoral economic concerns over climate risk mitigation.

However, Stoltenberg's government was short-lived. An October 2001 election brought Bondevik back to power, this time at the helm of a Conservative–Christian Democrat coalition.[52] Unlike Bondevik's previous minority government, his second government brought carbon polluters to the table through Conservative participation. This constrained Bondevik's efforts to champion climate causes.[53] For instance, the Conservatives made Bondevik accept Stoltenberg's gas-fired power plant permits. While Bondevik promised no additional concessions for fossil fuel plants, conflict over gas development created tension throughout the coalition's lifespan (Tjernshaugen 2011).[54]

Bondevik himself remained committed to pro-environment policies. He thus supported proposals from Conservative Party insider Børge Brende, who was appointed as Minister of the Environment. Brende lacked institutional ties to Norwegian industry and had been a strong environmental advocate as leader of the Young Conservatives in the late 1980s.[55] In Cabinet meetings, Brende often received his strongest support from Christian Democrats, particularly Bondevik.[56] The partnership between the Conservatives and Christian Democrats thus empowered environmental voices within the Conservative Party.[57]

Brende and Bondevik leveraged their position to push forward major environmental reforms, particularly related to land management. However, this influence did not extend to climate policymaking. Symbolically, Brende secured a few early successes. He withdrew Norway from the laggard JUSCANZ negotiating group at the Marrakesh climate conference.[58] He spearheaded a supplemental climate policy white paper that somewhat strengthened Labor's 2001 proposal.[59] However, Bondevik and Brende's climate ambitions stalled during consultative planning with economic stakeholders over cap and trade. The debate reprised divisions from the Quota Commission's report: which industries should receive free pollution permits? And how many permits should be distributed?[60] Mirroring carbon tax debates during the late 1990s, the peak business association HSH argued that industrial polluters had to start paying their fair share. However, HSH again found itself institutionally marginalized compared to peak associations with more established policy design access.[61]

Once again, the overlapping interests of the labor association LO and business association NHO played a critical role in obstructing any significant new climate policy. These groups now opposed the enactment of a domestic trading scheme, which they had previously pitched as an alternative to Bondevik's carbon tax reform. Industry argued the Norwegian economy was too small to support unilateral emissions trading without compromising industrial competitiveness.[62] Industry insisted emissions trading should only be pursued inside an integrated Europe-wide framework.[63] Because of LO's close ties to Labor and NHO's close ties to the Conservatives, their interventions received attention within both government and opposition ranks.

Both the Stoltenberg and Bondevik climate white papers envisioned unilateral Norwegian emissions trading. Norway's emissions trading debate thus preceded Europe-wide discussions. However, by mid-2002, EU policy negotiations were progressing faster than Norwegians had anticipated. Europe's progress actually helped to bridge internal Norwegian political tensions by making industry's demand to follow Europe's lead seem reasonable.[64] In effect, the emergence of an EU carbon pricing scheme solved many of the intractable distributional conflicts that had stalled Norwegian climate policy over the preceding decade.[65] Unlike carbon taxation, the European Union Emissions Trading System (EU ETS) proved an economic blessing for almost all Norwegian sectors since most domestic carbon tax losers became continental cap and trade winners.[66] Onshore process industries, with dedicated access to clean hydropower electricity, had a lower carbon-intensity than their fossil fuel-dependent European counterparts. Offshore oil and gas industries already faced a greater domestic burden than global peers. Further, European emissions trading could expand demand for Norwegian gas as central European countries switched from coal to gas.[67] For Norwegian industry, embracing EU policy was a way to permanently undermine pro-climate domestic factions working to impose domestic costs.[68] As one industry official remarked, when it comes to climate policy "we want to be Europeans—have a European focus on everything."[69] Throughout the early 2000s, pro-climate factions had not succeeded in imposing real costs on many Norwegian carbon polluters. A shift of climate policy action to Brussels reinforced the producer-friendly nature of Norwegian climate policymaking. It reinforced an approach to climate risk management that avoided domestic cost imposition and instead focused on climate mitigation in ways that protected the economic status quo.

By the time the EU agreed to its emissions trading scheme in October 2003, a consensus had emerged in Norway to support joining the EU scheme. Differences surfaced only around policy implementation details.[70] However, as a non-EU member state, Norway was not immediately eligible to join the EU ETS Phase 1 between 2005 and 2008.[71] Instead, Norway developed its own domestic emissions trading scheme to parallel Europe's first compliance period. This Norwegian Emissions Trading Scheme (Norwegian ETS) was explicitly modeled after the EU ETS. However, unlike the European policy, the Norwegian ETS excluded the offshore sector, which already faced a higher burden under Norway's existing carbon tax regime.

Climate policy proponents believed the Norwegian ETS might provide an opening to impose costs on onshore carbon pollution. Phase I of the EU ETS had coverage rules that, when applied to Norway, divided Norwegian onshore industries: some were included and others excluded. The Norwegian Ministry of Finance proposed to bring all onshore industries under the domestic cap, effectively expanding the scope of the Norwegian system relative to its EU counterpart.[72] Industrial stakeholders successfully resisted this approach. Ultimately, process industries covered by the EU ETS were brought into Norway's ETS with generous allowance allocations. Process industries outside the EU ETS were exempted, per industry's demands. With only a single oil refinery remaining as the scheme's market buyer, the carbon price effectively dropped to zero and there was virtually no trade.[73]

Instead, industrial stakeholders argued that uncapped onshore industries should establish voluntary emissions reduction agreements with the government. Voluntary agreements have a long history in many corporatist societies. However, Norway has largely avoided them, as Ministry of Finance economists have questioned their utility. According to critics, voluntary agreements put the government at a negotiating disadvantage because they generate information asymmetries between industry and bureaucrats.[74] By contrast, Norwegian industry insists that voluntary agreements have real teeth because they are negotiated under threat of regulatory action.[75] In the environmental domain, the Jagland government had begun negotiating voluntary environmental agreements with Norwegian industry in the aftermath of the GTC controversy, including on carbon pollution. However, only a sulfur hexafluoride (SF6) agreement with the aluminum sector was concluded before the government fell in 1997.[76]

During his first term, Bondevik rejected voluntary agreements as insufficient to mitigate climate risks.[77] However, his second government had deeper institutional ties to industry because the Conservative Party was a coalition partner with Bondevik's Christian Democrats. As a result, his government acquiesced to industrial demands for voluntary approaches. Under a 2004 voluntary agreement with Norsk Industri, onshore industry committed to a 30 percent reduction in carbon pollution below 1990 by 2007, a reduction of about 13.5 million tons.[78] Ultimately, industry reported reductions of 12.4 million tons by 2007; however, it remains ambiguous what part of this reduction was prompted by the agreement and what reflected secular trends in industrial efficiency and sectoral outputs.[79] The voluntary agreement thus reflected a level of costs that industry was "able to live with."[80]

Norwegian emissions trading rules further harmonized with the EU during Phase II of the EU ETS (2008–2012).[81] During Phase II, national governments outlined National Allocation Plans (NAPs) for domestic allowance distributions. Norwegian Environmental Directorate bureaucrats, supported by a pro-climate environment minister, attempted to use the NAP process to increase the cost levels faced by some Norwegian carbon polluters.[82] In its initial NAP proposal, the Directorate overruled objections from industry stakeholders to propose that some onshore businesses must purchase allowances.[83] Norwegian business fought these proposals in front of the European Free Trade Association (EFTA), which overturned the Norwegian plan and sided with industry in July 2008.[84]

Ultimately, Norway was only able to differentiate its domestic plan to impose costs on offshore actors, reaffirming *onshore* industry's belief that moving to a European system would weaken the policy threat they faced.[85] The situation was different for the offshore oil and gas industry. Again, Norway imposed higher costs on this sector. First, Norway made offshore industries buy all allowances through auctions. Second, the country lowered the total number of allowances it released into the market to create artificial scarcity. This forced offshore actors to purchase allowances from other European countries.[86] Yet, the government also lowered its offshore sector carbon tax so that the combined liability of domestic carbon taxation and trading allowances approximated previous carbon tax liabilities.

Curtailment of Norwegian policymaking autonomy accelerated under Phase III, with full EU-Norwegian harmonization of quotas, allocation rules, and offset rules.[87] From Phase III onward, Norwegian National Allocation Plans (NAPs) were set in Brussels, not Oslo.[88] This shift again served industrial interests. The Norwegian offshore industry received more free allowances during Phase III than they had under a more stringent Norwegian Phase II NAP. Between these free allowances and a collapsing EU carbon price, Norwegian oil and gas companies began to face carbon pollution liabilities well below historical averages.

To counteract these declining costs, the Norwegian government doubled its domestic carbon tax on the offshore sector in October 2012 to 410 NOK.[89] This policy was understood as a corrective to EU policy failures, rather than as a ratcheting up of domestic policy cost levels. Critically, the reform included an explicit guarantee to reduce domestic carbon taxes if the EU carbon price recovered.[90] The offshore industry was unhappy with the reform, but it saw the tax increase as insignificant relative to the benefits of doing business in Norway. Industry also appreciated that the new price was still low relative to the demands of environmental advocates, who wanted rates as high as 800 to 1000 NOK ($108.81 to $136.01 USD), which could have threatened investment profits.[91]

In sum, the introduction of emissions trading in Norway and its gradual harmonization with the European Union did little to upend patterns of climate policy conflict. Norwegian climate policy continued to be characterized by producer-friendly policies that preemptively addressed the distributive concerns of industrial labor and capital. Indeed, for many carbon polluters, the transition to a continentally integrated carbon market reduced the domestic policy threat by foreclosing advocates' policymaking autonomy. By 2005, a decade and a half after Norway introduced its carbon tax, double-represented carbon polluters still enjoyed an institutionalized shield from climate policy costs.

Climate Politics under a Norwegian "Red–Green" Coalition Government

In 2005, Norway shifted from a right-leaning Christian Democratic-Conservative coalition to a left-leaning red–green coalition government of Labor and the Socialist Left. Norway historically lacked a strong Green Party; instead, the Socialist Left was viewed as the natural home for left-leaning environmental voters. Yet, Socialist Left and green Labor voices still found themselves constrained by industrial factions within the "red" labor wing of the red–green coalition. Often, Socialist Left coalition partners were simply left out of climate-related decisions; LO and Labor would negotiate a policy that green voices would simply need to accept.[92] In some ways, green voices within this red–green coalition were even more curtailed than they had been under the previous blue–green government. While the Bondevik government had a climate advocate as prime minister, the red–green coalition, still led by Jens Stoltenberg, had a prime minister who was an ally of industry.[93]

As with the previous blue–green coalition, environmental advocates did find success in non-climate domains. For instance, the Labor–Socialist Left government strengthened the country's biodiversity laws.[94] However, on climate policy, government actors struggled to manage internal tensions between left-leaning industrial interests and green factions.[95] For example, Socialist Left members fought to slow expansion of oil and gas exploration. For a period, these voices pulled many Labor members towards a pro-climate position; after tense debates, Labor leaders gave in to pressure to delay oil exploration in the sensitive Lofoten area in March 2011.[96] However, this delay proved temporary and the government moved forward in 2013 against the Socialist Left's objections.[97] Tensions between government factions were also mirrored by cross-cutting differences in policy preferences among labor and business organizations. Such industrial unions as the Fellesforbundet supported the Lofoten expansion, while service and public sector unions opposed it.[98]

Similarly, the Stoltenberg government promoted investments in offshore extraction that counterbalanced the effectiveness of the sector's carbon tax. In principle, a carbon tax should reduce the profitability of extracting oil from marginal fields where extraction is most energy-intensive.[99] However, Norway simultaneously implemented a range of policies to mitigate the costs of mature field oil extraction, including investment rule changes to *promote* offshore development in a June 2011 energy policy white paper.[100] Debate made little reference to how such changes would undermine the offshore carbon price.[101] Discussions of avoided extraction proved even more controversial.[102]

Similar tensions structured carbon pricing specifically during the development of a 2007 climate policy white paper. When Labor was unwilling to endorse policy measures that posed costs on domestic carbon pollution, the Socialist Left threatened to leave the government.[103] In particular, negotiators could not agree on how to balance domestic versus international commitments in meeting Norwegian reduction targets.[104] These tensions rehashed old fault lines. During the Kyoto negotiations, Norway had joined the United States to lobby for "flexible mechanisms" that would allow Norway to meet emissions reduction targets by financing emissions reductions abroad. Flexible mechanisms were viewed as a way for Norway to make climate policy commitments without compromising domestic oil and gas development (Hovden and Lindseth 2002). Norwegian policymakers had long sought means to contribute to global climate reductions without imposing domestic costs on institutionally empowered polluters.[105] For instance, debates over the appropriate ratio between domestic and international commitments arose during the transition between the first Stoltenberg government and the Christian Democrat–Conservative coalition. While Stoltenberg's plan proposed that a "reasonable" share of Norway's emissions reductions should be achieved domestically, Bondevik's plan declared this share should be "considerable" (Hovden and Lindseth 2002).

As part of this new red–green coalition government, the Socialist Left wanted a stronger commitment to domestic carbon reduction in the form of meaningful reduction targets for industry. Ultimately, Labor's allegiance to carbon-intensive industry carried the day, and Stoltenberg's June 2007 white paper argued that up to one-third of Norway's carbon pollution commitments could be achieved by funding emissions reductions in the developing world. At the same time, the plan's domestic targets remained ambiguous.[106]

The opposition's response to the 2007 white paper underscored the persistence of cross-cutting climate policy cleavages. A trio of center-right opposition parties issued a public list of sixty-one environmental demands, criticizing the government's proposal as weak.[107] Among the demands was a proposal for an overall pollution reduction of 22.4 million tons, equivalent to a 27 percent reduction below 1990 levels.[108] Christian Democrats were particularly concerned that Norway not "buy" its way out of its climate obligations, emphasizing the need to couple international commitments with serious domestic cuts.[109] Environmental groups echoed this criticism.[110]

Norwegian Climate Settlements

The debates over the 2007 white paper exposed climate policy divides within the governing coalition. Socialist Left officials were upset that right-wing parties were proposing more ambitious climate policy than their left-leaning government coalition partners,[111] but Labor's industrial allies continued to block costly reforms. To solve this policy coordination problem, the government invited the opposition into cross-party talks. This move pre-empted the opposition's ability to criticize the government. Further, both Labor and Socialist Left party members believed their bargaining positions would be strengthened with new voices at the table.[112]

Cross-party negotiations have precedent in Norway, including on healthcare and pension planning.[113] Since climate policy was not a source of significant electoral conflict, partly because producers had not faced incentives to mobilize publicly against reforms, few political parties had incentive to oppose joint discussions.[114] Yet, willingness to negotiate did not bridge serious differences in policy preferences. In practice, the Labor–Socialist Left tension that had structured internal government debates spilled into the cross-party talks. At one end, Conservative and Labor politicians emphasized the need to maintain industrial competitiveness by reducing emissions at the point of least cost, including in developing countries. By contrast, the Socialist Left, Christian Democrat, and Liberals parties argued for costly domestic cuts.[115] The parties' key constituencies echoed this divide; again, peak business and labor organizations sided with their industrial members. For example, LO pushed to soften the ratio of climate policy measures undertaken domestically, even as other labor groups urged more domestic action.[116]

Negotiations culminated in January 2008 with a joint declaration by all political parties except the Progress Party, known as the First Climate Settlement (*Klimaforliket*). According to the agreement, Norway would cut emissions by about 15 million to 17 million tons of CO_2eq,[117] an extension from the 13 million to 16 million tons reductions in the original government White Paper.[118] Similar to the White Paper, two-thirds of these reductions were to be undertaken domestically, while the remaining third could be covered through international offsets. The agreement also included a range of subsidies for renewable energy, rail transit, and building efficiency. The only cost components were changes to more consumer-focused transport, gas, and diesel taxes.[119]

The value of international offsets brought Ministry of Finance officials into alignment with industrial interests, against the objections of climate reformers.[120] Neoliberal economists believed in efficient emissions reductions and supported international offsets when they presented lower marginal costs. Norwegian industrialists supported offsets to avoid domestic costs altogether. Consequently, the First Climate Settlement's centerpiece was the Norwegian Climate and Rainforest Initiative. This program helped countries in the global South avoid deforestation to offset domestic emissions. While a proposal of the Norwegian Society for the Conservation of Nature (*Naturnforbundet*), it was introduced into negotiations by opposition parties, most prominently the Conservatives, who wanted 6 billion NOK ($816 million USD) dedicated to rainforest conservation each year.[121] The final agreement reduced this figure to 3 billion NOK ($408 million USD).[122]

The cross-party agreement reflected a concerted effort by Norwegian actors to use consensual policymaking to navigate cross-cutting cleavages. The settlement focused on common policy denominators that could bridge green and industrial factions. It sidestepped policy costs for domestic carbon polluters, who were represented across the ideological spectrum inside the negotiations.

Carbon-intensive interests' efforts to restrict domestic costs continued in the settlement's aftermath when the government tasked bureaucrats to undertake sector-by-sector assessments of emissions reduction opportunities. This assessment was released as the ClimateCure (*Klimakur*) report in June 2010.[123] The report identified 160 concrete measures to reduce domestic emissions totaling 22 million tons of CO_2eq by 2020,[124] some of which involved substantial costs for domestic industry.[125] Importantly, the ClimateCure report was prepared outside usual corporatist consultative processes. Industry was upset by the report's findings, particularly its costly proposals, and by a preparatory process that largely shut economic stakeholders out from policy planning. Industry saw the ClimateCure as mapping the possible without deferring to industry on what was affordable.[126]

In response, Norway's industrial community accelerated a parallel effort to document emissions reductions potential from industry's perspective. Coordinated by NHO, a panel of industrial CEOs met from Spring 2008 through Fall 2009. In their final report, the NHO panel suggested a target of 12 million tons, lower than the government's proposal of 15 million to 17 million tons. Industry's plan focused largely on energy efficiency measures, renewable energy development, forestry practices, technology subsidies, transport taxes, and, crucially, global carbon offsets.[127] Few changes to industrial actors' domestic carbon pollution were discussed in the report.[128] Thus, industry's emissions reduction proposal was bounded by the status quo and what industry itself saw as economically viable.[129] Initially, politicians criticized the NHO report for its failure to outline a path to achieve the national goal of 15 million to 17 million tons.[130] However, the report's claim that the Climate Settlement's target was too expensive was gradually accepted by political actors.[131]

The Labor and Socialist Left parties renewed their red–green coalition after the Fall 2009 election. As part of their new coalition agreement, the Socialist Left demanded an increase in the country's emissions reduction commitment. A new target promised a reduction from the 2020 reference curve by 40 percent, rather than 30 percent, of the country's 1990 emissions.[132] However, the agreement did not provide a clear roadmap to achieving this, prompting a second climate policy white paper.[133]

This second white paper proposed a number of reforms, including doubling the offshore CO_2 tax by 200 NOK ($27.20 USD) to correct for the failures of the EU ETS (as described earlier). It also established a climate and energy technology fund, subsidies for low-carbon transport, building energy efficiency measures, and a proposal to strengthen the Rainforest Initiative.[134] Building on the perceived political success of the First Climate Settlement, the government invited opposition parties into analogous negotiations over this second white paper. As before, Labor and Conserva-tives negotiated as a block against Christian Democrats, the Liberals, and the Socialist Left.[135] Once again, the Progress Party did not participate.[136]

In contrast to the first, the Second Climate Settlement, announced in June 2012, did not initiate major new policies.[137] In fact, it *weakened* lan-guage describing the balance between domestic and international emis-sions reductions relative to the 2007 agreement. Economic stakeholders were thrilled; they interpreted the shift as belated recognition that previ-ous reduction targets were too costly.[138] As a whole, politics of both "cli-mate settlements" reveal how climate policy cooperation in Norway was achieved because policy content eschewed costs.

A Subsidy-Based Approach to Climate Policy

Neither multi-party climate agreement disrupted the producer-friendly ori-entation of Norwegian climate policymaking. No direct costs were imposed on onshore carbon polluters by either agreement, despite efforts by ideo-logically motivated reformers affiliated with the major political parties. Instead, the climate agreements reinforced a subsidy-based approach to climate reforms. These subsidies succeeded precisely because they bridged cross-cutting climate policy preferences within political parties and eco-nomic stakeholders.

The most dramatic subsidies were transport-related.[139] Electric vehicles were exempted from car purchasing taxes while biodiesel vehicles paid half the tax, an annual subsidy of 500 million to 1 billion NOK.[140] The Norwegian government also subsidized hydrogen-powered vehicles in the early 2000s by building hydrogen charging stations across Oslo. Annual support of 5 million NOK continued through 2013, even as fewer than ten hydrogen-powered cars remained on the road.[141] As Norwegian stakehold-ers noted, it was difficult to think of a subsidy that didn't exist for alterna-tive fuel vehicles.[142]

Government subsidies also supported carbon capture and sequestration technologies. During the early 2000s, CCS technologies emerged as a strat-egy for Norwegian Labor politicians to bridge the gap between the policy priorities of the party's climate reformers and industrial advocates (Tjern-shaugen and Langhelle 2009).[143] In 2000, senior union officials and Labor Party politicians jointly formed the Henriksen Committee to promote domestic gas consumption (Kasa and Underthun 2010). In a July 2001 report, the group recommended expanding gas use across Norway, including for electricity production and by households. Recognizing climate change, the report underscored the promise of "pollution-free" gas plants through CCS.[144] In other words, investment in CCS and related technologies papered over internal divisions over climate policy; the labor movement's grand bargain was expansion of gas use while committing to CCS investment (Tjernshaugen 2011).

Harkening back to Labor Party-led investments in onshore industry during the mid-twentieth century, Norsk Industri, LO, and the Norwegian Gas Forum business association lobbied successfully to establish Gassmaks, a ten-year program to research new petrochemical and gas products (Kasa and Underthun 2010).[145] The left's contradictory impulses to protect the climate while promoting industry also flared over a new gas-fired power plant to support offshore oil production. Mongstad was a marginally profitable refinery with poor efficiency. However, Statoil considered it to have strategic importance.[146] The refinery needed a heat and power plant, which Labor was prepared to approve. However, Socialist Left politicians insisted the plant require CCS technology to operate.[147]

Stoltenberg initially acceded to this demand, using CCS as a compromise to resolve intra-coalition tensions. He trumpeted investment in a full-scale gas CCS facility at Mongstad as Norway's "moon landing."[148] In return, the Socialist Left Party agreed that Mongstad could operate without carbon capture until the CCS plant was completed in 2014. The Mongstad power plant was built on schedule; however, CCS plans were abandoned in 2013 after the Norwegian Auditor General (*Riksrevisjonen*) criticized cost over-runs and the project's limited potential for success.[149] The end result was that Norway's red–green coalition authorized a new gas-fired power plant with massive but unsuccessful technology subsidies. Despite green voices' presence within government, the red wing of the red–green coalition ensured Mongstad would continue to produce carbon pollution for decades while accelerating fossil fuel extraction from the Norwegian continental shelf.

The Norwegian CO_2 Compensation Scheme

The producer-oriented nature of the Norwegian climate policy was again underscored in the late 2000s by debates over the Norwegian CO_2

Compensation Scheme. For two decades, onshore industry had justified policy exemptions by emphasizing risks to industrial competitiveness. If industry had to pay high prices for carbon, their products would be more expensive, and this could very well put them out of business. The argument then went even further, claiming that imposing costs on domestic carbon pollution would *increase* rather than decrease climate risks through "carbon leakage": reductions in Norwegian production would require foreign production increases to meet persistent industrial demand. Because Norwegian industry used clean electricity, it produced a given unit of industrial goods with less carbon pollution than foreign companies. While targeting onshore industries might reduce Norwegian carbon pollution, net global emissions could increase through industrial activity substitution to other parts of the world.

The true risk of Norwegian carbon leakage remains controversial. Union officials estimate that carbon leakage could increase net global emissions several times over.[150] Yet, many economists believe carbon leakage is a minor issue; while Norwegian industry uses low-carbon electricity, its plants also use inefficient older technologies. These economists believe potential carbon leakage is modest, closer to 10–30 percent.[151] They also note the marginal profitability of Norwegian energy-intensive industries, so Norway's economy would not suffer large economic consequences if they disappeared.[152]

Ultimately, however, the voices warning of carbon leakage were persuasive enough to launch a national CO_2 Compensation Scheme. Norwegian electricity grids are connected to continental Europe, so the marginal price of electricity in central Europe partially shapes Norwegian domestic prices.[153] Under a 2009 amendment to EU emissions trading rules, countries could compensate domestic economic sectors that cause carbon leakage as a result of ETS-linked electricity pricing shifts.[154] This EU rule change opened a domestic policymaking opportunity that Norsk Industri and LO took advantage of by pushing the government to implement a compensation scheme for onshore industry.[155] Norsk Hydro went so far as to threaten to move aluminum production abroad if the Norwegian government did not introduce the program.[156]

Negotiations took several years, in part because the government wanted to coordinate its policy with European authorities. However, both Labor and Conservative officials threw their weight behind industrial labor and business demands. Other interests were more divided, but they were ignored during the policymaking debate. Again, peak service organizations, prominently Virke, spoke against the policy, splitting from the LO and NHO coalition, but had little influence over the debate's contours.[157] In government, the Socialist Left Party struggled with how seriously to take the issue of carbon leakage. At least some party factions believed the compensation scheme was bad policy.[158] The environmental community was also divided: some advocates had serious reservations while others, notably industry-aligned Bellona, championed the policy.[159] Within the bureaucracy, the Ministry of Finance deeply opposed the policy, seeing it as defeating the ETS's purpose.[160]

Industrial polluters' interests were once again served. In September 2012, the Labor–Socialist Left coalition government introduced a 500-million NOK policy to reimburse industry for shifts in marginal electricity prices. Under this Norwegian CO_2 Compensation Scheme, a government agency calculates the fraction of the market price of electricity that can be attributed to the EU ETS and reimburses energy-intensive Norwegian industry for these indirect costs. As a senior political official in the Labor–Socialist Left coalition later reflected, the "most awful thing about it is that there are not many ways the ETS affects Norway actually—this was actually a very, very tiny bit that would have affected us. And the compensation scheme removes the last little bit."[161]

The compensation scheme became a political flash point again when the Conservative Party returned to government after the fall 2013 election under the leadership of Erna Solberg. It invited the Progress Party (*Fremskrittspartiet*) to participate in a coalition government for the first time. The Progress Party had not participated in either climate settlement because it actively expressed skepticism about climate change during the late 2000s. However, the party agreed to defer to Conservative climate change positions during coalition negotiations.[162]

Siv Jensen, leader of the Progress Party, was given the finance portfolio. Her October 2013 state budget proposal surprised almost everyone by introducing a 30 NOK ($4.08 USD) price floor into the CO_2 Compensation Scheme, which would have eliminated the compensation at low cost levels. The reform was likely the initiative of Ministry of Finance bureaucrats who remained uneasy with the compensation scheme.[163] These officials used Jensen's interest in budget cuts to push the proposal past an inexperienced minister.[164] It is, of course, ironic that a climate skeptic would be the political actor to push a pro-climate reform against the corporatist consensus.

Labor and business stakeholders reacted immediately. Working together, LO and Norsk Industri publicly decried the economic damage that would result from the potential reform,[165] and labor unions threatened to strike if the policy was enacted.[166] They were joined by some industry-friendly environmental NGOs, who objected to the price floor on carbon leakage grounds. This confused other environmental actors but generated industry goodwill.[167] Ultimately, the government relented to intense stakeholder pressure. While Finance officials continued to push for the price floor, the government removed the proposal from its final state budget in November 2013.[168]

Solberg's election also precipitated the convening of yet another Norwegian expert commission to evaluate prospects for national environmental taxes.[169] The Conservatives and Progress Party still depended on support from both the Christian Democrats and Liberals to pass legislation in parliament. While these smaller parties refused to enter into government with Progress, they used their voting power to force the government to revisit, yet again, environmental taxation. Unlike previous commissions, this new panel mostly comprised economists and finance officials. It did not include environmental groups or economic stakeholders. Removed from the corporatist policymaking process, it was viewed as performative.

The committee's final report, released in December 2015, once again argued that carbon pollution should be subject to undifferentiated costs, recommending a price of 420 NOK per ton. Except for the offshore sector, actors covered by the EU ETS were to be exempted. However, the report did not change the dynamics of Norwegian climate policymaking. Its recommendations were met with immediate opposition from many economic stakeholders and top politicians in both Conservative and Progress parties. As before, when pro-climate voices pushed policies onto the agenda, the persistent representation of carbon polluters ensured the failure of costly reforms.

The Trajectory of Norwegian Climate Policymaking

As a result of close ties to the largest Norwegian political parties and corporatist institutions that gave them access to policy design, Norwegian carbon polluters shielded themselves from substantial climate policy costs for almost two decades. Enough support for climate policy existed to continually add reforms onto the political agenda, but industrial labor and business associations were always powerful enough to make sure any resulting policy did little to actually impose costs on carbon polluters. Onshore industry was systematically exempted from Norway's carbon tax throughout the 1990s. Both Conservative and Labor party officials defended exemptions when a Christian Democratic coalition tried to expand the tax's sectoral coverage in 1997. As pressure to introduce onshore costs intensified, successive governments set up voluntary emissions reduction agreements to shield industry from formal obligations. These agreements provided industry a policy bridge until Norway could join the EU Emissions Trading System on favorable terms. In the late 2000s, Norway even introduced a compensation package to protect onshore industries from indirect costs associated with EU climate policy.

Unlike the onshore sector, the offshore oil sector faced a substantial carbon tax beginning in 1991. During the 2000s, the Norwegian government repeatedly preserved this price signal even as the country underwent a gradual transition from carbon taxes to emissions trading. However, this offshore carbon tax has not constrained carbon pollution growth in a highly profitable sector: between 1990 and 2011, carbon pollution from the energy extraction sector doubled, despite the country's carbon pricing regime.[170] Industry could pay the tax without compromising its economic viability.[171] While it may have driven some carbon pollution reductions at the margin, the tax did not cause not the economic shifts required for deep decarbonization (Riksrevisjonen 2010). Worse, costs generated by the tax were counterbalanced by pro-offshore development policies that subsidized the industry's continued expansion.

Nor has the Norwegian government used available regulatory levers to impose costs on producers. Expansive latent authority to control Norwegian carbon pollution under the country's Pollution Control Act has rarely been invoked. Instead, substantial domestic costs have been imposed only on consumers; during the 1990s, for every unit of cost imposed directly on Norwegian producers, consumers bore about two to three units of cost (Svendsen et al. 2001). Instead, climate risk mitigation emphasized subsidy-based approaches to transform the Norwegian economy without jeopardizing existing carbon-intensive jobs or capital. These investments bridged cross-cutting climate policy cleavages within both left- and right-leaning coalitions. Norwegian politicians also pushed carbon pollution reductions abroad, for instance as part of the Norwegian Forestry and Climate Initiative.

Overall, Norway presents a case of early but producer-friendly climate policy action. This trajectory reflects the privileged influence of carbon-dependent economic actors on Norwegian climate policy design. As in other advanced economies, the climate threat's emergence exposed cross-cutting climate policy preferences on both the left and right. This influence was institutionally reinforced by formal ties between carbon polluters and political parties (particularly on the left) and by the institutionalized access of carbon polluters to climate policy design through corporatist policymaking. On the left, close institutional ties between LO and the Labor Party gave industrial unions and their political allies direct access to policy debates within Labor-led governments. Similar ties between the Conservative Party and carbon-intensive business structured climate debates on the right. By contrast, smaller parties on the Norwegian right (Liberal), center (Christian Democrats), and left (Socialist Left) lacked strong ties to carbon-polluting constituencies and advocated for greater cost imposition on domestic carbon pollution. However, even during Norwegian blue–green and red–green government coalitions where these pro-environment parties sat in government, industrial wings of both the blue (Conservative) and red (Labor) parties persistently frustrated efforts by green coalition partners and pro-climate copartisans to ratchet up costs on carbon polluters.

The result is that Norwegian climate policymaking mostly proceeded through a causal pathway where policies shielded producers from costs; posing few economic threats, these policies were enacted, once proposed, with minimal controversy. Norwegian policy timing was thus a function of policy content. Norway enacted early climate reforms because carbon-dependent economic actors could reliably shield themselves from costs within consensual policymaking processes.

On rare occasions, climate policy advocates took control of the agenda from Labor or Conservative actors to press for stronger reforms. For example, in 1997, a Christian Democratic minority government proposed expanding the carbon tax to onshore industries. When this happened, carbon polluters mobilized quickly and publicly alongside their political allies on both sides of the ideological spectrum to beat back policy threats. However, as the following chapters will make clear, this mode of climate debate remained rare in Norway relative to the United States and Australia because carbon polluters were rarely excluded from Norwegian policy design.

Further, the continuous representation of carbon-dependent economic actors within the policymaking process tempered the strategic need for sustained public campaigns against domestic climate policy. Norwegian carbon polluters' ability to pre-empt climate policy threats during the design stage weakened incentives for these interests to politicize climate change more broadly or make climate cleavages a focus of national elections.[172] Instead, "quiet" management of the climate issue facilitated cross-party political agreements on climate policy. However, having agreed to policy goals, the country struggled to implement the costly policies necessary to achieve them without undermining the political conditions that made the agreements possible.

More generally, this chapter challenges the narrative that corporatist policymaking institutions facilitate environmental policy outcomes by providing an institutional context favorable for collective action. According to some scholars, corporatism helps build trust among governments and key economic stakeholders in business and labor communities, facilitates compensation to distributional losers, weakens the incentive to free-ride by nurturing collaborative decision-making, and promotes the management of long-term policy risks through stable government-interest group relationships (Scruggs 2003; Jacobs 2011). Put otherwise, the features of corporatism that make the system successful in providing social and economic protections may also facilitate environmental protections. This chapter finds the opposite in the climate domain. Consensual corporatist policy bargaining meant that policy losers always had influence over policy design. The careful, incremental approach to policy that characterizes corporatist decision-making was not able to address a policy threat like climate change, which demands non-incremental changes and cost imposition on powerful economic actors.

In Norway, the result was a domestic climate-policy trajectory that achieved significant incremental changes but never imposed sufficient costs on carbon pollution to disrupt the economic status quo. Norwegian government audits suggest that the carbon tax and early emissions trading efforts had little impact on the carbon intensity of the Norwegian economy (Riksrevisjonen 2010). Between 1990 and 2011, net Norwegian carbon pollution rose, driven by the offshore oil sector and transportation. Stand-alone evaluations of Norway's carbon pricing system suggest the country's carbon tax may have reduced Norwegian carbon emissions by only 2 percent over the 1990s (Bruvoll and Larsen 2002). Because the tax was selectively applied to the most inelastic pollution sources, it was more a fiscal than environmental measure.[173]

It remains unclear whether Norway can meet the reduction targets it set in its 2012 Second Climate Settlement, particularly its goal to achieve two-thirds of its reductions through domestic reforms.[174] For example, a Norwegian auditor's report from 2010 questioned Norway's likelihood of achieving its goals (Riksrevisjonen 2010). The report emphasized that Norwegian greenhouse gas emissions had continued to grow, particularly in the energy sector, reaching 8.4 percent above 1990 levels by 2008, despite a Kyoto commitment to limit growth to 1 percent above 1990 levels by 2012. Moreover, Norway's pollution reduction efforts to that point depended on offset purchases from other parts of the world. As a result, the collaborative nature of Norwegian climate policymaking that some scholars hope can facilitate collective climate action may instead undermine deep decarbonization by locking in the double representation of carbon polluters.

4 US Climate Policy Inaction, 1988–2006

While Norway enjoyed its green political reputation, the United States spent much of the 1990s and 2000s portrayed as a climate villain. On the international stage, the United States was one of only three countries that refused to ratify the Kyoto Protocol. At the time, the United States released the most carbon pollution in the world, a position it would hold until surpassed by China in 2007. And it was in the United States where climate denialism took root in a network of fossil fuel lobby groups and conservative think tanks (Oreskes and Conway 2010; Layzer 2012).

World leaders did little to hide their frustration with American policy inaction. After the Bush administration rejected Kyoto in 2001, incensed EU parliamentarians passed a resolution that the "long-term interests of the majority of the world population are being sacrificed for the sake of short-term corporate greed in the US."[1] Swedish leaders, who held the EU presidency at the time, condemned US inaction as "appalling and provocative."

What explains US climate policy inaction? For contemporary observers, partisan polarization looms large. Many Republican politicians, aligned with fossil fuel interests, vocally oppose climate reforms. By contrast, many Democratic actors embrace costly climate policies. However, these recent political alignments do not reflect the structure of past US climate policy conflict. Instead, support for and opposition to US climate policymaking historically cut across party lines and economic stakeholders. Moreover, narratives of US climate policy inaction obscure repeated efforts by powerful US actors to impose domestic costs on carbon polluters.

This chapter traces US climate policymaking conflict from the climate threat's emergence to the George W. Bush administration. It details how climate policy divided the executive branch under successive Republican and Democratic administrations and how cross-cutting congressional divisions systematically frustrated climate reform efforts. As was the case in Norway, numerous climate policy proposals made it onto the political agenda in the 1990s and early 2000s. The Clinton administration proposed an energy tax as part of its 1993 deficit reduction package that was explicitly framed as increasing carbon pollution costs. This tax passed the House but was rejected by the Senate after intense interest group mobilization. During Kyoto Protocol debates, Clinton administration officials debated an emissions trading scheme architecture, but the proposal never gained traction due to a lack of support outside the executive branch. Two years later, after both Bush and Gore expressed support for emissions trading during the 2000 presidential campaign, Senate legislators from both parties forged a bipartisan carbon pricing compromise. Yet, as president, Bush withdrew his support under pressure from fossil fuel interests inside his administration. Subsequent Senate efforts to enact an emissions trading scheme between 2003 and 2006 were stymied by opposition from labor, business, and political officials on both the left and right of the political spectrum.

As a result, early US efforts to enact climate reforms were persistently frustrated by climate policy opponents. This opposition reflected the presence of an ongoing policy threat; compared to Norway, climate policy advocates had fewer ties to carbon-intensive constituencies and were more willing to impose direct costs on major carbon polluters. This dynamic was reinforced by the country's pluralist policymaking institutions. Fearing these costs, carbon-dependent business and labor mobilized to reshape legislative and electoral incentives associated with policy enactment. Unlike Norway, where carbon polluters managed the policy threat quietly through corporatist institutions, US carbon polluters had to move debate into the public sphere.

Overall, opponents' success in blocking climate policy partially reflected US institutional biases against policy change. The country's balance of power system creates numerous veto points that make it easier to block policy reforms than to fashion viable pro-reform coalitions. However, the double representation of carbon polluters compounded this status quo bias. The cross-cutting distribution of climate policy preferences dispersed climate policy opponents within both left- and right-leaning political coalitions.

It ensured that opponents occupied at least one climate policymaking veto point whether pro-climate policy or anti-climate policy factions controlled the political agenda. Democrat-led efforts to enact climate reforms were persistently compromised by opponents inside the Democratic party. Correspondingly, pro-climate Republicans were repeatedly marginalized by their copartisans. The result was a systematic failure of all federal climate policy proposals between 1988 and 2006.

In chapter 5, I describe the United States' shift from climate policy laggard to climate policy leader between 2007 and 2015. While a complex set of interacting factors shaped this transition, this second period is best demarcated by the Supreme Court's landmark 2007 decision in *Massachusetts v. EPA*. That decision reshaped the US policy status quo to empower policy proponents. After 2007, business and labor stakeholders continued to exert influence on both the left and right. However, executive branch climate reformers were able to use new this regulatory power to threaten carbon polluter costs, sidestepping copartisan opponents in the legislative branch. US climate leadership was later undermined by Trump administration policies. Figure 4.1 charts milestones in US climate policymaking between 1988 and 2015.

Fossil Fuels, Political Power, and the US Economy

Fossil fuels play a central role in the US economy. Coal production and consumption were central to nineteenth-century US industrialization (Adams 2006). Today, coal is primarily used for energy production. Oil supported US transportation systems throughout the twentieth century; transport still accounts for two-thirds of US oil consumption. The remainder is used for diverse manufacturing and energy needs, from fertilizers to plastics. Natural gas is also consumed in significant quantities by households, industrial processes, and, increasingly, electricity generation.

While fossil fuels are entrenched across the US economy, their significance does vary by region. This creates differentiated state-level economic threats from climate policy (Rabe 2018). Further, the system of federal institutions in the United States makes some states more powerful at the federal level than their populations would suggest; many of these overrepresented states are particularly dependent on fossil fuels. A fossil fuel-dependent state like Wyoming, with a population of under six hundred thousand,

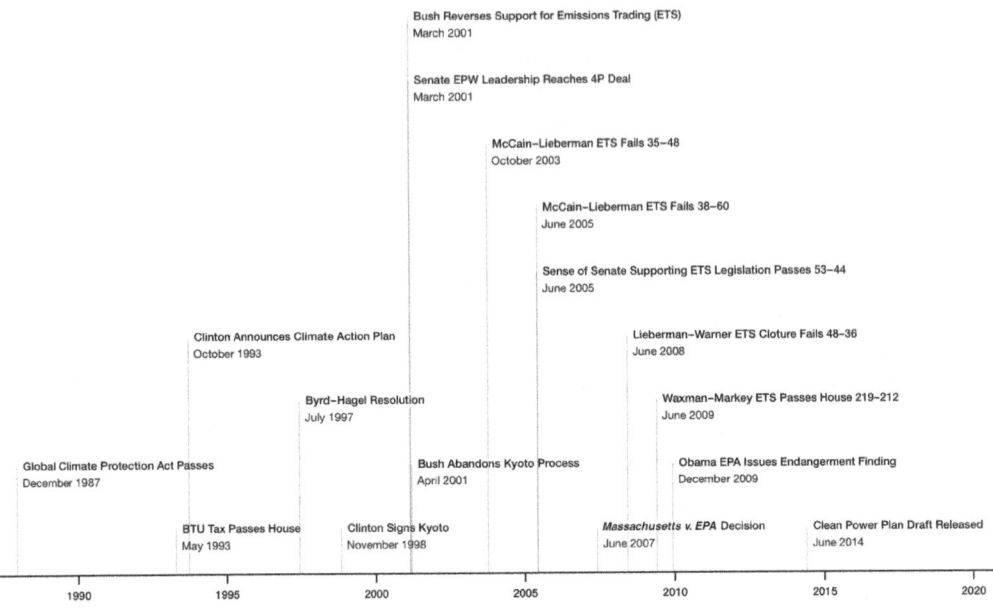

Figure 4.1
Milestones in US carbon pricing, 1988–2015

has the same number of US senators as California, an economically diverse state with almost forty million residents. In US presidential elections, Wyoming residents enjoy more than 3.5 more electoral college votes per voter than Californians. As a result of these federal institutions, fossil fuel interests receive outsized representation in the US Senate.

Fossil fuel interests are also well organized as political lobbies. Coal, oil, and gas-dependent businesses organized throughout the twentieth century to represent sectoral interests through such organizations as the Western Fuels Association (coal) and the American Petroleum Institute (oil). Fossil fuel interests also came to be represented through business lobbies. The National Association of Manufacturers (NAM) was established in 1895 to advocate for manufacturing interests, including in the environmental policy domain. The US Chamber of Commerce, founded in 1912 as an explicit counterforce to US labor, repeatedly intervened in environmental policy-making debates through the twentieth century. More recently, scholars have profiled the emergence of the Koch network as a US political force (Skocpol and Hertel-Fernandez 2016). Koch Industries, the country's second-largest private company, makes most of its revenue from fossil fuels.[2] In turn, such

Koch-funded organizations as Americans for Prosperity have emerged as major climate reform opponents.

Industrial unions are also well organized. Many carbon-intensive unions first organized under the Council of Industrial Organizations (CIO) banner. After the CIO rejoined the American Federation of Labor (AFL) in 1955, these labor actors found a voice through the AFL-CIO's Industrial Union Council (IUC), an internal labor forum to advocate for labor's energy and manufacturing interests.[3] However, unlike in other advanced economies, US labor organizing did not result in the establishment of a labor party. Class-based political coalitions emerged by the early twentieth century, the result of Great Depression party realignment that shifted industrial elites toward the Republican Party and working-class voters toward the Democratic Party (Silbey 2010). However, Democratic New Deal outreach to select farm and labor groups stymied momentum for a US labor party (Eidlin 2016). Correspondingly, formal institutional links between the Democrats and the US labor movement never developed as they did in Australia and Norway, where the dominant left-leaning political coalition emerged *as* the labor movement's political arm.

The Emergence of Climate Change on the US Policymaking Agenda

The earliest climate science grew out of the US military's atmospheric science research (Weart 2008). Senior US government officials were made aware of the climate threat due to human activity during the early Cold War.[4] However, the issue received only low-profile executive branch attention during the 1960s and 1970s.[5] This inattention reflected scientific uncertainty about the timing and extent of human-caused climate change. Compared to such high-profile challenges as acid rain, climate change remained an ambiguous future threat. Correspondingly, early US climate debates centered on support for additional climate science research.[6]

Conflict over climate research intensified during the early 1980s. An early attempt by the Reagan administration to cut climate research funding was abandoned after congressional hearings organized by then-Tennessee Representative Al Gore (Weart 2008). After failing to cut the budget for climate research, the Reagan administration turned instead to low-profile efforts to contest the climate threat's severity. For example, the administration criticized a 1983 US Environmental Protection Agency (EPA) climate report for purportedly exaggerating climate-related risks (Weart 2008; Layzer 2012).[7] At the same time, Democratic congressional representatives complained during the mid-1980s that other states' fossil fuel severance taxes inflated energy prices in their home states (Rabe 2018). Severance taxes are imposed on natural resources at the point of extraction. Democrats were arguing, in effect, *against* Republican states imposing costs on carbon pollution.

US political attention to climate change intensified in the late 1980s as the science of climate change became more certain. In part, climate change's increased profile stemmed from high-salience weather events, including a record 1988 drought in the American Midwest and West (Weart 2008). At the same time, June 1988 congressional testimony profiling the climate threat by scientist James Hansen to the Senate Energy and Environment Committee received widespread media coverage.[8] Hansen's testimony and the year's extreme weather were linked by the media, resulting in a spike in public attention (Kamieniecki 2006).

During this period, climate concerns cross-cut partisan divisions. Early proposals ranged from requests that the executive branch develop a national climate response strategy to new funding for energy conservation and climate-friendly technologies.[9] While these efforts did not gain legislative traction, they enjoyed support from Democrats and some Republicans. For example, climate policy proponents inserted a Global Climate Protection Act in the 1988 Foreign Relations Authorization Act, which passed both the House and Senate with large bipartisan majorities in late 1987. This statute gave the State Department and the EPA two years to report back to Congress with a national climate response strategy.[10] During his 1988 presidential campaign, George H. W. Bush sought to differentiate himself from Reagan by endorsing a pro-environment agenda. In an August 1988 speech, he declared that "those who think we're powerless to do anything about the 'greenhouse effect' are forgetting about the 'White House effect.' As President, I intend to do something about [climate change]."[11]

As climate policy gained traction on the legislative agenda, US carbon polluters mobilized against climate reforms.[12] A project of the National Association of Manufacturers in 1989, the Global Climate Coalition (GCC) brought together more than fifty companies and trade associations with ties to fossil fuel extraction or combustion (Kamieniecki 2006). Presenting themselves as the "voice of business" in the climate debate, the GCC attacked climate policy as scientifically unnecessary and economically unfeasible: it disseminated research that was skeptical of human-caused climate change and profiled the alleged economic costs of climate policy (Leggett 2001).[13] Faced with uncertainty over the scope of future climate reforms, fossil fuel interests sought to undermine legislative incentives to act.

Divided Climate Policy Preferences under President George H. W. Bush

Although George H. W. Bush campaigned on a "White House effect" to address climate change, his party and administration were torn apart by divergent policy preferences once in office. White House Chief of Staff John Sununu was a staunch climate policy opponent throughout the administration's early years. Sununu attempted to remove climate policy from the political agenda by dissuading US media from reporting on climate science (Gore 1993). Sununu and his fellow opponents were opposed by others in the administration, including Council of Economic Advisors Chair Michael Boskin and EPA Administrator William Reilly (Layzer 2012).[14] By early 1990, several administration officials were urging Bush to take some form of climate action.[15]

In his signature environmental policy effort, Bush sided with Reilly against administration critics to champion the 1990 Clean Air Act amendments, which addressed a variety of air pollutants but not carbon pollution (Layzer 2012).[16] While Bush sided with pro-environmental voices on some environmental issues such as acid rain, he supported the Sununu faction during debates over climate policy. During the campaign, Bush committed to hold a high-level global climate meeting during his first year in office.[17] This summit was eventually held in April 1990, against Sununu's wishes. However, Bush's position at the summit was muted, emphasizing a "no regrets" policy approach that would pose no economic costs for individual companies (Leggett 2001).[18] Bush also resisted climate policy reform at other international conferences. While the administration committed to the concept of an international climate treaty in May 1989, it frustrated efforts to develop emissions reduction targets at a November 1989 Netherlands conference.[19] By the 1990 World Climate Conference in Geneva, the United States was the only advanced economy not to have announced a carbon pollution reduction target (Leggett 2001).

Mirroring the divided executive branch, congressional climate advocates came from Democratic *and* Republican ranks. Bush's climate obstructionism at the 1989 Netherlands conference drew criticism from both Democrats and pro-climate Republicans including Pennsylvania Senator John Heinz.[20] Led by Rhode Island's John Chafee, five Republican senators wrote to Bush demanding the administration support an aggressive climate target, including a commitment to stabilize US carbon emissions at 1988 or 1989 levels.[21] Both the House and Senate passed bipartisan resolutions with large majorities urging Bush to attend the Rio Earth Summit.[22] Advocates also introduced climate provisions to the Clean Air Act amendments of 1990 and the Energy Policy Act of 1992. However, without strong support from environmentalists who were not yet prioritizing domestic climate policy advocacy, these efforts never gained traction.[23]

Sununu's eventual departure from the White House facilitated a more accommodating stance toward climate issues (Weart 2008). However, Bush's early pro-environment rhetoric did not translate into climate action; during his four years in office, no new policy was adopted that explicitly addressed carbon pollution. Instead, Bush doubled down on climate policy opposition during the 1992 election by attacking Clinton's "drastic" plan to impose a "punishing carbon tax" that would purportedly cost 600,000 jobs and $100 billion annually.[24] These attacks included swing-state radio ads opposing carbon taxation in the final days of the 1992 campaign, even though Clinton had not actually proposed such a policy.[25] In short, despite the presence of pro-climate voices within Congress and the Bush administration, policy opponents maintained control of the political agenda.

Divided Climate Policy Preferences under President Bill Clinton

With the election of Democrat Bill Clinton in 1992, pro-climate forces enjoyed unprecedented influence through Vice President Al Gore.[26] Gore had been a congressional climate policy leader, having engaged the issue personally since taking a class with climate scientist Roger Revelle during college (Gore 1993). In addition to spearheading the fight against Reagan's proposed climate research funding cuts, Gore had authored a number of aggressive climate bills in the early 1990s. Unlike contemporaneous proposals focused on subsidies and conservation measures, Gore's efforts threatened significant costs on carbon pollution. His World Environment

Policy Act of 1991 proposed that large, new carbon-polluting facilities must buy carbon offsets or pay a $250 per ton penalty.[27] His May 1992 Global Climate Protection Act tasked the EPA with regulating carbon pollution to stabilize carbon pollution at 1990 levels by 2000.[28] In his best-selling 1992 book, *Earth in the Balance*, Gore argued for an expansive carbon tax on all fossil fuels and carbon-intensive products.[29] However, Gore questioned emissions trading approaches to climate mitigation, seeing cap and trade as a conservative, Republican idea because Bush had embraced the policy instrument during acid rain debates (Pooley 2010). More generally, Gore's importance to climate politics during this period reflects the elite nature of attention to climate concerns. Climate advocacy was led by scientists and elected officials, not environmentalists.[30]

Gore's climate policy enthusiasm was not shared by everyone in the Clinton administration.[31] Just as the climate threat had divided senior Bush administration officials, climate change became the most divisive environmental issue during the early Clinton presidency.[32] Gore wanted the United States to commit to stabilize US carbon pollution at 1990 levels by 2000. However, Energy Secretary Hazel O'Leary and Treasury Secretary Lloyd Bentsen were both concerned about the economic impact of this proposal on US industry.[33] Clinton sat in the middle: while not opposed to climate policy, neither was it his key priority.[34] On stabilization targets, Gore's view would prevail in early 1993 when Clinton included the 1990 by 2000 target in an Earth Day address; however, Clinton offered few plans for achieving the goal, promising vaguely to "continue the trend of reduced emissions."[35] In an attempt to bridge the differences within his own political coalition, Clinton's commitments emphasized hypothetical win-win conditions under which environmental protections could be aligned with job gains and economic growth.

By contrast, pro-reform forces would push energy taxation during the Clinton administration's first year. However, that reform would still fail as the result of coordinated opposition from double-represented carbon polluters across the political divide.

The BTU Tax Debate

During his February 1993 State of the Union address, Clinton proposed a British Thermal Unit (BTU) tax as a revenue component of his deficit reduction package.[36] A BTU is the quantity of heat required to raise the temperature of one pound of water by one degree Fahrenheit. Clinton's BTU tax proposal would have taxed energy-intensive economic activity by individuals and companies to raise an estimated $22 billion annually.[37] The intense debate engendered by this proposal underscores the cross-cutting nature of US climate policy preferences. Unlike the Norwegian carbon tax enacted two years earlier, the BTU tax deliberately threatened major economic stakeholders with real costs. Facing this threat, policy opponents quickly mobilized to publicly undermine electoral and legislative incentives to pass the tax. Simultaneously, Democratic and Republican officials with ties to carbon-dependent economic interests weakened the policy's content and, later, blocked its passage.

During deficit reduction planning, the Clinton team had considered a range of policy instruments, including a carbon tax, a BTU tax, a clean fuels tax, an oil import fee, and an increased gasoline tax.[38] Ultimately, a carbon tax was rejected because it would burden Democratic constituencies, particularly workers and politicians in coal-dependent states. The administration was particularly concerned that West Virginian Senator Robert Byrd, chair of the Senate Appropriations Committee, would block any proposed carbon tax as his state was a leading domestic coal producer (Erlandson 1994). As Clinton argued in a speech to Congress: "Unlike a carbon tax, [the BTU tax] is not too hard on the coal states."[39] In this way, the BTU tax was more politically acceptable in the face of within-party differences in Democrats' material interests.[40]

Distributive concerns also shaped the tax's architecture. Provisions included lower rates on home heating oil, to avoid disadvantaging oil-dependent Northeastern consumers, and the inclusion of energy taxes on clean hydropower and nuclear energy, to include such regions as the Pacific Northwest as tax revenue sources.[41] In other words, the Clinton administration tried to balance the effect of the tax across regions of the country. As is typical of energy taxes, the proposal was regressive, as it would hurt low-income Americans more than the wealthy ones. However, its effects on low-income Americans were counterbalanced by other provisions in the deficit reduction package. These included expansion of the earned income tax credit (a tax provision benefiting low-income Americans), increased funding for the Low Income Home Energy Assistance Program (LIHEAP), and increased funding for food stamps. The administration argued these benefits exceeded the marginal energy cost increases that the BTU tax would impose on low-income Americans. In this way, the tax was far more sensitive to consumer costs than the contemporaneous Norwegian carbon tax, which burdened consumers while shielding producers.

It is also noteworthy that the tax exempted feedstock energy use, for instance crude oil or gas used to produce natural gas or oil, to minimize industry costs. In short, the BTU tax was far from the climate policy instrument that some administration advocates wanted; instead, it provided an uneven environmental signal that focused on bridging regional differences within the Democratic coalition.[42] Nevertheless, the tax was sold, in part, as an effort to support the US goal of cutting carbon emissions to 1990 levels by 2000.[43]

The BTU tax received mixed support from environmental stakeholders. Of the major green groups, the Environmental Defense Fund (EDF) and the World Resources Institute (WRI) were strong promoters of market-based policies; however, EDF was committed to emissions trading and was reluctant to shift toward tax-based instruments.[44] Instead, momentum for an energy tax came from Gore and other political officials, not environmentalists.[45] Leading up to the State of the Union, Gore and other senior officials held private meetings with environmental groups to update them on the administration's agenda. The groups then coordinated their efforts to lobby for the tax.

Similarly, labor unions were followers, not leaders, on the BTU tax. Perhaps the most engaged labor actors were coal miners who supported the decision to shield the coal industry from concentrated costs.[46] Yet, minimal evidence of labor involvement in BTU tax advocacy exists in archival materials, contemporaneous media reports, or stakeholder recollections.

By contrast, carbon-intensive businesses mobilized quickly and early against a potential carbon tax, starting in fall 1992.[47] In January 1993, fossil fuel companies formed an Alliance Against a Carbon Tax to catalyze a broad-based coalition against carbon taxation.[48] After Clinton's announcement of the BTU tax, manufacturing and energy-intensive industries criticized the plan, including refiners, utilities, oil and gas companies, manufacturers, and transport companies.[49] With the specific threat of a carbon tax dodged, the Alliance changed its name to the Energy Tax Policy Alliance (ETPA) to support industrial opposition to the BTU tax. ETPA was joined by a second umbrella organization, the Affordable Energy Alliance, a coalition of 900 organizations opposed to any energy tax. The Alliance represented a diverse membership base across almost every state. These groups emphasized the significant producer costs associated with the policy, alongside alleged costs to the US economy.[50] In early 1993, ETPA hired Bonner & Associates to incite grassroots opposition against the proposal, using mail, phone calls, and organized visits from "home-district influentials."[51] The idea was to "smother" Congress in letters and phone calls.[52] Starting in April, policy opponents paid Burston-Marsteller over $1 million to create grassroots opposition against wavering Democrats on the Senate Finance Committee, involving almost forty-five staff members in twenty-three states. The staff blanketed the country with local anti-BTU editorials, produced local economic impact reports nationwide, and even helped school boards estimate BTU tax impacts on their budgets.[53] The BTU tax was also opposed by farm interests who cited cost burdens to US agriculture in excess of $500 million annually.[54]

Business lobbies thus targeted legislative incentives to support the tax directly by attacking the tax's merits to legislators and indirectly by mobilizing the public against the idea. For instance, Mobil Oil placed a full-page advertisement in the *New York Times* comparing the energy tax to Ford's Edsel automobile, a famous 1950s commercial failure, noting "both looked good on paper."[55] Utilities sent their customers mailers critical of the policy and ran newspaper campaigns to frame the tax as a "job destroyer" (Erlandson 1994). The Affordable Energy Alliance disseminated a briefing book rallying businesses against the bill.[56] Other opponents prepared district-by-district breakdowns of projected job losses.[57] Farming interests prepared county-by-county estimates of tax-related agricultural costs increases.[58] The Koch-funded Citizens for a Sound Economy (CSE), representing itself as a consumer advocate, ran local ads and information sessions arguing that environmental pricing took away consumer freedoms.[59]

Other businesses supported the reform, previewing growing climate policy divisions within the business community. These included the Business Council for a Sustainable Energy (BCSE), an early coalition of renewable energy companies, energy efficiency actors, and the natural gas industry.[60] These coalition members all stood to profit from the reform. Similarly, while the American Petroleum Institute opposed the tax, some oil companies publicly endorsed the proposal, including Atlantic Richfield Company (Arco), Unocal, British Petroleum, and Shell.[61] These groups ran advertisements highlighting household benefits associated with the tax.[62]

Similar divisions split political officials. Even before the BTU tax announcement, a bipartisan coalition of congressional representatives from largely coal-dependent districts wrote to President Clinton in February 1993 to condemn carbon pricing.[63] Figure 4.2 charts the congressional districts of the letter's signatories. These districts mostly overlap the distribution of US fossil fuel deposits, and emphasize how both Republicans and Democrats with ties to carbon-dependent economic actors contested reform efforts.

Efforts to balance the BTU tax's distributive burden across states did not mollify these opponents.[64] Instead, elected officials allied with interest groups, sometimes parroting corporate opponents' materials. For example, in its criticisms of the BTU tax the House Republican Policy Committee used financial estimates taken directly from a National Association of Manufacturers–commissioned report.[65]

Industry hoped to turn public sentiment against the tax by making consumer costs particularly salient. Their goal was to emphasize "hidden" policy costs for consumers and drown out the Clinton team's efforts to

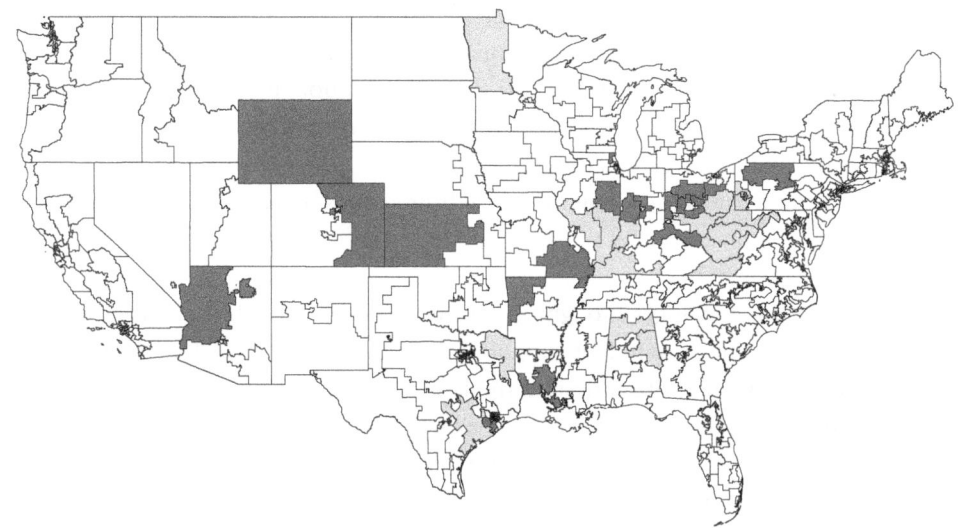

Figure 4.2
Cross-cutting opposition to carbon taxation in the 103rd Congress. Map displays the districts of congressional representatives who wrote to Clinton on February 11, 1993, opposing the introduction of a US carbon tax. Democrat-held districts appear in light gray, Republican-held districts in dark gray.

contextualize consumer impacts with the deficit package's compensatory policies.[66] Yet, interest groups' apocalyptic economic predictions about BTU tax costs were not borne out by independent analysis. A Congressional Research Service report compared the BTU tax's impact against other potential deficit-reduction measures, and found nothing specifically harmful about the policy.[67]

Facing opposition from Democrats representing fossil fuel-dependent states and districts, the Clinton administration tried to bridge within-party divisions. One point of tension was a demand by some Democratic senators, including Louisiana's Bennett Johnston and Oklahoma's David Boren, that tax collection points be closer to retail consumption, to minimize producer costs.[68] A member of the Senate Finance Committee, Boren threatened to stall the tax proposal entirely. Industry echoed these demands. To maintain sufficient Democratic Party support, the Clinton administration acceded to these demands in April 1993 and moved liability points closer to consumers.[69] It also expanded exemptions for some fuels championed by prominent Democrats, including home heating oil and ethanol, despite the fact that these changes undermined the policy's environmental objectives.[70]

Even with these concessions, some Democrats maintained reservations. The BTU tax proposal was thus amended again in mid-May. First, the Ways and Means Committee reversed the ethanol exemption, instead exempting farm gas and diesel use.[71] Second, it created new exemptions for energy used in aluminum production.[72] Most critically, the proposal bowed to Oklahoma Democratic Congressman Bill Brewster's demand for line-item BTU charges on consumer utility bills, which had been lobbied for by utility companies.[73]

While these changes brought more Democrats on board, they did not secure economic stakeholder support. For example, the Farm Bureau did not reverse its opposition despite the amendments cutting the industry's cost burden in half.[74] Instead, policy opponents redoubled their efforts through the Affordable Energy Alliance.[75] Still, with interests of key energy-dependent Democrats satisfied, the BTU tax passed the House on a narrow 219–213 vote in late May 1993 as part of the 1993 Omnibus Budget Reconciliation Act.[76]

Although the amended policy bridged sufficient distributive tensions to gain majority support from House Democrats, it still faced opposition from regional Democratic interests in the Senate, including Senator Boren and Louisiana Senator John Breaux. Meanwhile, the same interests that had lobbied the House to transparently target consumers, now argued the tax was unfair to consumers precisely because it *protected* producers (Erlandson 1994). For example, the Koch network's CSE saturated Oklahoma with radio and TV ads, asking the public to contact Boren and "spend some of your energy to stop the energy tax."[77] With both Boren and Breaux holding seats on the Senate Finance Committee, prospects for the BTU tax effectively died (Milne 2008).[78]

Acknowledging this political reality, the administration dropped the BTU tax from its deficit reduction package.[79] The Senate passed a version of the Omnibus Budget Reconciliation Act, minus the BTU tax, in June by a 50–49 margin.[80] During House-Senate conference negotiations, even Democrats who had voted for the House bill criticized the tax proposal (Erlandson 1994). Later, the failed BTU tax was seen by many Democrats as a key driver of their 1994 midterm losses. Over the following decade and a half, many Democrats remained reluctant to engage with climate policy for fear of being "BTUed" by an unsupportive electorate.[81]

Ultimately, the BTU tax faltered not only because of Republican and industrial opposition, but also because the Democratic Party could not reconcile its cross-cutting policy preferences. The climate threat's emergence had revealed latent differences in the material interests of different Democratic constituencies. From the beginning, a carbon tax was rejected as a way to accommodate coal-dependent Democrats. The Clinton administration instead hoped the BTU tax could advance climate policy objectives without imposing regional costs. However, after carbon-dependent economic actors mobilized against the tax, the administration was forced to moderate the policy's impacts on an expanding set of Democratic constituencies. While these amendments bridged internal House differences, the House bargain could not satisfy the distribution of regional interests in the Senate.

Domestic Divisions over International Climate Negotiations

Clinton's willingness to impose costs on carbon polluters faded with the BTU tax's failure. While Gore continued to press for a robust policy response to climate change, senior economic advisors such as Deputy Treasury Secretary Lawrence Summers and Council of Economic Advisors Chair Janet Yellen cautioned against early action to avoid other countries' free-riding off of US policies. (Layzer 2012). When Clinton released a National Climate Action plan in October 1993, he recommitted the country to adoption of a stabilization goal of 1990 levels by 2000. However, his plan mostly relied on voluntary measures.[82] Environmental groups doubted the package could meet its emissions reduction goals.

With the 1994 Republican takeover of Congress, domestic policymaking efforts stalled. As a consequence, the focus of pro-climate Democrats shifted to international climate negotiations. Despite domestic inaction, US negotiators endorsed efforts to limit global emissions at the Berlin Climate Change Conference (COP 1) and spoke in favor of a binding treaty at the Geneva Climate Change Conference (COP 2).[83] However, leading up to the December 1997 Kyoto conference (COP 3), Clinton's Cabinet reached a climate policy stalemate.[84] Clinton's senior economic advisors cautioned that the United States should not agree to an international treaty without full developing country participation; by contrast, the State Department, Department of Energy, and the EPA all advocated a "two-step" approach in which developed countries committed to stabilize carbon levels first and then brought developing economies on board later.[85]

Within the US Cabinet, there was consensus on several small measures to address climate change: tax cuts for energy efficiency and low-carbon technologies as well as low-carbon research and development. However, the Cabinet split over the appropriateness of further domestic policy measures. One camp, led by Lawrence Summers, Janet Yellen, Office of Management and Budget Director Frank Raines, Secretary of Commerce William Daley, and Secretary of Labor Alexis Herman cautioned against US action before developing countries accepted binding commitments; after the international community completed a binding agreement, the group advocated a global emissions trading scheme or a global carbon tax on the order of $22 per ton by 2015. A second approach was put forward by Director of the National Economic Council Gene Spirling and Chair of the Council on Environmental Quality Kathleen McGinty, supported by Secretary of Energy Federico Peña, Secretary of Transportation Rodney Slater, and then-White House Staff Secretary and senior White House climate negotiator Todd Stern. This camp advocated early action and a commitment to stabilize emissions at 1990 levels by 2015 or 2020. Their proposal included a domestic emissions trading scheme with a "safety valve" (e.g., a ceiling price) of about $25 per ton to protect the US economy from excessive costs. The scheme would have begun in 2005. A third camp also supported a cap and trade system with a safety valve, but its members believed that steeper carbon pollution reductions were necessary to assert US leadership and mitigate climate risks. This camp was led by EPA Administrator Carol Browner, Secretary of the Interior Bruce Babbitt, Secretary of Agriculture Dan Glickman, USAID Administrator J. Brian Atwood, Deputy National Security Advisor James Sternberg, Director of the Office of Science and Technology Policy Jack Gibbons, and Deputy Secretary of State Strobe Talbott; these administration voices wanted an emissions reduction goal of 1990 levels by 2010 and a ceiling price as high as $50 per ton.[86]

By this point, the administration faced anti-climate policy pressure from both business and labor stakeholders. Senior advisors warned that "aggressive approaches [to climate policy] would expose us to well-financed campaigns—*by major corporations and labor unions*—that demagogue our policies as excessively costly and as a large energy tax increase" [emphasis added].[87] Similarly, in advocating for his middle-ground position, Spirling argued that a future cap and trade scheme would need less aggressive reduction timetables—reaching 1990 levels by 2020 instead of 2015—to "attenuate corporate and union opposition."[88]

Ultimately, Clinton sided with the middle group: the United States would support early international action and a domestic trial emissions trading scheme for the 2005 to 2008 period.[89] Yet, Clinton was still reluctant to propose costly reforms. In an October 1997 climate policy speech, Clinton outlined energy efficiency and clean energy tax credits, electricity deregulation to increase market competition, and private-sector emissions reduction partnerships. He did call for the establishment of a "market system for reducing emissions ... that will draw on our successful experience with acid rain permit trading" but stopped short of a concrete proposal.[90] Instead, under the Clinton plan, short-term climate policymaking would consist of subsidies and voluntary efforts; after a first economic review, there would be a pilot market program, followed by a second economic review before broader expansion of US emissions trading. In short, this was a politically symbolic plan that sidestepped meaningful action (cf. Rabe 2018).

In the Kyoto lead-up, domestic economic losers stepped up their opposition to US international engagement. On the business side, the Global Climate Coalition ran a $13 million campaign to emphasize Kyoto's alleged catastrophic economic costs (Layzer 2012). The auto industry-funded Coalition for Vehicle Choice argued against any global treaty. And the coal-funded Centre for Energy and Economic Development allied with the United Mine Workers of America to oppose US action (Layzer 2007).[91]

Opposition also came from labor interests. The UMWA had been mobilized around climate policy since the early 1990s. Alarmed by the asymmetric commitments that developed countries made under COP 1's Berlin Mandate, an international agreement to establish binding climate reduction targets, the UMWA began to organize labor climate policy opposition. For instance, it arranged a meeting with the AFL-CIO to brief senior labor leaders on the emerging climate policy threat and encouraged them to join the debate.[92]

Industrial unions soon worked to frustrate US participation in a global climate treaty, led by the Industrial Union Division (IUD) of the AFL-CIO. In a November 1996 IUD resolution, the group criticized the Berlin Mandate for excluding developing countries. It called on Congress and the executive branch to reject any global deal that did not include developing country participation or that would cause US job losses.[93] Despite divisions within the AFL-CIO, its executive council would eventually side with industrial labor to endorse this position.[94]

The AFL-CIO brought the resolution to Congress, working with legislators, including both Senator Byrd and Senator Hagel, to craft what would become the Senate's Byrd–Hagel Resolution that the United States should reject any international climate agreement that did not set pollution limits on developing countries or that would harm the US economy.[95] Labor voices were joined by energy industry officials in pushing Senate action against US climate treaty participation (Royden 2002).[96]

With the Byrd–Hagel Resolution accumulating sixty-five cosponsors from both parties, Senate climate advocates, including Democratic Senator John Kerry and Republican Senator John Chafee, decided to throw their weight behind the resolution in an attempt to muddy its significance. They hoped that, by supporting the resolution, its political symbolism would be diluted.[97] Labor interests take credit for this, claiming their support was what "caused [Byrd–Hagel] to be adopted unanimously 95–0, rather than along partisan lines, which could have produced a 65–35 vote."[98] Regardless, climate policy proponents did not anticipate that their unanimous endorsement of the Byrd–Hagel Resolution would become a powerful rallying cry for policy opponents over the coming decades.

At Kyoto, US negotiators eventually agreed to a 7 percent reduction below 1990 levels by 2010; they believed this was nearly equivalent to Clinton's previous goal once carbon sinks and synthetic gases were factored into compliance requirements (Royden 2002). Yet, the absence of binding commitments for developing countries stunted congressional support. Thus while Clinton would sign the treaty, he never sent it for Senate ratification. Meanwhile, industry continued to oppose the Kyoto Protocol, while environmentalists worried the agreement was too weak (Royden 2002). Similarly, the AFL-CIO remained opposed to Kyoto through the late 1990s. Even an effort to organize an internal Climate Change Working Group to define climate change principles the federation could support fell apart when member unions could not find sufficient common ground (Obach 2004a). As before, while some high-level officials supported costly climate change policy, opposition from industry and labor was influential enough within both the Republican and Democratic parties to stymie meaningful reform.

Toward the end of his presidency, Clinton did take a few executive actions to reduce carbon pollution. For instance, a June 1999 directive instructed federal agencies to reduce building energy use emissions by 30 percent below 1990 levels by 2010.[99] During this period, some EPA voices also suggested Clinton go further by using the Clean Air Act to directly regulate carbon pollution. The idea first surfaced publicly during a March 1998 appropriations hearing when Texas Republican Tom Delay grilled EPA chief Browner on a leaked internal EPA brief.[100] In an April 1998 legal memorandum that responded to Delay's inquiry, the EPA general counsel suggested carbon pollution regulation was within EPA's scope provided the administrator found that "CO_2 emissions are reasonably anticipated to cause or contribute to adverse effects on public health, welfare, or the environment."[101] The EPA's legal opinion drew opposition from some elected congressional officials. Browner herself testified in February 2000 that regulating carbon dioxide was not yet within the EPA's mandate, though the EPA's general counsel maintained that carbon dioxide could become subject to EPA authority if a future administration so desired (Royden 2002).

Notwithstanding these efforts, Clinton's term in office was characterized by limited climate action. Policy opponents distributed across the US government within both the Democratic and Republican parties were able to systematically block even small efforts to impose costs on carbon pollution. In moments when pro-climate factions moved climate policymaking onto the political agenda, as with the BTU tax and US Kyoto engagement, dispersed policy opponents in government, allied with industrial business and labor interests, effectively undermined climate reform efforts.

Climate Policy Inaction under President George W. Bush

After George W. Bush won the presidency over Gore in 2000, pro-climate voices lost executive branch influence.[102] However, the climate policy opposition that would define the Bush administration's later years was not apparent at first. In fact, pro-climate Republicans advocated carbon pricing during the administration's early months. These efforts drew from legislative momentum to manage the threat of carbon pollution as part of a four pollutant (4P) strategy. In a September 2000 campaign speech, Bush promised to implement a 4P strategy if elected, requiring the power sector to reduce sulfur oxide, nitrous oxide, mercury, and carbon pollution. Bush also suggested "market-based incentives, such as emissions trading, to help industry achieve the required reductions."[103] A mandatory carbon pollution cap was listed in Bush administration transition planning documents that were prepared for the incoming cabinet in late 2000 (Whitman 2005).

Anticipating that, no matter the election result, there would be momentum on a 4P strategy, senators and staff on the Environment and Public Works Committee explored the basis for a 4P policy beginning in fall 2000. These negotiations were boosted by ongoing federal lawsuits against coal-fired power plants.[104] As part of the 1977 Clean Air Act amendments, Congress established that existing industrial air pollution sources would need to undergo a "New Source Review" (NSR) after substantial modifications. Between 1998 and 2000, the Clinton administration filed almost two dozen lawsuits against utility companies, claiming they had made serious facility changes without requesting an NSR, under the cover of routine maintenance.[105] Senate 4P negotiations explored a bargain to legislate restrictions on the four pollutants in return for Clean Air Act regulatory relief, including from NSR provisions. By March 2001, the contours of a deal in principle had been agreed to by the top two Republicans and top two Democrats on the Senate Environment Committee.[106] Under the deal, carbon pollution would be managed through an emissions trading scheme, with a cap declining over twenty years.[107]

The Senate deal was negotiated without stakeholder consultation; it was set to be announced in early March 2001 when, surprising principals in both parties, Bush reversed his support for carbon pollution's inclusion in the bill.[108] As late as the first week of March, EPA administrator Christine Whitman had recommitted the US administration to establishing a cap and trade system at a G8 environment ministers meeting in Trieste, Italy, with approval from Bush's National Security Advisor and chief of staff (Whitman 2005). However, senior business officials and anti-climate Republicans became concerned with climate policymaking momentum and forced an internal review of the proposal. Outside the administration, four Senate Republicans also wrote Bush to declare their opposition.

On March 13, 2001, Bush abruptly informed Whitman that he was reversing his position (Whitman 2005). In a letter to Republican senators, Bush announced he no longer supported efforts to impose mandatory reductions on carbon pollution, noting that it was "not a pollutant under the Clean Air Act."[109] Bush justified his reversal with Department of Energy analysis that carbon pollution inclusion in a multipollutant strategy would increase the price of coal and, consequently, consumer electricity costs. With Bush's announcement, the Senate bargain fell apart before it could be announced. The administration's shift angered both Democrats and pro-environment Republicans, including Oregon's Gordon Smith, who had strongly supported the deal.[110] Legislative proposals for a multipollutant deal that included carbon would continue in Congress.[111] For instance, a 4P bill that included an emissions trading scheme for carbon pollution was proposed by Independent Senator Jim Jeffords. The bill made it out of committee in June 2002 on a 10–9 vote, with Democrat Max Baucus voting against the bill but Republican Lincoln Chafee supporting the bill as a cosponsor.

As time went on, the Bush administration distanced itself further from climate policy as carbon-intensive constituencies flexed their influence within the administration. In April 2001, the Bush administration announced its "withdrawal" from the Kyoto Protocol. When, in February 2002, Bush proposed the US economy reduce its carbon intensity by 18 percent by 2012, he offered only voluntary policies centered on $4.6 billion in tax incentives for alternative vehicles and renewable energy.[112] Bush also established the voluntary Climate Leaders program, which allowed businesses to work with the EPA to track their carbon pollution reductions. Because his climate measures imposed no costs on carbon pollution, they were harshly criticized by environmentalists while being applauded by many carbon polluters.[113]

In another win for fossil fuel interests, the Bush administration wound down efforts to prosecute coal-fired power plants for New Source Review violations.[114] Instead, the Bush EPA enacted new NSR rules in December 2002 and October 2003 that expanded the activities that utilities could undertake without triggering a permit.[115] The Department of Justice had to continue prosecuting cases initiated during the Clinton administration, many of which remained staffed by lawyers committed to the file. However, the administration did its best to hamstring its legal team; for example, in some NSR cases, lawyers were prohibited from mentioning climate change during arguments.[116]

The administration also contested climate science in a bid to further undermine policymaking incentives. In June 2001, Bush questioned whether scientists could differentiate natural from anthropogenic global warming.[117] The administration also purportedly banned federal officials from referencing a climate science report mandated by the Global Change Research Act of 1990 (Piltz 2005).[118] Internally, administration officials attempted to disrupt statutory obligations for climate science reporting, despite pushback from both Republican and Democratic legislators and legal conflict with environmental groups.[119] Officials also weakened government statements about the climate threat. For example, Phillip Cooney, the chief of staff for the White House Council on Environmental Quality, edited official reports to eliminate climate threat descriptions and references to National Academy, IPCC, and National Assessment climate reports.[120] Cooney had formerly worked as an American Petroleum Institute lobbyist.[121] In 2003, the EPA decided to avoid mentioning climate change in its annual report because the only text acceptable to the White House would, in the agency's eyes, have compromised EPA integrity (Shulman 2008).

Then, in May 2003, EPA Administrator Whitman resigned, leaving anti-climate Republicans in near control of the executive branch. By this time, congressional reformers had moved on from 4P proposals to advocate standalone carbon pricing. Between 2003 and 2007, diverse Senate and House legislators introduced climate-related bills, ranging from fully realized emissions trading schemes to simple statutory authorization for the EPA to undertake all measures necessary to achieve specified levels of carbon pollution reductions.[122]

The first emissions trading bill to receive serious political consideration was the Climate Stewardship Act of 2003, a joint effort by Republican John McCain and Democrat Joe Lieberman. McCain and Lieberman began work in August 2001, supported by emissions trading proponents at the Environmental Defense Fund (Pooley 2010).[123] Their bill proposed a cap and trade scheme covering about 85 percent of US carbon pollution, including electricity, transportation, industry, and commercial sectors. It aimed to stabilize US greenhouse gas emissions at 2000 levels by 2010, with further targets set by the Commerce Department on a biennial basis.[124] Yet, by design, the McCain–Lieberman bill sketched the symbolic contours of a future deal, rather than a fully specified climate policy architecture. For example, the bill did not make allowance distribution decisions.[125]

The McCain–Lieberman bill faced a challenging political environment. Senate Republican climate opponents had blocked even minor climate policy provisions in the Energy Policy Bill of 2003 (Layzer 2012). Senator Jim Inhofe, chairman of the Environment and Public Works Committee, refused to consider the McCain–Lieberman proposal. And Bush's EPA refused to perform customary economic and environmental analysis of the draft bill, leading McCain and Lieberman to accuse the administration of discrimination against legislation that did not parrot its climate policy opposition.[126]

McCain and Lieberman forced floor time for their bill by threatening to add their proposal as an amendment to the 2003 energy policy bill.[127] Among interest groups, EDF ran an ad campaign in support of the bill,[128] but other environmental groups were surprised by the bill's sudden profile and organized only informal campaigns in support.[129] A few industrial actors, including Alcoa and BP, began to signal some emissions trading acceptance.[130] However, the bill was mostly opposed by lobbies representing oil, gas, coal, manufacturing, and metal industries.[131]

Against this steep opposition, the bill's modest 43–55 defeat in late October 2003 was viewed as a legislative milestone. It revealed a significant reserve of Republican support for climate action as well as a critical mass of Democratic opponents from fossil fuel-dependent states. This partisan gradient—broad Republican opposition tempered by a supportive Republican minority and broad Democratic support tempered by a vocal Democratic minority—would continue to define emissions trading debates throughout the remainder of the George W. Bush administration.[132]

Among Republicans, McCain continued to criticize Bush's "terribly dis-appointing" climate policies.[133] Along with Lieberman, he reintroduced a mostly unchanged Climate Stewardship and Innovation Act in 2005.[134] The bill was defeated in June 2005 by a vote of 38–60, losing senators who had previously supported the 2003 version. These senators switched both because climate policy had become increasingly controversial and because the new bill included pro-nuclear provisions (Layzer 2007).[135] Yet, immedi-ately after, the Senate passed a "Sense of the Senate on Climate Change" as an amendment to the Energy Policy Act of 2005. The resolution affirmed the existence of climate change and made a symbolic commitment to, by the "end of the first session of the 109th Congress, ... enact a comprehen-sive and effective national program of mandatory, market-based limits on emissions of greenhouse gases." The amendment collected five Republican cosponsors and eleven Republican votes.[136] In short, even at the height of the Bush administration's climate policy recalcitrance, a quiet faction of pro-climate Republicans waited in the US Senate. Climate reformers hoped this group would join in a bipartisan climate reform coalition after Bush left office.

The Trajectory of US Climate Policymaking, 1988–2006

By the final years of the Bush administration, US climate reformers had little to show for their two-decade effort to enact federal climate policies. Early attempts to impose costs on carbon polluters culminated in the Clinton administration's BTU tax proposal. However, that effort collapsed under pressure from carbon-dependent economic actors, relegating climate change to the bottom of the federal political agenda throughout the 1990s. While the Clinton administration discussed emissions trading toward the end of Clinton's second term, these conversations were largely symbolic. A second climate policymaking window peaked with a 2001 Senate deal-in-principle for a carbon price as part of a 4P multipollutant strategy. That deal collapsed after the Bush administration reversed its support. Subsequent Senate efforts to negotiate emissions trading schemes acted more as trial runs than serious legislative efforts. As a result, despite the United States' Kyoto promise to cut carbon pollution by 7 percent relative to 1990 levels, US carbon pollution by 2005 was more than 16 percent higher than it had been in 1990.[137]

However, dismissing the United States as a climate policy laggard during this period obscures the persistent efforts of climate reformers to propose and enact costly climate policies. While Norway was implementing a carbon tax, the United States also considered a climate-flavored energy tax in 1993. Just as in Europe, US policymakers began to discuss a domestic emissions trading scheme in the late 1990s and early 2000s. Throughout this period, climate reformers held important congressional and executive branch posts, including the vice presidency under Clinton. What set the United States apart from Norway was that, once proposed, climate reforms failed. Divergent climate policy outcomes did not reflect differences in policy proposal timing, but instead differences in enactment success rate.

Policy enactment failures were, in turn, a function of policy content. Compared to the Norwegian carbon tax, the BTU tax threatened producers with stronger costs and was designed with less sensitivity to their economic needs. At the same time, US climate policies were more sensitive to consumer costs. For example, the BTU tax design deemphasized consumer costs to buttress public support. In response, policy opponents worked to increase cost transparency to undermine the proposal's political viability.

US climate policy proposals thus involved a relatively stronger policy threat against producers. This threat led to coordinated opposition by carbon-dependent economic actors. Lacking a guaranteed voice in policy design, these opponents mobilized conflict in the public arena in an effort to undermine the electoral and legislative incentives associated with climate reforms. Their efforts included political attacks on pro-climate politicians and campaigns to depress the public's climate policy beliefs and attitudes. Partly as a result of these efforts, climate policy became an object of contentious public debate from the early 1990s forward.

Similarly, US climate policy inaction during the 1988–2006 time period cannot be fully explained by party polarization. From the 1990s forward, a partisan gradient in climate policy preferences did exist: Democrats were more likely than Republicans to propose and support costly climate reforms. However, this gradient obscured significant anti-climate policy factions within the Democratic Party and significant pro-reform factions within the Republican Party. In both parties, climate opponents represented constituencies that faced economic threats from carbon pricing. These constituencies included both carbon-dependent businesses and industrial unions. The AFL-CIO, for example, was a driving force behind the Senate's Byrd–Hagel

Resolution that opposed Kyoto. Carbon polluters, in short, enjoyed double representation.

Climate policy proponents were also present within both parties. George H. W. Bush promised to unleash a "White House effect" during his 1988 campaign. In office, his administration was torn between pro-climate and anti-climate voices. Vice President Gore advocated for ambitious climate action during the Clinton years, but faced pushback from more reluctant Democrats. And George W. Bush initially supported a cap and trade scheme before reversing course under pressure from carbon-aligned administration actors.

As these examples show, neither climate policy opponents nor climate policy advocates had a complete stranglehold on US policymaking. Instead, just as in Norway, carbon polluters were dispersed across political and economic coalitions as a result of cross-cutting climate policy preferences. This guaranteed that double-represented opponents occupied at least one policymaking veto point no matter which party controlled the political agenda. However, because carbon-dependent economic actors did not enjoy guaranteed access to policy design, they could not reliably preempt costly reforms. This forced climate debates out into the public arena to manage future policy threats.

By the mid-2000s, carbon polluters seemed invincible to some observers. Anti-climate factions within the Republican executive branch opposed all climate mitigation. Democratic and Republican climate policy opponents repeatedly marginalized pro-climate voices within all branches of government. Yet, just as environmental groups found themselves shut out of climate policy design during the early 2000s, carbon polluters would find their influence abruptly diminished when reformers proposed threatening climate policies in the aftermath of the Supreme Court's landmark 2007 decision in *Massachusetts v. EPA*. In the next chapter, I trace these renewed efforts to impose costs on domestic carbon polluters from 2007 to 2015 as the strategic context for climate policymaking shifted. I then reflect more broadly on the broader dynamics of US climate policy conflict.

5 US Climate Policy Action, 2007–2015

Even as climate opponents' influence intensified under the George W. Bush administration, the strategic policymaking context shifted. The Supreme Court's landmark *Massachusetts v. EPA* decision empowered the EPA to regulate US carbon pollution under an existing statute, creating an executive branch pathway for climate reforms. At the same time, state-level policies threatened businesses with a regionally fragmented policy environment. Coupled with accelerating media attention to climate change and heightened public climate concerns, many stakeholders decided that climate reforms were imminent.

This chapter traces distributive conflict over US climate policy from 2007 through 2015. Initially, policy proponents leveraged the climate threat's salience and the prospect of EPA regulation to assemble a broad reform coalition within business and labor communities. These efforts culminated in the House passing the American Clean Energy Security Act in June 2009. However, parallel efforts stalled in the Senate, undermined by blanket Republican opposition and reservations from Democrats with carbon-dependent constituencies.

With the failure of congressional climate reforms, the Obama administration instead used its new executive authority to propose far-reaching costs on diverse US carbon polluters. Between 2010 and 2016, the administration advanced a series of escalating regulations, seeking to constrain carbon pollution from motor vehicles, new industrial facilities, new power plants, and, eventually, existing power plants. After two decades as one of the planet's policy laggards, the United States transformed abruptly into a climate policy leader. Yet, this leadership would prove ephemeral: by 2017, the Trump administration had launched aggressive efforts to retrench and repeal Obama-era reforms.

Changes to the Strategic Climate Policymaking Context

During the 1980s and early 1990s, climate policy opposition was coordinated by oil companies, coal companies, and energy-intensive manufacturers through such organizations as the Global Climate Coalition. Analogously, among labor organizations, coal-dependent mine workers first engaged climate policy. This pattern of asymmetric mobilization conforms to now-standard theories of collective action (Olson 1965): concentrated losers have the easiest time organizing against policy threats.

Yet, as climate change's salience increased throughout the 1990s, crosscutting policy preferences came into sharper relief. Diverse stakeholders who stood to benefit from climate policy also emerged to contest climate reforms. For example, the Business Council on Sustainable Energy (BCSE) was founded in 1992 to represent natural gas, energy efficiency, and renewable energy interests.[1] Over time, insurance and financial service companies joined this pro-climate coalition.[2]

Other businesses shifted their strategic climate demands in response to increased scientific certainty about climate risks. Beginning in the late 1990s, some multinational companies in carbon-dependent industries became concerned about long-term climate policy costs. These companies came to view their strategic interests as best served by proactively shaping a climate response (Meckling 2011; Jones and Levy 2007). For instance, through the International Climate Change Partnership, such US corporations as Dupont and General Electric (GE) backed the Clinton administration's efforts to promote market-based climate policies (Layzer 2007; Meckling 2011).[3] These corporate shifts were nurtured by environmentalists. In May 1998, the Pew Center for Global Climate Change established the Business Environmental Leadership Council to support dialogue between environmental groups and carbon polluters. Similarly, in 2000, the Environmental Defense Fund brought seven carbon-intensive companies into a Partnership for Climate Action program (Meckling 2011).

Shifting business interests in turn weakened some of the earliest anti-climate lobbies. Beginning with British Petroleum (BP) in 1996, a steady stream of carbon polluters left the Global Climate Coalition.[4] While many US corporate lobbies, from the US Chamber of Commerce to the National Association of Manufacturers, continued to oppose climate policy, business interventions in climate debates increasingly reflected cross-cutting preferences. By 2005, utilities such as Cinergy and Exelon, oil companies such as Shell and BP, and corporations such as General Electric had all positioned themselves as climate policy advocates (Layzer 2007).

The climate threat's profile increased again between 2005 and 2007, with sharp increases in TV and newspaper coverage (Boykoff 2011). Public climate concerns swelled over the same time period (Brulle, Carmichael, and Jenkins 2012), the product of multiple, interacting factors. Al Gore's Oscar-winning documentary on climate change, *An Inconvenient Truth*, became a box-office hit after its May 2006 release. In October 2006, Nicholas Stern released a major report on climate change economics commissioned by the British government to widespread coverage, known as the *Stern Review*. Stern argued that the economic costs of policy inaction outweighed the costs of early action. Then, in 2007, the IPCC released its Fourth Assessment Report, containing strident warnings about the dangers of unmitigated global warming. Both Gore and the IPCC would receive the 2007 Nobel Peace Prize for their efforts to profile global climate risks. Responding to these developments, US environmental groups refocused their advocacy efforts on climate change. After the Democrats won back the Senate and House in 2006, climate legislation emerged as the environmental community's coordinated policy priority (Pooley 2010).

State-level developments also boosted national climate policymaking. A coalition of Northeastern states formed the Regional Greenhouse Gas Initiative (RGGI) in December 2005 to develop a utility sector emissions-trading scheme (Meckling 2011; Raymond 2016; Rabe 2018).[5] RGGI began allowance auctions in September 2008 for a January 2009 policy start date. California also pushed state-level climate reforms beginning in 2005 after Republican Governor Arnold Schwarzenegger issued an executive order setting substantial state reduction targets.[6] In 2006, California adopted the California Global Warming Solutions Act of 2006 (AB-32). AB-32 empowered the California Air Resource Board to promulgate any regulations necessary to meet a 1990 levels by the 2020 target, including through carbon pricing. California then recruited four other states (Arizona, New Mexico, Oregon, and Washington) to discuss regional emissions trading in February 2007.[7] Meanwhile, the governors of Minnesota, Wisconsin, Illinois, Iowa, Michigan, and Kansas signaled interest in yet another regional scheme through the exploratory November 2007 Midwestern Greenhouse Gas Accord. These state-level climate policies would have a mixed political history over subsequent years (Rabe 2018). However, they contributed to a perception of significant momentum behind state-level climate policymaking. This perception, in turn, induced strategic support for federal policies from businesses looking to forestall an inconsistent patchwork of state-level policies (Jones and Levy 2007).[8]

This shifting climate policymaking context further exposed cross-cutting climate policy preferences within business and labor communities. On the business side, the US Climate Action Partnership (USCAP) emerged from conversations between GE CEO Jeff Immelt and the Environmental Defense Fund (Pooley 2010).[9] By early 2006, Pew Center on Global Climate Change, EDF, NRDC, and ten major US companies had begun confidentially discussing climate legislation.[10] This dialogue included Duke Energy, the third-largest carbon polluter among US utilities.[11]

USCAP conversations culminated in a public statement of principles in January 2007, endorsing "prompt enactment" of a national emissions trading scheme.[12] The scheme was envisioned as reducing carbon pollution by 60 to 80 percent below 2007 levels by 2050.[13] Congressional advocates immediately seized on the USCAP statement. By mid-February, California Democrat Barbara Boxer had organized an entire Senate Environment and Public Works Committee hearing dedicated to the group's report.[14]

By contrast, fossil fuel-dependent business and labor groups attacked the USCAP statement. The Competitive Enterprise Institute cautioned that the "risk of global warming must be set off against the risk of global warming policies."[15] Coal utilities were privately furious with other utilities who participated (Pooley 2010).[16] The United Mine Workers of America protested Caterpillar over its USCAP membership (Pooley 2010).

Yet, business support for climate action went beyond USCAP. The Edison Electric Institute (EEI), the largest utility trade association, had helped found the GCC and labeled US support for the Kyoto Protocol as "economic suicide" (Layzer 2007). EEI executives, however, were split on US climate policy.[17] Such coal-dependent utilities as Southern Company and American Electric Power (AEP) objected to any reform. However, other members believed some form of climate policy was inevitable and that the sector should adopt a constructive approach (Pooley 2010). In February 2007, EEI released a set of climate policy principles that, despite their abstract nature, marked a new openness to federal action (Meckling 2011). This was followed, in March, by a statement from sixty-five investors and companies affiliated with the Coalition for Environmentally Responsible Economies (CERES) and the Investor Network on Climate Risk. That statement called for a national reduction targets of 60 to 90 percent below 1990 levels by 2050.[18]

During this period, labor voices spoke about climate change with more volume.[19] The Bush administration's climate intransigence pushed unions into dialogue with environmentalists because they shared a common political enemy.[20] In 2003, the two communities founded the Apollo Alliance to support clean energy jobs. As a joint coalition with environmentalists, the Apollo Alliance represented a significant departure for US labor (Obach 2004b). Previous labor–environmentalist contact had been more transactional.[21]

However, strategic incentives for labor groups and environmentalists to align against a common opponent did not generate convergent climate policy preferences. While some unions expressed tentative support for carbon pollution reduction targets, others rejected policies beyond climate-friendly technology investments (Stevis 2013). The Apollo Alliance serviced the least common denominator among industrial unions and environmental groups. Member organizations who supported more stringent policy commitments remained frustrated.[22]

Seeking a stronger platform, the United Steel Workers and the Sierra Club founded the BlueGreen Alliance (BGA) in 2006.[23] On the environment side, the BGA grew to include the NRDC, EDF, the National Wildlife Federation, and the Union of Concerned Scientists. On the labor side, it would include diverse service and industrial unions, from the Service Employees International Union (SEIU) and the American Federation of Teachers, to the United Automobile Workers (UAW) and Utility Workers Union of America. Yet, even the BGA was circumscribed by the material interests of its labor members. Its lobbying was limited to issues where environmental and labor groups' preferences intersected, from domestic renewable investments and construction jobs to climate adaptation spending.[24]

Judicial Changes to the Policy Status Quo

These shifts in stakeholder interests would then interact with a dramatic institutional shift. In June 2007, the Supreme Court surprised many observers with its decision in *Massachusetts v. EPA*, which created a viable non-legislative path for climate policy. *Massachusetts v. EPA* transformed the policy status quo by shifting the onus onto policy opponents to block reforms.

Controversy over whether carbon pollution was covered by the Clean Air Act began late in the Clinton administration's second term (see chapter 4). With Clinton reluctant to assert this authority, environmental advocates tried to force the government's hands. In October 1999, a coalition of environmental advocates and clean energy business groups petitioned the EPA to regulate carbon dioxide, methane, nitrous oxide, and hydrofluorocarbons from new motor vehicles under Section 202(a)(1) of the Clean Air Act.[25] They argued that carbon pollution and other greenhouse gases were Clean Air Act pollutants because—under the statute's language—they could be "reasonably anticipated to endanger public health or welfare."[26]

The petition's response fell to the George W. Bush EPA. After numerous delays, the agency rejected the challenge in August 2003 arguing, first, that it lacked statutory authority under the Clean Air Act to regulate carbon pollution and, second, that it did not believe greenhouse gas standards for motor vehicles were necessary in either case.[27] Environmental groups contested the EPA's determination through the courts, beginning in late 2003. Lawsuits brought by fourteen environmental groups, twelve states, and three cities became consolidated into *Massachusetts et al. v. EPA*. In a split decision, the DC circuit upheld the EPA's denial in July 2005. However, the Supreme Court agreed to review the decision and, in April 2007, found for environmental advocates in a landmark 5–4 decision. The court agreed that greenhouse gases were air pollutants under the Clean Air Act, and that the EPA must regulate carbon unless it "determines that greenhouse gases do not contribute to climate change or if it provides some reasonable explanation as to why it cannot or will not exercise its dissertation to determine whether they do."[28]

The ruling reversed the status quo: climate policy opponents suddenly had to justify *not acting* to manage climate risks. Carbon-intensive industries immediately fretted about the EPA's new regulatory power;[29] stripping the EPA of this new authority became an immediate legislative priority. Regulatory relief would become a significant bargaining chip for proponents building a reform coalition. Further, senior Democrats who had opposed carbon pollution regulation using the Clean Air Act became more open to climate action. For example, Michigan Democrat and Congressman John

Dingell suggested the Court's decision forced Congress to enact far-reaching climate legislation to replace the new status quo.[30]

The Bush administration struggled to respond to the Supreme Court decision. Internal debate centered on whether to issue an "endangerment finding" for carbon pollution, a formal statement linking greenhouse gas emissions to US public welfare. In November 2007, the administration approved the preparation of an endangerment finding at the level of deputy chief of staff for policy; the EPA was to propose initial regulations in Spring 2008 for final approval in Fall 2008. This decision was supported by Energy Secretary Samuel Bodman, Treasury Secretary Henry Paulson, Council on Economic Advisors Chairman Edward Lazer, and Council on Environmental Quality Chairman Jim Connaughton.[31] Reflecting this consensus, officials finished a draft endangerment finding by December 2007.[32] However, climate policy opponents pushed back, including in Vice President Cheney's office, the Office of Management and Budget (OMB), and the Department of Transport. OMB officials simply refused to open the EPA's email with the endangerment finding attached, deciding that if they did not read the document, they were not legally required to respond.[33] Then, on December 5, 2007, the EPA was directed to "cancel" its proposed endangerment finding.[34] Instead, the Bush administration initiated a broader effort to solicit stakeholder comments, deferring action to the winner of the 2008 presidential election.[35]

Despite these efforts to stall a response, the Supreme Court's decision radically transformed US climate politics. Until that point, policy opponents could block climate reforms by occupying at least one legislative veto point. However, the Supreme Court's decision upended the policy status quo. So long as policy proponents controlled the executive branch, climate policy was a credible threat.

Senate Conflict over Emissions Trading, 2007–2008

In the midst of this increased political attention to climate change, Senator Lieberman took a third crack at climate legislation during the 110th Congress. Initially, Lieberman worked with Senator McCain to reintroduce a version of the duo's previous bill in January 2007.[36] By this time, even President Bush was signaling growing acceptance of the climate threat, making a passing reference in his January 2007 State of the Union. However,

McCain was preoccupied with his 2008 Republican presidential primary race and the bill never gained traction.[37]

Instead, Lieberman turned to Virginia Republican Senator John Warner to cosponsor a new legislative effort. Warner accepted that climate change was happening and had supported the pro-climate "Sense of the Senate on Climate Change" amendment to the 2005 energy policy bill. He was an avid hunter and angler, and was influenced by military warnings that climate change was an emerging national security concern.[38] Although he represented Virginia, Warner was largely unresponsive to his state's coal interests when it came to climate change, particularly after announcing his retirement in August 2007.[39] After Democrats regained Senate control in 2007, Senator Boxer structured the organization of the Environment and Public Works Committee to facilitate conversations between Warner and Lieberman, creating the awkwardly titled Private Sector and Consumer Solutions to Global Warming and Wildlife Protection subcommittee with Lieberman as its chair and Warner as ranking minority member.[40]

Lieberman and Warner began by negotiating a set of common principles for federal climate policy, including that it should cover carbon pollution across the economy.[41] Staffers then canvassed diverse economic constituencies to understand the features each wanted from any climate bill. They solicited input from senators who represented economic interests of particular regions.[42] While USCAP played an influential symbolic role in shaping these discussions, its members' January 2007 statement proved too vague to influence the specifics of legislative design.[43]

Lieberman and Warner released their policy's proposed structure in August 2007 and a draft bill, the Climate Security Act, in October 2007. The duo approached bill design as a revenue allocation exercise.[44] Emissions allowances created value, and this value could be distributed to induce support.[45] The Lieberman–Warner bill initially proposed to stabilize US greenhouse gases by 2012, cut them 15 percent by 2020, 33 percent by 2030, 52 percent by 2040, and 70 percent by 2050.[46] The final 70 percent by 2050 target bridged political differences between such senators as Boxer and Vermont Independent Bernie Sanders, who both wanted 80 percent reductions, and other senators more comfortable with 60 percent.[47] The 70 percent target also split the 60 to 80 percent range introduced in USCAP's statement. The Lieberman–Warner bill auctioned 24 percent of its total allowances, using revenues to fund climate technology and adaptation measures, consumer energy assistance, and worker retraining.[48] The Environment and Public Works (EPW) committee voted the bill out in December 2007, just before the Bali Climate Change Conference.[49] Only Warner voted with the Democrats, though Republican senators Susan Collins, Norm Coleman, and Elizabeth Dole were all non-committee member cosponsors.[50]

During this period, a competing bipartisan bill emerged from Senators Jeff Bingaman and Arlen Specter through the Senate Committee on Energy and Natural Resources. Bingaman and Specter released their discussion draft in January 2007 and a full bill in July.[51] The Bingaman–Specter proposal, the Low Carbon Economy Act of 2007, was more sensitive to the demands of carbon-intensive business and labor; it drew from the National Commission on Energy Policy (NCEP) (Meckling 2011).[52] NCEP was a 2002 Hewlett Foundation initiative that brought diverse economic stakeholders together to develop long-term energy policy recommendations.[53] The Bingaman–Specter proposal was to cap carbon pollution at 2006 levels by 2020, and 1990 levels by 2030. The bill included generous allowances that grandfathered in the coal industry, and more circumscribed sectoral coverage than Lieberman–Warner. It also included a "safety valve" to cap the carbon price at $12 per ton, increasing at 5 percent over inflation annually.[54] Program revenues flowed into an Energy Technology Deployment Fund. In sum, Bingaman–Specter protected incumbent economic interests from serious policy costs.

Bingaman–Specter enjoyed broad industrial labor and business support, from the Edison Electrical Institute to the AFL-CIO and United Steelworkers.[55] The UMWA opposed Lieberman–Warner but supported Bingaman–Specter.[56] However, both Lieberman and Warner opposed the safety valve proposal, viewing it as ineffective.[57] They instead wanted to contain costs through a Carbon Reserve Efficiency Board that could temporarily adjust the policy's cap and allowance distribution to moderate unexpected economic pressures.[58]

After Lieberman–Warner passed through the EPW, Lieberman's staff consulted with any willing interest groups, trying to gauge what amendments could get the bill sixty Senate votes (Pooley 2010). This included efforts to broker a compromise with the Bingaman–Specter team on the safety valve.[59] Yet, Boxer simultaneously postured to strengthen, not weaken, the bill's green credentials (Pooley 2010). In March 2008, she threatened to kill the bill if it weakened before a final Senate vote.[60]

Boxer was also trying to accommodate emerging opposition from environmental advocates. After Lieberman–Warner's introduction, environmental groups jointly called for the bill to be strengthened.[61] Yet, this nonspecific call obscured internal divisions. USCAP-aligned groups such as EDF and the NRDC were strong supporters. EDF lobbied aggressively for Lieberman–Warner, including disseminating its own economic impact models to blunt pessimistic industry reports (Pooley 2010).[62] Yet other environmental advocates were skeptical about market-based approaches to climate risk mitigation; they objected to free allowance distributions for industrial carbon polluters and insufficiently ambitious targets. Friends of the Earth called Lieberman–Warner a "wholly inadequate response to the greatest environmental crisis of our time" and called for the agreement to be jettisoned if its ambition was not increased.[63] The Sierra Club declined to endorse the bill, arguing that an effective climate bill needed fully auctioned allowances.[64]

Boxer would unsuccessfully attempt to bridge these competing demands. However, her mid-May manager's amendment did little to grow Lieberman–Warner's support coalition. Substantively, her amendment added a Strategic Allowance Reserve to the policy, described as an "emergency off-ramp" that could sell allowances from the 2031 to 2050 period in the 2012 to 2017 period. It also reallocated $800 billion toward consumer tax relief, delegating the details to the Finance Committee.[65] Yet, Boxer's efforts created tension among the bill's sponsors, particularly Warner who disagreed with Boxer's efforts to channel auction revenues toward green subsidies.[66] Warner ultimately compromised on many of his concerns, but only because he knew the bill had no prospect of becoming law.[67] In a realistic legislative setting, it is less likely Boxer and Warner would have bridged their divide.

Still, even as Warner acquiesced to Boxer's amendments, Lieberman–Warner divided the Democratic caucus. Several senators were uninterested in anything more ambitious than Bingaman–Specter, if that stringent.[68] Lieberman–Warner was ultimately "political theatre," a rehearsal for post-Bush legislative action; yet, coal-state Democrats became frustrated when Boxer publicly described the bill as more than a symbolic vote.[69] And while Senate Majority Leader Harry Reid had promised to bring a completed deal to the Senate floor, he faced caucus pressure to block the bill's consideration.[70] In June, ten Democratic senators wrote to Reid and Boxer warning that they could not support the final Boxer-amended bill. These senators all came from states with fossil fuel industries or carbon-intensive electricity sectors, and included labor-aligned Ohio progressive Sherrod Brown. Collectively, they demanded that any cap and trade bill must equitably distribute costs across states, invest aggressively in new technologies to soften regional economic impacts, and protect industrial and manufacturing jobs.[71]

Business and labor communities echoed this Democratic split. The AFL-CIO had come to support the Bingaman–Specter proposal but raised concerns over the Lieberman–Warner bill's impacts on US workers. The union worried the bill's strong 2020 target would not provide time for CCS technology commercialization.[72] In a letter to Boxer, the AFL-CIO cautioned against the bill's "overly aggressive Phase 1 emissions reduction target" while praising the bill's numerous producer carrots.[73] The United Mine Workers worried the bill would undermine the coal industry's viability.[74] The UAW demanded the bill preempt EPA authority to regulate carbon pollution under the Clean Air Act, and more clearly protect against offshoring of manufacturing jobs as a result of new low-carbon standards.[75]

Other labor groups were more supportive. The International Brotherhood of Boilermakers and various building trades unions endorsed the bill, believing that it would create fifty years of construction work. These groups emphasized expansion opportunities for the coal industry through CCS technology support. Steelworkers straddled both sides. Emphasizing their general support for cap and trade, they withheld endorsement on Lieberman–Warner because it lacked the Bingaman–Specter safety valve, and did not provide enough free allowances to steelworker-supported industries. In a joint letter through the BlueGreen Alliance, the Sierra Club and the Steelworkers chastised the bill for insufficient attention to competitiveness concerns, and demanded stronger provisions to prevent clean energy manufacturing from developing mostly abroad.[76]

Among businesses, key carbon polluters continued to oppose Lieberman–Warner. As before, opposition focused on building narratives about economic harms from climate policy, often supported by dramatic business group-commissioned economic models.[77] These efforts attempted to indirectly shape public opinion as a means of pressuring elected officials. For example, the US Chamber of Commerce cited economic costs and US bureaucratic expansion in its bill opposition;[78] it ran an ad campaign featuring unwitting US consumers waking up cold and running to work because they couldn't afford to drive in a world devasted by the legislation.[79] It also coordinated local informational events across the country to publicize policy costs (Pooley 2010). The National Association of Manufacturers declared the bill a "Key Manufacturing Vote," citing insufficient provisions to preserve US economic competitiveness.[80] The Farm Bureau felt the bill did not create an adequate market for agricultural offsets and would increase the cost of US agricultural inputs.[81] The coal industry, through a newly established American Coalition for Clean Coal Electricity (ACCCE), ran ads warning the bill threatened the public's way of life.[82] Even USCAP member Duke Energy actively lobbied against the bill, spending over $1.5 million to argue that, while it supported cap and trade generally, it did not support Lieberman–Warner specifically.[83] Duke Energy's position reflected an internal split within USCAP; only six of the organization's corporate members supported Lieberman–Warner with the rest joining Duke in opposition (Pooley 2010). Instead, pro-climate USCAP members, including GE, Pacific Gas & Electric (PG&E), and Exelon signed a separate endorsement letter in coalition with environmental groups.[84]

Echoing the heaviest carbon polluters, most Republicans also opposed the bill, citing consumer costs for low-income and middle-class Americans.[85] At the beginning of the 110th Congress, Boxer had arranged an EPW hearing for senators to testify on their global warming beliefs. This had revealed a distribution among Republicans from Inhofe's skepticism to Maine Senator Snowe and Tennessee Senator Alexander's sincere belief in anthropogenic climate change.[86] Yet Republican moderates refused to commit; for instance, Specter balked after it became clear that both steel businesses and steel unions opposed Lieberman–Warner (Pooley 2010). At the same time, Bush threatened to veto Lieberman–Warner in early June, arguing it would shrink the economy by $10 trillion dollars and reduce household income by almost $4,500 annually.[87]

In this way, cross-cutting policy preferences again stymied policy action. The Lieberman–Warner bill failed to break a filibuster on a 48–36 vote in June 2008.[88] Six absent senators later offered written statements saying they would have voted yes if present.[89] However, nine of the senators who had written to Reid objecting to the Lieberman–Warner bill voted for cloture. These senators were unlikely to have supported the bill in a final vote. Even the cloture bill was opposed altogether by Democrats Sherrod Brown, Byron

Dorgan, Tim Johnson, and Mary Landrieu. In short, apparent support levels for Lieberman–Warner reflected partisan posturing more than a broad Democratic coalition in favor of the legislative package.[90]

Climate Policymaking during the 111th Congress

In sharp contrast to George W. Bush, both 2008 presidential nominees took climate change seriously. John McCain maintained his climate policy commitment despite growing criticism from Republican primary opponents.[91] Both Senators Obama and McCain had skipped the Lieberman–Warner vote while campaigning, but both sent letters declaring their support. Further, Obama had joined only Republican Olympia Snowe in cosponsoring the McCain–Lieberman Climate Stewardship and Innovation Act of 2005.

On the campaign trail, Obama promised to introduce a cap and trade scheme that auctioned 100 percent of available allowances.[92] Obama also promised to pursue executive climate policy actions including issuing the long-delayed EPA Endangerment Finding.[93] Yet, asked about climate policy during the presidential debates, neither candidate profiled these commitments. McCain emphasized nuclear energy investments while Obama emphasized clean energy technologies.[94]

After Obama won, the president-elect and his transition team signaled continued cap and trade interest.[95] President Obama reiterated this support in his first State of the Union address. In order to "save our planet from the ravages of climate change," he argued, "I ask this Congress to send me legislation that places a market-based cap on carbon pollution and drives the production of more renewable energy in America."[96]

Climate advocates were cautiously optimistic that a policymaking window had opened. To leverage this window, pro-reform officials moved to seize control of congressional leadership positions. Importantly, California Democrat Henry Waxman deposed Michigan's John Dingell as chair of the House Energy and Commerce Committee. Dingell had long advocated a cautious approach to climate policymaking, reflecting his deep ties to Michigan's automotive industry. After Democrats took control of the House in 2007, Dingell did not signal that climate change was a personal priority.[97] House Speaker Nancy Pelosi consequently tempered his role by establishing a select committee on climate change under Massachusetts Democrat Edward Markey.[98]

For most of the 110th congress, Dingell's climate efforts centered on collaboration with Virginia Democrat Rick Boucher, chair of the Energy and Air Quality Subcommittee. While Dingell had deep ties to automotive interests, Boucher's district was located in the heart of Virginian coal country. Throughout the 110th Congress, Dingell and Boucher held twenty-seven hearings, released numerous white papers, and engaged in extensive stakeholder consultation. In October 2008, they released a draft bill for debate by the next Congress.[99] The Dingell-Boucher plan took climate seriously, but was constrained by the status quo economic interests of carbon polluters. It proposed an economy-wide cap covering about 88 percent of US carbon pollution, including a 6 percent reduction below 2005 levels by 2020, 44 percent below 2005 levels by 2030, and 80 percent below 2005 levels by 2050.[100] It also included generous offsets provisions and bonus allowances for CCS deployment.[101]

The Waxman versus Dingell contest exposed active conflict between Democratic factions preferring high and low climate policy costs; the leadership fight was explicitly understood as an effort by pro-environment factions to replace a more senior committee chairman less sympathetic to the climate agenda.[102] In contrast to Dingell, Waxman had previously proposed costly climate bills. For instance, Waxman's Safe Climate Act of 2007 offered reduction targets of 1990 levels by 2020 and 80 percent below 1990 levels by 2050.[103]

In a narrow 137–122 vote, Waxman became head of the House Energy and Commerce Committee in November 2008. Having deposed Dingell, he immediately restructured relevant subcommittees. Most critically, Waxman merged the Energy and Air Quality Subcommittee (EAQ) and the Environment and Hazardous Materials Subcommittee (EHM) into a single new Energy and Environment Subcommittee. The EAQ had been chaired by Boucher. The EHM had been chaired by Gene Green, a Texas Democrat with close ties to the oil industry. Instead, Waxman appointed Markey as chair of the new committee, completing the leadership coup by green Democrats. Waxman and Markey's ascendance thus displaced a leadership team with strong ties to carbon-intensive Democratic constituencies, bringing into power a team of committed green legislators. Coupled with President Obama replacing Bush, climate proponents suddenly enjoyed an unprecedented level of federal policymaking control.

Newly empowered pro-climate factions moved quickly. Initially, legislators found it easy to form political coalitions to distribute benefits to diverse Democratic constituencies. Climate-related provisions were central to the massive stimulus bill responding to the global financial crisis, the American Recovery and Reinvestment Act of 2009, which included over $50 billion in green jobs support, electricity grid investments, renewable energy loan guarantees, energy efficiency measures, and clean energy research funds. The stimulus bill's green jobs focus aligned with the Apollo Alliance and BlueGreen Alliance efforts to use clean technology investment to bridge labor and environmental interests. For example, a labor–environment model bill, the Apollo Economic Recovery Act, contained prototypes of the stimulus bill's low-interest loans.

Parallel efforts to pass a comprehensive climate reform bill that included costs would prove more challenging. On this front, House leadership decided to begin immediate work on an energy and climate bill, hoping that the Senate would pass its own climate bill in advance of the Copenhagen Climate Change Conference in December 2009. They planned for the House and Senate to conference the bill in Copenhagen's aftermath.[104] BTU tax memories largely killed residual appetite for carbon taxation.[105] Instead, Waxman began work on an emissions trading system. While Lieberman–Warner had been conceived as a dress rehearsal for future climate reforms, House legislators believed the climate bill would be so intricate and far-reaching that it should be built from scratch.[106]

USCAP re-engaged this new debate with more detail than during Lieberman–Warner drafting, but still struggled to bridge its members' differential needs. On targets, industry preferred Obama's campaign target of 14 percent reductions, but environmentalists wanted more aggressive targets, arguing that Obama's plan had not factored in the distribution of free allowances included in the USCAP proposal. This difference could not be bridged, so the group instead agreed to lobby for a target range (Pooley 2010).[107] Under the USCAP proposal, the target boundaries—within which members agreed to lobby—were 97–102 percent of 2005 levels by 2012, 80–86 percent of 2005 levels by 2020, 58 percent of 2005 levels by 2030, and 20 percent of 2005 levels by 2050. This represented a general increase in carbon pollution reduction levels relative to the 2007 USCAP statement. Targets were conditional on offset availability.[108] USCAP members also were divided over allowance distributions, with some industrial members demanding as much as 40 percent free allowances; ultimately, members bridged their differences by using ambiguous language to force agreement (Pooley 2010).

USCAP support gave congressional action symbolic legitimacy from corporate America.[109] Waxman's first hearing as head of the House Energy and Commerce Committee invited USCAP leaders to testify in support of emissions trading. However, USCAP was not the only industry group preemptively working to shape potential climate legislation. The Edison Electric Institute also began internal discussions on a framework for utility participation in setting a federal carbon price.[110] In January 2009, EEI released its plan, supported by eighty utility CEOs.[111] The proposal called for a long-term carbon pollution reduction target of 80 percent by 2050, but weaker short-term targets to facilitate CCS and nuclear technological breakthroughs. It also included a "cost collar" of ceiling and floor allowance prices. EEI requested 40 percent of total allowances, proportional to the sector's contribution to US carbon pollution.[112]

These business proposals shaped policy design.[113] For example, Waxman and Markey integrated supplemental performance standards for coal-fired power plants (to incentivize CCS adoption) from the USCAP plan.[114] They also signaled privately to EEI that they would consider its proposed utility allocation formula (Pooley 2010). Meanwhile, the Obama administration backed away from its election commitments to auction 100 percent of the emissions trading scheme's pollution permits, partly in reaction to opposition from Senate Democrats representing fossil fuel-dependent states.[115]

Waxman and Markey introduced their American Clean Energy Security (ACES) Act discussion draft to the House Energy and Commerce Committee in late March 2009. Known as the Waxman–Markey bill, it proposed capping US carbon pollution at 3 percent below 2005 levels by 2012, 17 percent by 2020, 42 percent by 2030, and 83 percent by 2050.[116] This scheme covered all US emissions except from agriculture. Covered entities could use offsets to cover some of their liabilities under a complex formula.[117]

This emissions trading scheme was only one part of the Waxman–Markey bill. It also proposed clean energy and energy efficiency requirements, including a clean energy standard of 25 percent by 2025.[118] It offered transition assistance to regulated sectors, particularly coal.[119] It contained consumer support measures and funds for worker retraining.[120] Finally, as a carrot to induce industrial support, the bill partially preempted carbon pollution regulation through the Clean Air Act.[121] However, this was not the blanket preemption sought by some industry stakeholders. Instead, absolute preemption from Clean Air Act regulations was withheld until conference bargaining with a prospective Senate bill.[122]

The Waxman–Markey discussion draft deliberately sidestepped allowance allocations, which the drafters wanted to use to build voting coalitions over subsequent months (Pooley 2010). Waxman and Markey believed the cap was the critical design feature. Allowances could then be used to generate support for the strongest possible targets, including to mollify fossil fuel-intensive Democratic constituencies.[123]

With this strategy, Waxman and Markey moved to secure support from House Energy and Commerce Committee members. They sequentially bargained with senior Democrats closely aligned with various carbon-intensive constituencies, amending the draft to respond to these constituencies' needs. Critical votes included Rick Boucher (coal), Gene Green (oil), John Dingell (automotive), and Mike Doyle (heavy industry). With Republicans largely absent from negotiations, business leaned on these Democratic legislators to represent their interests.[124] By contrast, it was assumed that the green edge of the Democratic coalition would support any comprehensive climate reform bill, no matter its contents.[125]

Negotiations began with coal interests, represented by Boucher. The bill's advocates believed finding middle ground with Boucher would position the bill as attractive to a broad House coalition (Pooley 2010). Boucher made a variety of demands to alleviate coal-industry costs, including free utility allowances.[126] As Boucher proclaimed in a TV interview: "I personally think coal's golden age lies ahead. We are structuring this legislation to accommodate growth in coal sales and production and use."[127] The Boucher compromises also satisfied industrial labor constituencies who had opposed previous Senate reforms. Working with Boucher, coal unions focused their lobbying on CCS subsidies to mitigate long-term sectoral policy costs.[128] With Waxman's coal concessions, the UMWA believed Waxman–Markey offered as favorable a set of climate policy terms as the union was ever likely to achieve.[129] However, coal businesses still rejected the deal. They felt the bill's compensation was insufficient relative to their long-term climate policy exposure; it was better to gamble on their ability to stall EPA regulations in court.[130] In general, coal unions and coal businesses did not coordinate their climate lobbying efforts.[131] Just as, in the absence of engaged Republicans, Democratic congressional representatives stood in as access points for regional business interests, unions represented their sectors' economic interests within policy design debates.

The utility sector emerged satisfied with the draft bill, once EEI wrote their preferred allowance distribution formula into the discussion draft.[132] Some major utilities stood to benefit from the deal's structure, given their low-carbon generating mix.[133] More generally, utilities believed their sector had many of the cheapest abatement opportunities; Waxman–Markey would provide a business opportunity when they profited from excess allowance sales.[134] However, EEI's coal-dependent utilities were more sanguine. A group of coal-dependent Midwestern utilities formed the Midwest Climate Coalition, arguing the bill would compromise their economic viability. Some of these companies had not contested the January 2009 EEI proposal, thinking prospects for an emissions trading bill were low. Others, such as MidAmerican, rejoined EEI only to protest the organization's climate policymaking efforts.[135]

With support from coal-aligned Boucher and many utilities, Waxman began secondary negotiations with other carbon-intensive industrial interests. Indiana Democrat Mike Doyle demanded 15 percent of allowances on behalf of trade-exposed heavy industries. Michigan's Dingell wanted the bill to expand loans for car manufacturers, and direct 1 percent of allowances to car companies for fuel efficiency support.[136] Oil interests were championed by Texas Democrat Gene Green. The oil industry had remained relatively disengaged from negotiations.[137] By the time oil interests shifted from blanket opposition to serious lobbying, the available allowance pool had dropped.[138] The oil industry received few of their overall demands.[139] However, Green represented a Houston district with oil refineries. He demanded protections for refineries by having them defined as trade-exposed, and giving them privileged access to some offsets.[140] These demands were seconded by industrial unions, whose members worked in small refineries with less capacity to adapt to climate policy.[141] The 2 to 3 percent of total allowances allocated to refineries was viewed as a decision made entirely for political, not economic, reasons.[142]

With allowance distributions bridging intra-Democratic committee tensions, the bill received an uneventful markup in late May. It passed out of the committee on a 33–25 vote, with Republican Mary Bono Mack (CA) voting for the bill, and four Democrats voting against.[143] However, Democratic bargaining was not complete: other House committees also claimed jurisdiction over parts of the complex, economy-wide bill.

The largest roadblock came from agricultural interests, led by House Agricultural Committee Chairman Colin Peterson. Agriculture was not included under the cap, but was engaged by a crediting scheme where farmers could sell offsets into the system. Agricultural interests wanted their offset program managed by the US Department of Agriculture, separately from other bill offsets. The USDA would then control judgments around the climate impacts of such practices such as methane burning or till management.[144] Environmental groups were skeptical the department would structure offset programs to meet climate goals.

Peterson had long been concerned with climate policy costs for agriculture, having already lobbied against a carbon tax in the early 1990s. First, Peterson threatened to block Waxman–Markey unless the EPA changed its framework for calculating ethanol-linked carbon emissions. Peterson then attacked the EPA's proposed jurisdiction over agricultural offsets, declaring that "a lot of us on the Committee don't want the EPA near our farmers."[145] Peterson even rejected a compromise proposal to allocate a set amount of crediting value to the Secretary of Agriculture to coordinate across the farming community.[146] With Peterson controlling more than enough rural and agricultural votes to block House approval, Waxman conceded control over agricultural offsets to the USDA. He also inserted a provision blocking the EPA from considering the indirect land-use effects from ethanol for five years.

While Democrats worked to bridge cross-cutting climate policy preferences, Republicans mostly united in opposition. House opposition was boosted by anti-climate Republican senators, who coordinated with House allies. Again, policy opponents' strategy was to raise the salience of consumer costs in an effort to undermine legislative support.[147]

This coordinated Republican opposition reflected growing party polarization around climate issues (McCright and Dunlap 2011b; Guber 2012). In part, this trend mirrored general US party polarization (Abramowitz and Saunders 2008).[148] However, it also responded to interest group mobilization. Movement conservatism and anti-climate positioning became linked over the course of the 1990s and 2000s (Layzer 2012), as anti-climate Republican leaders echoed carbon polluters' climate science misinformation.

Such figures as Oklahoma Senator James Inhofe became persistent climate skeptics. And intellectual and financial ties deepened between conservative leaders and fossil fuel companies. For instance, Koch Industries and such Koch-funded conservative organizations as Citizens for a Sound Economy actively contested climate policy as far back as Clinton's BTU tax.

This Republican shift toward climate skepticism occurred even as Democratic elites remained divided over climate policy. Waxman's legislative efforts involved major distributive compromises to accommodate energy- and carbon-intensive Democratic constituencies, from agriculture to coal. Even so, these bargains were still insufficient for a large number of Democrats from states with significant carbon polluters who refused to support the final Waxman–Markey draft. The Waxman–Markey bill only passed the House in late June by a 219–212 vote, gaining the support of 8 Republicans but losing 44 Democrats.[149] EPA analysis of the final bill suggested allowance prices of $13 in 2015 and $16 by 2020, a lower carbon price than anticipated because generous offset provisions reduced anticipated costs. Consumer costs were anticipated around $80–$111 annually, including household benefits from bill provisions but not factoring in climate mitigation benefits.[150]

Interest Group Mobilization against the House Package

Even as Waxman bridged internal Democratic divisions to assemble a pro-reform coalition, the business community mobilized against his deal. Estimates suggest at least 1,150 individual business interests maintained a lobbying presence during negotiations, including almost five hundred that became active in the weeks immediately before the bill's passage.[151] Many companies spent significant amounts to influence the bill, hiring dozens of lobbyists and spending millions.[152] These efforts' intensity underscored the policy's transformative impact on the US economy. Emissions trading schemes quite literally create new markets and institutional rules that shape most businesses' profits and opportunities.[153]

Corporate lobbying over Waxman–Markey highlighted just how split business preferences had become. Among major oil companies, Shell and BP lobbied heavily for the bill while ExxonMobil and Chevron lobbied against, alongside the American Petroleum Institute. Similarly, while chemical interests maintained their opposition, the American Chemistry

Council supported cap and trade in principle but rejected Waxman–Markey specifically.[154] EEI leadership pressured legislators to endorse the legislation even as its renegade Midwest coal members opposed the effort.[155]

These climate policy differences spilled into the public domain. Prominent companies, from Apple to Exelon, quit the US Chamber of Commerce to protest its climate policy obstruction, including a controversial appeal by Chamber of Commerce leadership to contest climate science through a "Scopes monkey trial of the 21st century."[156] At the same time, Alcoa and Duke Energy withdrew from the American Coalition for Clean Coal Electricity (ACCCE).[157] Industry thus split between factions who wanted a rearguard attack on climate science and those who preferred constructive engagement.[158]

After major carbon polluters failed to stop House passage of Waxman–Markey, policy opponents regrouped to increase the scope of climate policy conflict. Over the 2009 summer recess, diverse industry actors worked to sour Senate support. These efforts all served to construct elite perceptions of the public's preferences and, by extension, electoral incentives to support climate reforms. For example, the ACCCE subcontracted Bonner & Associates, the same firm that had coordinated astroturfing during BTU tax debates, to recruit 225,000 "volunteers" to advocate for coal energy at local political events.[159] Bonner & Associates also sent forged letters to Congress, fraudulently pretending to write on behalf of such groups as the Dunmore Pennsylvania Senior Citizen's Center, the Jefferson Area Board for Aging, and the American Association of University Women.[160]

Meanwhile, API launched an "Energy Citizen" campaign to organize rallies against climate policy in twenty states.[161] These rallies sought to "put a human face on the impacts of unsound energy policy" by emphasizing jobs and consumer costs. API recruited employees from member companies, as well the US Chamber of Commerce, NAM, the US Farm Bureau and the trucking industry.[162] NAM and the National Federation of Independent Business (NFIB) launched a multimillion-dollar ad campaign.[163] The Farm Bureau also engaged. Whereas EDF had presented Senator Reid with baseball caps emblazoned with "Just Cap It" in 2007, the Farm Bureau instead encouraged farmers to send elected officials caps warning: "Don't Cap Our Future."[164]

These efforts played out simultaneously with the Tea Party's emergence, which adopted climate policy opposition as a central commitment

(Bartosiewicz and Miley 2013; Skocpol 2013).[165] For instance, the Koch-funded CSE had split into two organizations in the mid-2000s: Americans for Prosperity (AFP) and Freedom Works. Both organizations worked to embed climate policy opposition within the Tea Party movement.[166] Throughout 2009 and 2010, these efforts shifted Republican political incentives as anti-climate Republicans enjoyed electoral success. For example, staunchly conservative South Carolina Republican Bob Inglis of the House was primaried in June 2010 for his conservative climate policymaking support. Likewise, nineteen of twenty Republican candidates in contested 2016 Senate races expressed some degree of climate skepticism.[167]

In the House vote's aftermath, Waxman–Markey proponents found themselves outmatched by opponents. Environmental groups remained divided on the legislation's merits. While some groups hailed the landmark bill, including EDF, others such as Greenpeace believed it was compromised by industry giveaways.[168] James Hansen, who had assumed a double role as climate scientist and policy advocate, denounced the effort as a "counterfeit climate bill."[169] Still others, including the Sierra Club, were unsatisfied with the final deal but hoped Senate negotiations could fix its weaknesses (Pooley 2010).

Divided among themselves, these advocates allowed opponents at least one uncontested week to frame bill debate.[170] Similarly, pro-climate business actors did not step in to fill the communication gap. For instance, USCAP issued a congratulatory press release, but noted the organization would push Senate amendments to better align the bill with its January 2009 statement.[171] In other words, diverse policy supporters hedged in their effort to continue shaping the bill's distributive bargain. There was no political constituency to defend the Waxman–Markey compromise from coordinated opponent attacks.

Climate Policymaking in the US Senate

Senate climate proponents struggled to pick up where the House left off. While House actors report substantial strategic interaction between the two bodies, Senate staffers felt isolated from House decision-making.[172] The result was that Senate climate debates moved along an independent trajectory before stalling entirely.

In early 2009, Lieberman and McCain discussed yet another iteration of their joint cap and trade bill. By this point, Lieberman had lost his chairmanship of the Global Warming subcommittee after endorsing McCain during the presidential campaign. However, McCain was disinterested and faced a competitive primary challenge from the right in Arizona.[173] Instead, Lieberman sought new climate allies. His team decided to begin by drafting a nuclear energy provision in a bid to entice Republicans who were otherwise skeptical about climate policy but saw expanding nuclear energy generation as a policy priority.[174] Through this effort, Lieberman eventually brought South Carolina Republican Senator Lindsey Graham into a conversation that also included Massachusetts Senator John Kerry.[175]

The Senate remained occupied with other legislative matters throughout the first half of 2009, delaying serious climate debates until after ACES Act passage.[176] Then, in late September, Boxer and Kerry introduced a modified Waxman–Markey bill into the Senate. Compared to the House bill, the Senate version departed mostly by strengthening the 2020 target to a 20 percent reduction (Bartosiewicz and Miley 2013). Yet, Boxer struggled to form a Democratic support coalition, stymied by Democrats from fossil fuel-intensive states. Montana Democrat Max Baucus wanted the target weakened to below Waxman–Markey's 17 percent.[177] West Virginian Democrat Jay Rockefeller attacked the Boxer bill for providing insufficient incentives for coal.[178] Ohio's Sherrod Brown remained deeply concerned about policy costs for US manufacturing and labor. He wanted additional allowances for energy-intensive manufacturers and more transition assistance.[179] By December 2009, senators from states with carbon-intensive economies, including Senators Kent Conrad (ND), Evan Bayh (IN), and Mary Landrieu (LA), began pushing to abandon emission trading, at least until after the midterms.[180] The same distributive bargain that pushed the bill over the House line couldn't satisfy the political geography of the US Senate.

Waxman–Markey's failure in a Senate dominated by regional carbon-intensive interests provided an opening for Kerry, Graham, and Lieberman to broker a new deal. After finding no additional Republicans willing to support a potential climate bill, the trio opted to negotiate directly with industry. They reasoned that a bill supported by such organizations as the Chamber of Commerce and the American Petroleum Institute might allow some Republicans to endorse climate reforms.[181] Senate negotiators thus invited carbon-dependent economic actors to collaboratively design reforms. For example, the American Petroleum Institute demanded a separate pricing mechanism as a precondition for its participation: a "linked fee." The oil industry believed they had missed the allowance-allocation bandwagon during House negotiations. The linked fee proposal exempted the industry from cap and trade altogether; instead, the sector would be subjected to a standalone carbon tax with revenues recycled toward consumer rebates and transportation investments. This fee was envisioned as a line item charge paid by consumers at the pumps.[182] In return for the linked fee proposal, API still would not agree to endorse a climate deal but instead agreed to *not criticize* a climate bill for six weeks after its introduction. In separate negotiations, EEI demanded additional allowances and a policy start delay from 2012 to 2015. The Senate trio met EEI's demands.[183] After consultation with Texas industrialist T. Boone Pickens, the bill folded in generous natural gas support.[184]

By negotiating directly with industry, the Kerry–Graham–Lieberman proposal became a lowest-common-denominator policy that satisfied the most significant policy opponents' needs. Corporate involvement in the Waxman–Markey bill was significantly shaped by business; indirectly through uptake of USCAP and EEI proposals and directly through internal negotiations between Waxman and Democrats with carbon-intensive constituencies. The Senate efforts heightened this involvement by bringing business directly into policy design.[185]

Even these concessions to business were insufficient to expand the bill's support coalition. No new Republicans wanted to engage. Meanwhile Boxer, still shepherding her own bill, was not supportive.[186] Other Democrats rehashed the same distributive concerns that had roiled previous efforts: Michigan's Senator Stabenow demanded agricultural offsets be overseen by the USDA; Louisiana's Blanche Lincoln intervened to represent the economic interests of refiners.[187] Labor unions wanted a guarantee that only unionized labor be used for climate infrastructure projects, creating splits between Kerry and Graham.[188]

Despite these tensions, the three senators arrived at a tentative agreement in March 2010. However, support for their bill, the American Power Act of 2010, collapsed days before its unveiling. The collapse was apparently the result of distrust within the Democratic leadership and the badly timed BP oil spill that made the bill's concessions to offshore drilling unpalatable to Democrats.[189] Slighted after someone in the White House leaked false reports of a Graham-supported gas tax, the lone Republican bailed. While Kerry and Lieberman would still introduce the American Power Act, it could not attract any Republican support after Graham's departure. Kerry and Lieberman would pivot to discussions of a utility-sector-only bill, not dissimilar from the approach that such Republicans as Lamar Alexander had advocated three years earlier during Lieberman–Warner debates. Yet, this gained little traction from potential Republican supporters and economic stakeholders. If the utility sector was going to shoulder a disproportionate carbon pollution reduction burden, EEI demanded more expansive regulatory relief from mercury and non-climate-related EPA measures. Kerry was unwilling to make these concessions.[190] By the end of July, senior Senate Democrats formally abandoned climate reform efforts, with the Obama administration's tacit support.[191]

With Republicans recapturing the House in late 2010 and making gains in the Senate, the climate policymaking window closed. With it, the business coalition assembled to boost federal cap and trade also wound down. For example, many companies began to leave USCAP from January 2010 onward and the organization soon went dormant.[192] For many of these companies, incentives to strategically accommodate US climate reform efforts faded as the policy threat weakened.

The Era of Robust Executive Climate Action

Even as legislative efforts floundered, the Obama administration intensified efforts to directly regulate carbon pollution through executive action. Empowered by the Supreme Court's decision in *Massachusetts v. EPA*, executive branch climate proponents worked to impose costs on carbon polluters through a policymaking process that largely sidelined carbon-intensive economic stakeholders

Executive action after Obama's election was swift. In December 2007, the Bush EPA had refused California a waiver to implement stronger motor vehicle emissions standards. On taking office, Obama immediately reversed this decision.[193] The Obama administration then reinitiated efforts to prosecute coal-fired power plants for failure to comply with New Source Review requirements.[194]

At the same time, incoming EPA Administrator Lisa Jackson moved to issue an Endangerment Finding for greenhouse gases, asserting the agency's capacity to directly regulate carbon pollution through the Clean Air Act. The EPA released a draft in April 2009 for public comment, in the midst of Waxman–Markey debates. Simultaneously, the agency released a Cause and Contributes Finding, linking emissions from motor vehicles to greenhouse gases, and thus subject to the Endangerment Finding. Together, these documents created a legal obligation for the EPA to issue new carbon pollution standards from motor vehicles under Section 202(a) of the Clean Air Act.

Carbon polluters reacted swiftly to the development. The API called the action an "endangerment to the American economy and to every American family."[195] NAM warned that EPA actions would "result in serious risks to our nation's short-term economic recovery and long-term international competitiveness."[196] Even companies that supported the Waxman–Markey bill pushed back against the prospects of direct carbon pollution regulation. For example, Duke Energy suggested EPA efforts would be "a blunt instrument that would have unintended effects."[197] Undeterred, the administration issued its final Endangerment Finding in December 2009. Even as congressional efforts faltered, the Obama administration then announced new climate targets in the lead-up to the 2009 Copenhagen climate conference, adopting the Waxman–Markey bill's stipulation of 17 percent below 2005 levels by 2020 as a provisional goal.[198]

In response, policy opponents intensified effort to legislatively preempt Clean Air Act regulation. In the Senate, Alaska's Republican Senator Lisa Murkowski led a charge to strip the EPA of its authority, but her resolution failed 47–53 in June 2010, despite garnering all Republican votes plus six Democrats from carbon-intensive states.[199] A bipartisan coalition led by Democrats Collin Peterson, Ike Skelton, and Jo Ann Emerson introduced a parallel House bill.[200] Concurrently, Democratic Senator Rockefeller proposed a two-year moratorium on EPA action with his Stationary Source Regulations Delay Act of 2010.[201] In the House, West Virginia Democrats Nick Rahall and Allan Mollohan, alongside Boucher, mirrored this effort.[202]

With EPA authority established as the policy status quo, it was policy proponents who now only needed a single policymaking veto point to preserve executive climate policy authority. Climate policy opponents continued to introduce EPA preemption statutes between 2010 and 2015. While some measures passed the House, none reached the sixty-vote threshold in the Senate. While Republicans made significant Senate gains over this period, the regional distribution of interests remained more constant. As of the 2014 midterms, only Sherrod Brown remained of the Senate Democrats who had actively contested climate policy during the 2000s. In this way, Republican gains disproportionately came from reclaiming seats occupied by Democrats with weak climate policy preferences. The result was an erosion of cross-cutting policy preferences within the Democratic caucus, but not a major shift in the distribution of overall Senate preferences.

Instead, executive action proceeded unhindered by legislative preemption. In March 2010, under its "timing" rule, the EPA found that carbon pollution from motor vehicles would be actively subject to the Clean Air Act as of January 2011. This, in turn, triggered a new set of EPA actions. Once a pollutant is covered by any part of the Clean Air Act, it automatically becomes subject to a Prevention of Significant Deterioration (PSD) provision, which imposes obligations on new or modified stationary pollution sources that emit more than 100–250 tons per year of the regulated air pollutant. In this way, regulating one source of carbon pollution (from motor vehicles) opened the door for the EPA to regulate nearly every source of carbon pollution.

An immediate issue, however, was the unusual number of actors who emit carbon pollution in the 100–250 ton range, including many small businesses that would never have otherwise been subject to EPA regulations. To manage this issue, the EPA proposed a "tailoring" rule in May 2010, restricting PSD requirements to new sources emitting over 100,000 tons annually or modified sources increasing their carbon pollution by 75,000 tons or more. The EPA justified this departure from the Clean Air Act by emphasizing the agency's discretion to phase in regulations in an economically reasonable fashion. In April 2012, the administration then announced the intention to develop greenhouse gas emissions standards for new power plants.

Industry contested this "tailoring" rule, arguing that the EPA was overstepping its authority in departing from the statutory air pollution thresholds specified by the Clean Air Act. Thus, even as the EPA chose not to exercise its regulatory discretion on small-scale polluters, climate policy opponents criticized these efforts as arbitrary. Carbon polluters thus paradoxically attacked the EPA's efforts as insufficiently ambitious. By arguing that the EPA—if it wanted to regulate carbon pollution—would need to regulate all sources above the statutory 100 tons per year, opponents hoped to make these regulations administratively and politically unviable. In a June

2012 decision, the DC Circuit Court of Appeals upheld the EPA's timing and tailoring rules. The Supreme Court eventually agreed to hear a part of this decision, though it rejected efforts to reconsider either the EPA's Endangerment Finding or its motor vehicle greenhouse gas standards. The Supreme Court's final decision in June 2014 invalidated the tailoring rule, finding for industry that the EPA did not have the authority to arbitrarily relax the pollution levels at which an entity became subject to regulation. However, this caveat only applied to carbon pollution point sources that were not already subject to the Clean Air Act. The Supreme Court simultaneously found that any existing stationary source already subject to the Clean Air Act because of co-pollutants could have their carbon pollution regulated at the EPA's discretion. Only a small fraction of new stationary sources emit greenhouse gases without other co-pollutants, so the final Court decision largely entrenched the EPA's power despite invalidating the tailoring rule specifically.

Executive action to manage carbon pollution further intensified during President Obama's second term. Most dramatically, the administration expanded its rulemaking under the EPA to include existing utility sector sources of carbon pollution in June 2013.[203] As part these efforts, the EPA announced eleven "listening sessions" beginning in September 2013 to give agency input.[204] This tour was criticized by Republicans for skipping such states as Ohio and West Virginia that were most coal dependent and most likely to be impacted by the proposed rule.[205] Yet, the EPA consulted widely in its rulemaking design. For instance, the EPA held over two hundred meetings with affected parties from environmental groups to industry officials, perceived by industry as an above-average level of agency outreach.[206]

As a culmination of these efforts, Obama announced the Clean Power Plan (CPP) in June 2014.[207] The administration saw the CPP as part of a "three-pronged" policy that included expanded support for renewable energy and climate adaptation investments.[208] A number of key personnel involved in CPP design had previously worked on designing the Clean Air Act Amendments of 1990. They remained interested in market-based mechanisms despite the regulatory basis of their efforts.[209] Consequently, the Clean Power Plan set strict emissions standards for individual states, but offered states flexibility on how best to comply with these standards.[210]

Overall, the draft Clean Power Plan proposed cuts to utility sector carbon pollution by 30 percent below 2005 levels by 2030. The proposed regulation set an intensity-based carbon pollution standard for each state through a complex formula that estimated how easy or difficult it would be to decarbonize a state's electricity system. Because the EPA made no instrument choice requirements, advocates hoped the plan would induce coalitions of state-level actors to band together into regional emissions trading schemes.

During this period of strong executive action, climate policy opponents felt under siege. For some senior industry officials, the Clean Power Plan was an indication that environmentalists were now getting what they wanted.[211] The Koch-funded American Energy Alliance had already geared up to contest the CPP in the beginning of 2014 as the rule was being designed.[212] The American Coalition for Clean Coal Electricity argued that it was "a complete disregard for our country's most vital fuel sources, like American coal."[213] The CPP proposed significant costs on carbon-intensive sectors, in particular compromising the economic viability of coal combustion. It imposed these costs without any of the technological subsidies or policy carrots that had been previously inserted into the various emissions-trading schemes designed in collaboration with carbon-dependent economic actors.

Frustrated opponents attempted to mobilize the scope of conflict over the Clean Power Plan. Senior Republicans from coal-dependent states appealed directly to state governments to ignore the EPA. Most prominently, Republican Majority Leader Mitch McConnell, from coal-dependent Kentucky, argued that states should fight back against the administration by refusing to comply with the regulations.[214] Judicially, climate opponents contested the CPP through the courts. These opponents included carbon-dependent businesses, unions, and state-level government from carbon-intensive states. The crux of their opposition was the EPA's calibration of reduction targets according to a "best available system of emissions reductions." The EPA interpreted this system as being the broader "electricity system," which allowed the agency to pursue more ambitious targets linked to system-wide energy efficiency measures. By contrast, industry argued that "system" must be interpreted as bounded to the individual facility, which would weaken the feasible levels of carbon pollution reductions with available technology.

Still, business and labor climate policy preferences continued to cross-cut existing coalitions. In June 2014, a number of prominent member companies publicly distanced themselves from a Chamber of Commerce report preemptively attacking the policy.[215] Other small and medium-sized businesses with renewable energy or low-carbon business interests came out in support of the rules.[216] Among the labor community, the Clean Power Plan fractured any convergent climate policy preferences nurtured by Waxman–Markey negotiations.[217] Senior union leaders lamented the loss of the "semi-rational" cap and trade bill as the politics of the issue "went to hell."[218] These leaders expressed skepticism about using the Clean Air Act to regulate carbon pollution. They complained they were "stuck with all sticks and no carrots."[219] In this way, the same unions that threw their weight behind the Waxman–Markey bill, where subsidies and technological supports bridged internal labor tensions, saw the CPP as threatening.[220] As a senior labor official remarked: "Now we are screwed ... we have regulation without investment ... and it will divide us in many ways, internally as well as externally."[221]

The result was a return to outright climate policy opposition among many unions, contesting the legal basis for the EPA's efforts and redividing the AFL-CIO.[222] For instance, such unions as the International Brotherhood of Electrical Workers (IBEW) and the International Brotherhood of Boilermakers (IBB) joined the United Mine Workers to oppose the regulation.[223] However, these industrial unions had relatively little influence with the Obama administration, which only weakly depended electorally on the regions in which these industries were concentrated.[224]

After substantial stakeholder consultation, President Obama unveiled the final version of the CPP in August 2015. Despite the intensity of opposition from carbon-dependent economic actors, the plan's cost levels increased from its draft form with a reduction target moving from 30 to 32 percent. At the same time, the EPA created additional flexibility in the rules states could use to define and track their overtime emissions reduction targets. Under the final rule, states were required to issue a State Implementation Plan (SIP) describing how they would comply with the regulation no later than September 2016.

As before, Obama's announcement highlighted persistent divisions among economic stakeholders. Within days of the rule's finalization, opponents moved to sue the government, arguing the EPA had overstepped its constitutional bounds. This effort delivered early success when, in an unusual move, the Supreme Court stayed implementation of the rule in January 2016, before lower courts had judged the merits of these lawsuits.[225]

This stay remained in effect throughout the remainder of Obama's term and framed the Trump administration's subsequent efforts to unwind the CPP.

Clean Power Plan divisions were also reflected in debates over the Keystone XL pipeline, the subject of substantial mobilization by climate advocates. On the one hand, a large number of labor groups supported pipeline construction, including Plumbers and Pipefitters, Laborers' International Union of North America (LIUNA), the Teamsters, and the IBEW. On the other hand, diverse unions such as the Transport Workers Union, the Amalgamated Transit Union, the National Domestic Workers Alliance, and Domestic Workers United all opposed the project (Sweeney 2013). At the AFL-CIO annual winter meeting in February 2013, the organization issued a strong endorsement of pipeline construction. The Building and Construction Trades department of the AFL-CIO joined in this support, suggesting Keystone could be a valued source of jobs, overcoming the opposition of such unions as the National Nurses United.[226] These divisions, in turn, created tensions within the BlueGreen Alliance.[227] For instance, LIUNA members walked indignantly away from the BlueGreen Alliance in January 2012, declaring that they were "repulsed by some of our supposed brothers and sisters lining up with job killers like the Sierra Club and the Natural Resources Defense Council to destroy the lives of working men and women."[228] The result was a fragmentation of the labor–environmental coalition, as cross-cutting climate policy preferences again frustrated attempts to build a broad coalition of traditionally Democratic constituencies in support of climate reforms.

The Trajectory of US Climate Policymaking, 2007–2015

The Supreme Court ruling in *Massachusetts v. EPA* of 2007 represented a seismic shift in the strategic context of US climate policymaking. It transformed the policy status quo through the courts and tilted the climate policymaking field toward proponents. Before 2007, these proponents had to assemble a pro-reform coalition, which proved difficult within the US separation of powers system. Instead, after President Obama took office in 2008, advocates successfully used the threat of executive action to push climate reforms through the House. Yet, these efforts ultimately stalled in the

Senate where Republicans and fossil fuel-dependent Democrats controlled sufficient votes to stymie action.

However, the EPA moved forward with climate regulations all the same, and policy opponents found themselves unable to assemble a coalition to preempt these efforts. While opponents achieved limited success with the Supreme Court's 2015 stay of the Clean Power Plan's implementation pending legal review, efforts to push back against Obama's proposals only bore fruit after the 2016 election. With control of the executive branch flipping to Republicans, US leadership oscillated once more to laggardship. President Trump's disavowal of the Paris Agreement in June 2017 also underscored this whiplash in US climate posturing. Trump's first EPA administrator, Scott Pruitt, previously fought EPA climate policy as Oklahoma's attorney general; in office, he immediately began efforts to retrench or repeal EPA climate regulations. Trump's second EPA administrator, Andrew Wheeler, a former coal lobbyist, continued Pruitt's work. In a replacement for the Clean Power Plan, Wheeler pushed through an Affordable Clean Energy rule to cut carbon pollution by 0.7–1.5 percent (rather than the CPP's 32 percent). This rule, finalized in June 2019, required power plants to lower their carbon intensity only in limited ways that posed few economic threats to individual plants. At the same time, Trumps deregulatory efforts became mired in legal conflict. Without overturning the Endangerment Finding or *Massachusetts v. EPA*, the administration still faces a legal obligation to protect public health from carbon pollution. It remains unclear whether policies like the Affordable Clean Energy rule satisfy the courts' interpretation of this obligation.

Variance in US climate policy outcomes has never been more dramatic. But this volatility is not surprising once we consider both the distribution of US climate policy preferences and the ways these preferences interact with US policymaking institutions. Over the 2000s, cross-cutting climate policy preferences within business and labor communities assumed a higher profile. Most prominently, a group of major US businesses joined with the environmental community through USCAP to shape emissions trading legislation. This pro-reform coalition included both businesses that stood to benefit from climate policy as well as policy losers hoping to forestall even costlier reforms. Around the same time, partnerships between environmental groups and some labor unions resulted in such groups as the BlueGreen Alliance to jointly advocate for labor-friendly climate policies.

Yet, facing a substantial policy threat, policy opponents mobilized to expand the scope of climate policy conflict. The 2007 Lieberman–Warner bill faced stiff opposition from carbon-dependent economic actors and a veto threat from President Bush. Other labor and business interests preferred a more industry-friendly legislative package from Senators Bingaman and Specter, designed in partnership with carbon-dependent economic actors. Eventually, Democrats and Republicans with carbon-intensive constituencies in the Senate blocked the Lieberman–Warner effort.

President Obama's election opened a new policymaking window of opportunity. In the House, pro-climate legislators led by Henry Waxman moved to take control of the House Energy and Commerce Committee, displacing a group of carbon-intensive Democrats from policy leadership positions. Yet, even after gaining control of the policymaking agenda, pro-climate legislators were forced to bridge intra-party tensions in their effort to build support for climate reforms. These policy advocates sequentially bargained with Democrats representing diverse carbon-intensive industries to accommodate the economic concerns of each, from coal to heavy industry to agriculture. Advocates also leaned heavily on climate policy blueprints from USCAP and the utility sector to win over carbon-aligned copartisans. This negotiating dynamic gave major US carbon polluters a substantial voice in climate policy design. The result was a policy that imposed few costs on the actors who were most actively engaged in climate policy debates.

As had been the case during the BTU tax debate in 1993, the distributive bargains sufficient to bridge intra-Democratic differences in the House did not hold in the US Senate. Senators with carbon-dependent economic constituencies, including some Democrats, undermined the House deal. Unable to find broad support for emissions trading, climate policy proponents instead began negotiating directly with major carbon polluters. These efforts resulted in an extremely producer-friendly Senate policy package. But even this package could not find Republican support.

Instead, with congressional reforms stalled, the executive branch acted on climate policy through the Clean Air Act. Its efforts included major regulations on carbon pollution from new and existing power plants. Unlike legislative emissions trading schemes designed in collaboration with major carbon polluters, these Clean Air Act actions did not include significant technological subsidies or industry financial support. This made US climate reforms somewhat unique among advanced economies for their degree of uncompensated producer costs.

Nor did the Obama administration accommodate carbon-dependent labor actors. Assertive opposition to the Clean Power Plan from such groups as the United Mine Workers and the Boilermakers did not change the EPA's course or temper projected costs for US coal. In this way, the trajectory of US climate policymaking was defined, for a time, by late action that imposed relatively high costs on carbon-dependent economic actors shut out of the climate policy design process.

As they had during debates over the BTU tax and the Kyoto Protocol, carbon-dependent economic actors worked to politicize climate policy to undermine the political incentives associated with climate policy enactment. These efforts bore fruit. In the early 1990s, there was an ideological gradient in climate policy preferences, with stronger support among Democrats. However, over the course of the 2000s, climate opponents reshaped political incentives for Republicans. For example, previously stalwart climate policy proponent John McCain turned away from climate policy when confronted with a Tea Party challenger during a Senate reelection campaign. At the same time, many anti-climate Democrats lost their reelection bids during the 2010 and 2014 midterms. This re-sorted the distribution of climate policy preferences among federal elected officials. Increasingly, Republicans represent the most carbon-intensive US regions in Congress. At the same time, the Democratic presidential coalition has become increasingly decoupled from the most carbon-intensive regions of the country.[229] With US climate policy increasingly tied to executive branch control, policy opponents succeeded in reversing US climate leadership after their political allies won leadership positions in the Trump administration. Costly Obama-era policies were stalled or repealed.

Nonetheless, this issue realignment stemmed from the same climate policy dynamics that drove climate policy inaction for the first two decades of US climate policymaking. As in most other advanced economies, cross-cutting climate policy preferences privileged the voice of climate policy opponents during early US climate policy debates. However, double-represented carbon polluters in the United States never enjoyed the types of institutionally reinforced access to climate policy design that their peers in other economies had. This dynamic created incentives for carbon polluters to mobilize the scope of climate policy conflict. In turn, it was this politicization of climate policy that contributed to partisan sorting around climate policy issues, and that eventually gave Democratic reformers the latitude to propose substantial costs on carbon-intensive constituencies. These dynamics also teed up the Trump administration's subsequent efforts to reverse high regulatory costs.

Ultimately, carbon-polluter double representation was more disrupted in the United States than in either Norway or Australia. Yet, this disruption came alongside the development of an increasingly homogenous anti-climate policy political coalition on the right, with concentrated economic losers as a core constituency. The success of US efforts to mitigate climate change will depend, in large part, on whether this distribution of preferences empowers or weakens reformers' capacity to enact climate policies over the coming decade.

6   Climate Policy Conflict in Australia

Few countries have experienced more dramatic climate policy conflict than Australia. Over the past decade, internecine factional conflicts over climate policymaking consumed Australia's left-leaning Labor Party and right-leaning Liberal Party, contributing to the downfall of three prime ministers and an opposition leader. While Australia enacted an economy-wide carbon price in 2011, it repealed this policy in 2014.

Yet, in this chapter I argue that climate policymaking volatility obscures substantial continuity in the dynamics of Australian climate policy conflict. Mirroring the pattern previously described in Norway and the United States, the climate threat's emergence exposed cross-cutting policy preferences among both economic and political actors. Policy opponents used their structural position to frustrate both Labor and Liberal reform attempts throughout the 1990s and early 2000s. However, as the climate threat intensified during the mid-2000s, pro-climate factions within both parties partnered with bureaucratic, labor, and business allies to champion carbon pricing. By Australia's 2007 federal election, both parties' platforms included national emissions trading.

Labor efforts to enact an emissions trading scheme initially faltered after the Liberals replaced their pro-climate leader. Yet, after the 2010 election returned a Labor minority government supported by the Australian Greens, the country passed a comprehensive climate reform that coupled a national carbon price with extensive consumer and producer support programs. This reform was subsequently repealed in mid-2014 by the Liberals and replaced with a policy to pay carbon polluters for voluntary pollution reductions. Figure 6.1 charts Australian climate policymaking milestones between 1988 and 2015.

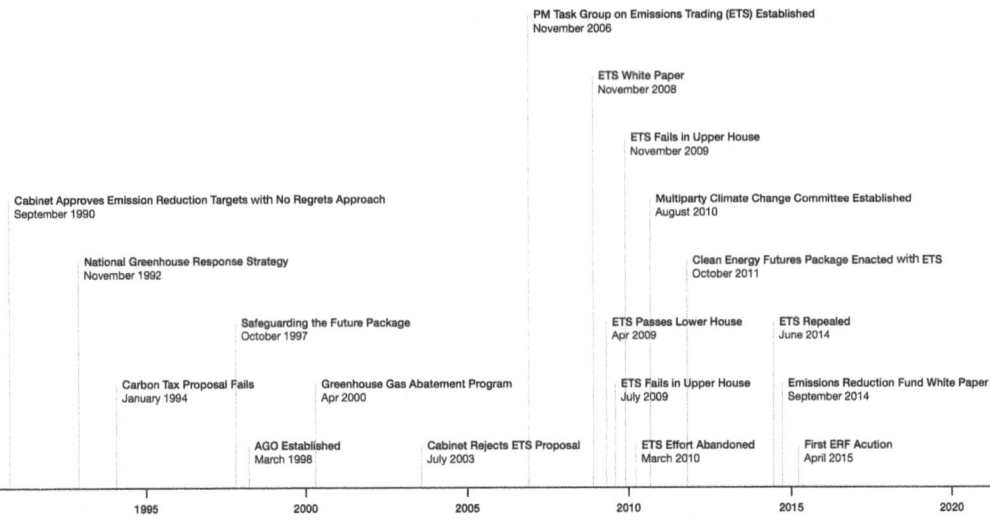

Figure 6.1
Milestones in Australian carbon pricing, 1988–2015

In this way, Australian climate policy action from 2011 to 2014 was characterized by the same patterns of distributive conflict that stalled climate reforms before 2011 and drove policy repeal after 2014. Early periods of climate policy inaction reflected double-represented carbon polluters' structural influence on Australian policy design, not the absence of strong pro-climate voices. In the context of an adversarial policymaking system, these pro-climate factions occasionally controlled the political agenda. However, advocates were still constrained by the political need to accommodate political constituencies with institutionalized ties to both left- and right-leaning political coalitions.

Situating Carbon Polluters in the Australian Economy

Despite gold-fueled economic expansion in the mid-nineteenth century, agricultural and pastoral goods dominated the Australian economy until the late 1800s (Walker 2002). Australia's carbon-intensive economic sectors only emerged during the twentieth century, including through wartime investments in such energy-intensive industries as steel, cement, and aluminum production. Over this period, the economic importance of Australia's extractive sector also grew. While domestic coal mining dated to the late 1800s,[1] coal production shifted toward export markets over the twentieth century. Today, Australia is the world's second-largest coal exporter and the third-largest gas exporter.[2] Industry economists argue that coal provides over $100 billion AUD in economic benefits.[3] By contrast, environmentalists deride the country as the "Saudi Arabia of coal."[4] Coal's abundance also spurred the development of a coal-dependent electricity system. As of 2013, coal provided 64 percent of Australia's power with natural gas supplying an additional 20 percent.[5] This stationary-source electricity generation produces over half of Australia's carbon pollution.[6] Today, Australia has the highest per-capita carbon emissions of any Western country.

Australian industry is politically represented by three major business organizations. The Australia Industry Group (AiG) conducts lobbying and member services on behalf of the broader business community.[7] The Business Council of Australia (BCA) represents the largest Australian corporations through lobbying and policy research. The Australian Chamber of Commerce and Industry (ACCI), closely aligned with the Liberal Party, represents small business interests.

In Australia, the Liberal Party anchors the dominant right-leaning political coalition. This coalition is a formal partnership between the Liberal Party and the rural, populist National Party.[8] In Australian politics, this semi-permanent Liberal–National alliance is described as the "Coalition." Here, I describe Coalition governments as Liberal governments, recognizing that the Liberal Party dominates the Coalition. This also provides clarity for scholars less familiar with Australian politics since it reserves the phase "political coalition" to unambiguously describe sets of interacting political actors, rather than the Australia-specific Coalition alliance.

The Australian labor movement enjoys institutionalized representation within Australian politics through the Australian Labor Party (ALP), founded in 1891. The ALP represents diverse working-class economic interests, balancing a working-class "blue-collar" constituency with a progressive, urban constituency. These constituencies have historically pulled in different directions on environmental issues. For example, Tasmanian Labor members, deeply invested in the forestry industry, are at odds with urban Labor voters elsewhere in the country.[9] Reflecting this ideological breadth, the party includes Left, Right, and Center-Left factions. Party factions are formal membership groups, streamlined in the early 1980s, that shape parliamentary and committee appointments (Leigh 2000). Traditionally, individuals chose factional alignment on economic grounds, with Right-aligned politicians supporting free trade and Left-aligned politicians supporting protectionism.[10]

Australian unions themselves are organized under a single peak organization, the Australian Council of Trade Unions (ACTU), in addition to sectoral unions.[11] Much like Labor politicians, unions also declare factional commitments. Crucially, carbon-intensive unions are affiliated with both Left and Right ALP factions. For instance the industrial Australian Workers Union (AWU) is the most important Right-aligned union, while both the Construction, Forestry, Mining and Energy Union (CFMEU) and the Australian Manufacturing Workers Union (AMWU) are Left-aligned.

The Emergence of Climate Change on the Australian Policymaking Agenda

Climate change emerged on the Australian policymaking agenda during the late 1970s, as Australian scientists began profiling the threat.[12] As early as 1981, an Australian Office of National Assessments report warned the Liberal Fraser government about climate risks (Hamilton 2007).[13] However, until the late 1980s, environmental conflicts centered on forest conservation and biodiversity protection (Toyne and Balderstone 2003).[14]

Environmentalists enjoyed significant influence during the Bob Hawke-led Labor government (1983–1991) (Hawke 1994; Toyne and Balderstone 2003).[15] Hawke placed significant value on environmental protection and viewed environment policies as electorally beneficial. For example, Labor used environmental commitments during the 1990 federal election to secure ballot preference flows from Green Party voters.[16] Yet, Hawke's environmentalism split the ALP. Within Australia's Cabinet, environmental decisions pitted Hawke and environment minister Graham Richardson against industrial-minded Labor politicians linked to blue-collar unions, who rejected the party's green agenda (Toyne and Balderstone 2003). These divisions cross-cut party factions. For instance, the green Richardson was a Right-faction power broker. By contrast, Resource Minister Peter Cook was affiliated with the Center-Left faction and routinely opposed environmental reforms (Economou 1996). Hawke's finance minister, Peter Walsh, would later cofound the Lavoisier Group, a coalition of Australian climate skeptics.[17]

Australia's first prominent climate debate centered on establishing a national carbon pollution reduction target.[18] Responding to the 1988 Toronto climate change conference, Hawke established a scientific advisory committee in April 1989 (Hamilton 2007). When Richardson proposed a 20 percent below 1988 levels by 2005 target to the Cabinet, unsympathetic senior ministers blocked the proposal over concerns it would limit industrial growth (Staples 2012).[19] This reflected a general Hawke-era pattern: despite pro-environmental politicians' success in driving conservation policies, industrial-aligned copartisans stymied all climate reforms. For example, Labor's 1990 pre-election "Our Country, Our Future" statement offered sweeping land use reforms but circumscribed climate commitments with an emphasis on preserving Australian economic competitiveness.[20]

Australia's Cabinet reconsidered emissions reduction targets in October 1990. After contentious negotiations with Labor's industrial block, the Cabinet would only endorse a 20 percent below 1990 target with a disclaimer that resulting policies could not adversely impact the national economy.[21] Moreover, the agreement positioned the Minister for Primary Industries and Energy at the center of climate policy planning to empower climate opponents in the policymaking process. This Labor decision helped establish the "no regrets" principle that shaped national climate policymaking through the mid-2000s.[22] Under this framework, only climate policies with positive economic co-benefits should be pursued. The "no regrets" policy bridged tensions between advocates of early climate action and political allies unwilling to impose costs on their carbon-intensive constituencies.[23]

After Paul Keating replaced Hawke as Labor prime minister in December 1991, the party become, in Richardson's words, "browner and browner."[24] Keating advocated environmental policies only when they fit his neoliberal economic worldview. Keating often downplayed the climate threat and sided with Labor's pro-industry camp to oppose policies that imposed costs on Australian business (Staples 2012).[25] Under Keating, Australia's 1992 National Greenhouse Response Strategy (NGRS) marginalized environmental voices and catered to industry and resource voices.[26] Similarly, Labor's 1993 election strategy departed from 1990 and 1987 to downplay environmentalism.[27] In short, the Australian left was not inherently pro-climate. Instead, climate change surfaced latent cleavages between otherwise aligned party members that reflected sectoral ties to labor and capital. Even when pro-environment leaders achieved environmental reforms, their efforts to drive climate policy specifically were routinely stymied by copartisans.

Climate Policy Inaction during the 1990s

While industrial-minded Labor leaders reversed Hawke's environmentalism during the Keating years, other Labor actors continued to push climate reforms. Carbon pricing policies had been discussed among environmental policymaking communities since the late 1980s.[28] However, the idea received its first substantial debate in June 1994 after Environment Minister John Faulkner floated a carbon tax.[29] Proposal details leaked to the press in February 1995.[30] The tax would have imposed a $1.25 AUD price per ton of CO_2 from stationary energy sources. It was expected to raise $960 million AUD over three years.[31]

Environmental groups pitched more aggressive tax levels to the government, worried the modest proposal would do little to help Australia meet its 20 percent reduction target.[32] By contrast, carbon-intensive industries opposed the proposal during a single round of business consultations in early February.[33] For instance, the Australian Mining Industrial Council derided the proposal as Australia taxing its "comparative advantage."[34] These industrial criticisms were echoed by industry-aligned Labor politicians, some of whom publicly attacked the proposal before their government formalized a position on the policy's merits.[35] Labor politicians were joined by carbon polluter allies within the opposition Liberal Party, who criticized the tax as "economic suicide."[36] Facing cross-cutting stakeholder opposition and divided Cabinet support, Faulkner shelved the policy.[37]

The Keating government instead turned toward voluntary abatement programs. Jointly developed by the environment and industry ministries, and enthusiastically endorsed by carbon-intensive companies, the Greenhouse Challenge program created a voluntary emissions abatement framework (Bulkeley 2000). Within this framework, climate policy costs were evaluated at the individual company level, eliminating any possibility that climate policy would impose distributive costs on major carbon polluters (Bulkeley 2001).[38] Only measures with individual economic benefits would be pursued.

The Liberal Party, led by John Howard, won the 1996 national election. Howard would continue to champion the "no regrets" approach to climate policy articulated by Labor Cabinet ministers and reflected in Labor's Green-house Challenge program.[39] Howard government debates initially centered on international climate negotiations. While Hawke had signaled interest in constructive engagement with global negotiations, Keating distanced his government from the international climate process.[40] For instance, Keating's Cabinet was reluctant to endorse the Berlin Mandate and only agreed after US endorsement.[41] Howard also intensified Australian obstructionism within multilateral climate fora. At COP 2 in Geneva, Australia resisted the EU-US compromise that all Annex 1 countries accept a uniform 1 percent reduction target. Australia's economy was already highly carbon-intensive, and undifferentiated reduction commitments would disadvantage the country's fossil fuel production.[42] After Geneva, Australia grew isolated as the country's insistence on special treatment generated international chastisement, including from the United States (Hamilton 2007).

Howard's climate reluctance stemmed from his view that Australian self-interest, shared by many senior Liberals, required maximizing fossil fuel-led economic growth (Hamilton 2007). Many Liberal officials within the Howard government had deep ties to Australian fossil fuel industries.[43] These close connections between such lobbying organizations as the Australian Industry Greenhouse Network (AIGN) and senior Liberal Party members created what some Australians described as a "GHG Mafia" in control of policy design during the late 1990s and early 2000s (Pearse 2007).

However, climate policy opposition was not homogenous within Liberal ranks. As with the Labor Party, cross-cutting climate policy preferences still structured the Australian right. During the late 1980s, the Liberals lagged the Hawke government in their commitment to environmental policies.[44] However, pro-climate factions occasionally championed more costly commitments. For instance, the Liberals took a more aggressive carbon reduction commitment than Labor into the 1990 election, proposing a 20 percent reduction below 1988 levels by 2000 (instead of 2005) (Staples 2012). Similarly, in the lead-up to Rio, pro-climate Liberals urged strong global action.[45] Moreover, even as Howard and his ministers signaled their climate skepticism, officials in Howard's local Liberal organization tried to persuade the prime minister that stronger climate actions were justified (Bulkeley 2001).

Australia's antagonistic climate posture left the country isolated. Hoping to temper domestic and international criticism, the Howard government defensively announced a domestic climate policy package in November 1997.[46] Formally abandoning Hawke's emissions reductions target, Howard's package aimed to reduce "net emissions growth" by a third between 1990 and 2010 at a five-year cost of $180 AUD million. The package included $65 AUD million in increased renewable energy support, a marginal increase in the country's Renewable Energy Target by 2 percent, new efficiency standards for electrical generators and cars, energy efficiency measures and expanded support for the voluntary Greenhouse Challenge program.[47] Howard also announced the establishment of the Australian Greenhouse Office (AGO) to coordinate Australian climate policy.

At Kyoto, Australian negotiators used the threat of withdrawal to secure concessions (Hamilton 2007). The final agreement included differentiated reduction targets and allowed Australia to cap emissions *growth* by 8 percent above 1990 levels by 2012. Australia also pushed for land-clearing accounting formulas so it could meet its commitment largely through natural shifts in land-use practices. In short, Australia got what it wanted.[48] Kyoto Protocol targets imposed few if any constraints on Australian carbon polluters.

Post-Kyoto, domestic climate policymaking centered on the AGO. The world's first integrated climate policy agency, the AGO faced significant institutional constraints. It was not an independent agency, instead reporting quarterly to a council of ministers representing agriculture, industry and environment.[49] Environment Minister Robert Hill held a number of Liberal Party leadership roles.[50] He used his influence to increase environmental issues' political profile within the government, developing strong relationships with conservation-minded environmental advocacy groups.[51] Yet, the council of ministers gave policy opponents the ability to continuously block AGO proposals.[52] Rather than facilitating climate policy action, the AGO's institutional structure enhanced policy opponents' veto powers.[53]

With these constraints, the agency focused on such voluntary measures as the Greenhouse Friendly program that helped companies confidentially track corporate carbon pollution using government-designed software.[54] Lacking policymaking sway, the AGO also focused on research. In 1999, the office prepared a series of emissions trading discussion papers.[55] Yet, even blue-sky carbon pricing discussions generated significant opposition from carbon-intensive constituencies. Bowing to industry pressure, the government reaffirmed in August 2000 that there would be no domestic emissions trading before an international trading system was in place (Hamilton

2007).[56] By 2000, intellectual momentum among government bureaucracy around emissions trading had stalled, despite the AGO's reports.[57]

Instead, the AGO's most consequential reform was an accidental by-product of Australian Senate politics. In 1999, the Howard government passed an indirect tax reform that required Senate votes from the Democrat Party. Among the Democrats' demands was additional climate policy funding. In response, the Liberals agreed to appropriate $400 million for the AGO, but neither party gave the AGO clear guidance on how to spend these funds.[58] AGO bureaucrats decided to use the money on abatement and established the Greenhouse Gas Abatement Program (GGAP).[59] GGAP paid companies to reduce carbon pollution using a competitive tender process. However, information asymmetries between government and industry soon compromised the program as officials struggled to evaluate the value of private tenders.[60] By mid-2003, only eighteen tenders had been approved, corresponding to a release of only $50 million AUD, 12.5 percent of the program's budget.[61] A final 2009 audit found GGAP failed to meet its abatement goals because it struggled to locate good projects and supported projects that failed outright or did not deliver promised abatement.[62] If GGAP had any positive effect, it was simply that it signaled to Australian carbon polluters that more serious abatement policies, including emissions trading, could reach the political agenda.[63] More generally, the GGAP program highlights the ability of small green-leaning parties to shape climate policy timing through strategic leverage; yet, the GGAP program's failure also highlights the persistent difficulty of enacting meaningful climate reforms under institutional conditions where climate policy opponents can veto policy design.

The Move toward Emissions Trading in Australia

As in the United States, cross-cutting preferences within the business community became more public during the early 2000s. Divisions within the business community surfaced during debates over Australian Kyoto Protocol ratification. Major carbon polluters assertively claimed that even Australia's 108 percent of 1990 target would still compromise Australian economic competitiveness.[64] Senior members of the Howard government began to echo business arguments for delaying Kyoto ratification, alleging the treaty would trigger major job losses.

However, the broader business community held divided preferences. Unlike a coal industry that rejected climate science outright, gas interests believed global action served their competitive interests. Such multinational companies as British Petroleum and mining giant BHP also called for constructive policy engagement. In turn, peak business organizations became paralyzed by their membership's divided preferences. When the Business Council of Australia's leadership signaled Kyoto ratification support in 2002, energy sector-member companies pushed back. Unable to resolve internal conflict, the BCA refrained from taking a Kyoto ratification stand (Hamilton 2007). Australia's broad-based business organizations similarly struggled to negotiate domestic policies on behalf of a divided membership.[65] The BCA's initial support for emissions trading fractured as financial sector advocates clashed with resource and metal processing sector opponents.[66]

Pro-climate business voices also emerged within Liberal Party ranks. On the one hand, carbon-intensive economic interests maintained their influence on Howard. For example, Howard decided not to ratify Kyoto, siding with emissions-intensive constituencies in June 2002. Yet, on the other hand, under Environment Minister Hill, pro-climate interests rallied during the early 2000s. These interests found a new ally in the Australian Treasury.[67] Treasury officials had not paid attention to climate change during the 1990s.[68] However, they became increasingly concerned about climate-linked economic risks beginning in 2000 under the leadership of Secretary Ken Henry. By 2001, Treasury officials were openly discussing market-based instruments to manage these climate risks.[69] Given Treasury's centrality in the Australian bureaucracy, climate attention by senior Treasury officials pushed climate change high onto the Australian economic planning agenda.

The early 2000s thus saw the emergence of a nascent coalition between environmentalists and economists interested in market-based threat responses. Internal debates culminated in an August 2003 Cabinet submission to evaluate emissions trading, jointly prepared by Hill and, critically, Minister of the Treasury Peter Costello.[70] While the proposal found additional support from other sympathetic Liberal Cabinet members, it was opposed by key politicians with ties to mining and resource interests.[71] Ultimately, emissions trading stalled as an idea because these policy opponents, chief among them Howard, controlled the party's agenda. Yet, even with the Cabinet submission's failure, many bureaucrats continued to discuss carbon pricing.[72]

State government actions also boosted carbon pricing's profile. In New South Wales, a state Labor government enacted the New South Wales Greenhouse Gas Abatement Scheme (GGAS) in 2003.[73] Then, in 2004, Labor-led state governments across Australia began to discuss a coordinated climate policy framework in an effort to force the federal Liberal Party's hand.[74] To do so, state governments established the National Emissions Trading Taskforce (NETT) to design and coordinate a *federal* climate plan.[75]

NETT discussions exposed the same cross-cutting climate policy preferences that had structured Labor climate politics a decade earlier. Labor Party officials affiliated with state-level industry or resource departments sat in one corner, facing off against officials affiliated with state-level environment departments. Premiers' offices and state-level treasury officials oscillated between the two camps.[76] Unable to bridge significant differences in target preferences between industrial and environmental groups, NETT participants left the target decision unresolved. Labor leaders instead punted target calibration to the federal government.[77]

While state-level Labor politicians discussed emissions trading design, federal bureaucrats and sympathetic federal Liberals championed the issue in Canberra.[78] Growing policy support also emerged among business actors through the Investor's Group on Climate Change (IGCC).[79] Likewise, the renewable energy sector became more professionalized, with the Australian Council for Sustainable Energy emerging as a serious policy player in 2002 (Hamilton 2007). Other businesses joined pro-climate ranks because they felt some form of carbon pricing was inevitable.[80] These businesses shifted from their blanket opposition to a defensive posture focused on shaping policy design to moderate potential costs.[81]

Meanwhile, the environmental movement began prioritizing climate policymaking in the early 2000s.[82] Some of these groups worked with business to promote domestic action. For instance, the Australian Conservation Foundation set up a Business Leaders Roundtable in May 2006 to raise the climate issue's salience in economic debates. This roundtable would later issue a joint statement advocating a "long, loud, and legal" carbon price.[83]

Business realignment and environmental advocate mobilization coincided with a confluence of political events that pushed climate change to the top of the Australian policymaking agenda. An intensification of a multiyear drought in 2006 precipitated serious water shortages in many parts of the country. Around the same time, the *Stern Review*'s publication, which highlighted the economic costs of unmitigated climate change, and the high profile given to Al Gore's documentary *An Inconvenient Truth* generated widespread media and public discussion about climate dangers.[84]

Buoyed by this attention, pro-climate Liberal factions began to flex their muscles, led by then–Minister for the Environment and Water Malcolm Turnbull.[85] These actors felt the Howard government's blanket opposition to climate policy was neither politically or economically sustainable. Howard began to find himself on the political defensive over climate change, and perceived a political need to manage growing public climate concerns.[86] Senior bureaucrats persuaded Howard that the Cabinet should, at a minimum, reconsider domestic carbon pricing. These bureaucrats saw the issue as one of economic risk management. Recognizing growing, if passive support from the business community, Howard agreed this was politically expedient. With the prime minister onside and backed by Treasury officials, Cabinet allowed senior bureaucrats to explore the possibility of establishing an Australian emissions trading scheme. In December 2006, the Howard government appointed a Prime Ministerial Task Group on Emissions Trading (PMTGET). The PMTGET gave major business actors a direct voice in climate policy design: it included carbon-dependent business actors, from mining (BHP Billiton, XStrata), airlines (QANTAS), electricity (International Power), gas (Australian Pipeline Trust), and aluminum (Alumina Limited).[87]

Cautious that the PMTGET should not be too costly, Howard not only appointed carbon-intensive industrial leaders to the group but also circumscribed its terms of reference to emphasize that Australia's comparative advantage in fossil fuel production not be compromised.[88] The group's discussions were also limited to "advise on the nature and design of a workable global emissions trading system in which Australia would be able to participate ... report on additional steps that might be taken, in Australia, consistent with the goal of establishing such a system."[89] However, the group immediately reinterpreted its mandate to include the design of a standalone domestic scheme if this was in Australia's economic interest. In doing so, PMTGET intentionally extended Howard's remit, exploiting reference term ambiguity to facilitate more expansive policy discussions.[90]

With few political incentives for federal Liberals to engage with state-level Labor, the NETT and PMTGET processes barely interacted. However, PMTGET participants, particularly within the bureaucracy, built their understanding of emissions trading from NETT materials.[91] In the end, PMTGET's process "shamelessly plagiarized" the NETT, and can be understood as its intellectual continuation.[92] In this way, both Liberal and Labor efforts to design an emissions trading scheme reflected important continuities. In practice, the PMTGET was a secretariat-led process, with uneven intellectual leadership from economic stakeholders.[93] Businesses wanted policy stability but did not possess nuanced understandings of emissions trading scheme (ETS) policies or their business impacts.[94] In public submissions, most economic stakeholders adopted a constructive tone, accepting the emissions trading premise but making specific policy demands to manage the policy's distributive burden. For instance, the Australian Coal Association signaled its desire for generous permit allocations and complementary measures to fund R&D and technological investments.[95] By contrast, environmental groups championed aggressive targets and costs. Meanwhile, Labor unions remained disengaged from the process.[96]

The PMTGET submitted its final report to Howard in May 2007. In almost direct contradiction to its terms of reference, it recommended a domestic policy in advance of meaningful global action, arguing that waiting for a global cap would "place costs on Australia by increasing business uncertainty and delaying or losing investment."[97] The report recommended comprehensive action with broad sectoral coverage, beginning as early as 2012.[98] Yet, mirroring the NETT process, the group punted on setting an emissions reduction target faced with internal divisions. It instead focused on emissions trading architecture more narrowly.[99] The report gave particular importance to compensating energy-intensive trade-exposed (EITE) industries.

Howard surprised observers by accepting the report's recommendations, gaining majority Cabinet support to continue emissions trading development.[100] The Liberal view became that some form of climate policy had become a "political necessity"; emissions trading seemed the conservative way to manage this need.[101] Howard received private reassurances from business leaders that they supported his decision.[102] The PMTGET offered divided Australian industry space to design a policy instrument that moderated costs on the largest carbon polluters.

Still, while Howard's motivations in setting the group up may have been defensive, it developed into a serious policymaking bid. With Liberal buy-in, the Australian civil service began preparing for an ETS in earnest. Marginal to the political agenda five years earlier, the idea of emissions trading had become the "political mainstream."[103]

The Carbon Pollution Reduction Scheme

The PMTGET released its report in advance of the Fall 2007 election. Alongside industrial relations, climate change defined that campaign. Both Liberal and Labor pro-climate factions enjoyed increased power. On the Labor side, a desire to defeat Prime Minister Howard united the party around Leader Kevin Rudd even as industrial factions remained skeptical of environmental policymaking.[104] Labor believed emissions trading could split the Liberal party by highlighting an unstable cleavage on the Australian right.[105] To differentiate the Labor plan, Rudd argued that any policy should start one year earlier, in 2011.

On the Liberal side, Howard's sudden emissions trading embrace attempted to sidestep the issue.[106] During the election, the Liberals promised to establish the "world's most comprehensive emissions trading scheme" no later than 2012.[107] However, this promise created space for such pro-climate voices as Malcolm Turnbull to advocate even more costly policies. These conversations generated serious tensions with National Party members inside the Coalition.[108] Turnbull then pressed the Howard Cabinet to reverse course and ratify Kyoto in September 2007.[109]

However, neither party made emissions reduction target commitments, echoing both the NETT's and PMTGET's reluctance to specify policy costs. Despite Labor's carbon pricing commitments, shadow cabinet divisions made a target agreement too difficult.[110] Nor did environmental groups press either party to make reduction commitments, focusing instead on Kyoto ratification.[111] Consequently, broad agreement over a policy instrument across the left and right did not resolve the deeper distributive issues implicit in climate policy design.

Labor won the 2007 election, elevating Kevin Rudd to prime minister. Rudd was aligned with the Right faction of the ALP but held unusually weak ties to the party's labor movement.[112] His commitment to climate action appeared sincere. While running for office, he repeatedly described climate change as the "great moral challenge of our generation."[113] Fulfilling a campaign pledge, he ratified Kyoto as a first act in office. Rudd also initiated work on Australian emissions trading development. He tasked Australian economist Ross Garnaut to review the economics of climate change, with the intent of producing an Australia-specific version of the Stern Review. To buy time for resolving target-setting tensions, he also tasked the bureaucracy with undertaking fresh emissions trading consultations.[114]

To coordinate policy development, Rudd established a new Department of Climate Change and Water, staffed mostly by Department of Environment bureaucrats; yet, the new department's leadership was largely drawn from Treasury.[115] These macroeconomists were seen as having policy control rather than—in the words of a relieved Labor Party insider—"touchy-feely" environmental folk.[116] This staffing arrangement underscored the emerging belief that Australian climate reforms were, first and foremost, a class of economic reforms.[117]

Green Paper

Labor policy consultations proceeded over the first half of 2008.[118] Then, in July 2008, Labor released a Green Paper outlining a proposed Australian Carbon Pollution Reduction Scheme (CPRS).[119] Given bureaucratic personnel continuity, the content of the CPRS mirrored the PMTGET, despite Rudd's new planning process. The proposal outlined an emissions trading scheme with broad sectoral coverage beginning in 2010 and covering companies producing more than 25,000 tons of carbon pollution annually.[120] Policy revenues were directed toward household and industry adjustment costs. The most emissions-intensive industries received 90 percent free permits while medium-intensive industries received 60 percent.[121]

Senior Liberal politicians found few differences between the PMTGET and Labor's proposal.[122] Nonetheless, Prime Minister Rudd's policy differed in several respects. Treasury official voices reduced carbon-intensive industry influence. These bureaucrats with strong neoliberal commitment to market-based mechanisms felt comfortable discussing the phaseout of certain economic activities in Australia, even in frank exchanges with business stakeholders.[123] While EITE assistance was present, the CPRS Green Paper capped assistance levels for the first time. This decision created tensions with carbon-intensive stakeholders, who believed the Green Paper reneged on their PMTGET bargain for 100 percent compensation.[124]

The package also offered a different relative balance of consumer versus producer costs. The Rudd plan instead focused more on cost mitigation for low- and middle-income Australians.[125] Policy planners familiar with discussions during both the Howard and Rudd governments believed the Howard plan would have instead directed income tax cuts toward upper middle-income individuals.[126]

The Green Paper also described emissions trading architecture in more specific terms than either NETT or PMTGET. It involved a "sharpening of the trade-offs" that made policy costs visible to stakeholders. As one official recalled, the Green Paper "forced decision makers to really understand the trade-off that they were facing for the first time. And what we observed was like a form of grieving process."[127] While large carbon polluters had actively followed climate debates for years, smaller and regional carbon polluters still found carbon pricing a strange and confusing new policy instrument.[128] Many businesses accepted climate policy's importance and endorsed a flexible, market-based approach to risk management. Yet, they had not developed clear mental models of such a policy's effects on their daily operations. The Green Paper abruptly made salient these costs.[129]

It also publicly surfaced business splits. It became common for IGCC representatives, trying to provide a pro-climate business voice, to clash publicly with Mineral Council representatives or the ACCI.[130] Peak business associations, including the Business Council of Australia, also struggled to bridge internal community tensions.[131]

Similar splits shaped labor response to the CPRS Green Paper. Before 2007, Australian labor leaders were mostly disengaged from climate policy.[132] The Australian Workers Union (AWU), which represents workers in a range of carbon-intensive sectors, grew alarmed by carbon pricing and aggressively attacked the government, a rarity for a labor union against a Labor government.[133] The AWU feared insufficient compensation to carbon-intensive industries, particularly for aluminum production, would kill union jobs.[134] Partnering with sectoral business leaders, union actors "game-planned it with various companies and industry associations so we were locked in together [in opposition to the CPRS]—they'd do Libs, [we'd] do Labor."[135]

By contrast, public sector and service unions supported aggressive climate policy action, splitting Australia's peak labor association, the Australian Council of Trade Unions (ACTU), which struggled to coordinate a response, much as its Business Council of Australia counterpart had struggled under Howard.[136]

ACTU efforts to negotiate compromise between green public sector unions and brown industrial unions predated the 2007 election, when it established a climate change subcommittee. However, the CPRS exacerbated a serious split between teachers and health workers who supported aggressive climate policy action, and industrial unions who viewed the policy as economically damaging.[137] ACTU leaders tried to accommodate industrial unions by elevating a CFMEU official as ACTU's climate point person.[138] Nonetheless, AWU still distanced itself from ACTU leadership. By the June 2009, AWU pressures had eroded ACTU's climate ambitions (Burgmann 2013).[139]

Even as a divided union movement debated carbon pricing, some of its leaders began engaging the environmental community. The Southern Cross Climate Coalition (SCCC), which grew out of a preexisting relationship between ACTU head Sharon Burrow and then head of the Australian Conservation Foundation (ACF) Don Henry, attempted to build climate policy consensus among core ALP constituencies, including ACTU, the ACF, the Australian Council of Social Services (ACOSS), and the Climate Institute.[140] Other environmental groups saw value in the process but weren't comfortable compromising with industrial unions to coordinate climate policy.[141]

SCCC deliberations worked to locate the strongest common policy denominator among its diverse member groups. For given climate policy design debates, each member organization declared its preferred policy positions, with junior staff collating recommendations to assess where there were similar policy preferences. Where policy differences could not be bridged, the SCCC described divergent preferences in its official recommendations to the Labor government.[142] On the hardest issues, unions and environmentalists could not reach agreement, including emissions reduction targets and compensation levels for carbon-intensive industries.[143] As a result, the SCCC boosted policy proposals where union and environment interests converged but discounted policies that disrupted the carbon-intensive economic status quo.[144] At the extreme, discussions about fossil fuel extraction, including coal mining patterns, remained outside the negotiating space.

White Paper

While the CPRS Green Paper only described emissions trading architecture, the December 2008 CPRS White Paper also proposed reduction targets.[145] These targets drew from Treasury macroeconomic modelling and the Garnaut review.[146] Australia would use the CPRS to achieve 5 percent reductions below 2000 levels unconditionally, and 15 percent below 2000 levels if a global climate agreement included comparable reductions from all major economies. The 5 percent unconditional target reflected a reduction of about 25 percent from business as usual, which saw significant Australian emissions increases by 2020. Framing a 25 percent reduction as 5 percent was viewed as soothing business concerns.[147] While the Green Paper suggested the rate of free permit allocation would decline over time, the White Paper also specified this decay rate at 1.3 percent annually.

Broadly, the CPRS core architecture remained unchanged, including the scheme's liability points, its timeline, and its general reporting and compliance features. Where differences existed, these included additional compensation to both business and households in response to stakeholder comments.[148] This included household compensation adjustments for pensioners and low- and middle-income families.[149]

Labor's proposed CPRS targets pleased no one. Green constituencies criticized the target as unambitious, arguing Australia had to make deep cuts to do the country's part in keeping climate change within safe levels. Union allies who might have been satisfied by the compromise proposal, particularly the ACTU, avoided an endorsement under pressure from pro-climate unions who wanted reductions in the 40 percent range.

During this period, the global financial crisis intensified, further complicating reform efforts. Labor officials wanted a broad-based political coalition to support the reform, not a narrow environmental or labor base.[150] In private negotiations, they brought together Australia Industry Group, the BCA, the WWF, and the Southern Cross Climate Coalition to agree on a "grand bargain."[151] Using this insider strategy, the government sought a compromise package major stakeholders could publicly endorse. Labor also wanted business support to provide cover for potential Liberal Party buy-in.[152]

Labor's insider bargain was negotiated confidentially, and the announcement of a deal in May 2009 took uninvolved stakeholders by surprise.[153] The compromise package differed from the CPRS White Paper in three respects.

First, it increased the *conditional* target from 15 to 25 percent if "the world agrees to an ambitious global deal to stabilize levels of CO_2 equivalent at 450 parts per million or lower by mid-century."[154] This increase was well below the 40 percent that environmental groups demanded but proved the maximum reduction business would tolerate. This increase was coupled to provisions allowing for international carbon credit use to reduce business compliance costs.[155] Second, it delayed policy onset by three years in recognition of financial crisis-linked economic challenges. Third, it introduced a one-year fixed price of $10 AUD per ton CO_2, to create a transition period of price certainty.

With the deal, begrudging carbon pricing acceptance emerged among many labor and business stakeholders.[156] However, the business community remained split. For Australia Industry Group, the national executive's decision to endorse masked a vocal membership minority protesting climate science outright.[157] However, organization leadership saw bipartisan appetite for emissions trading and wanted to shape what seemed an inevitable policy to their liking.[158] The Australia Industry Group and the Business Council of Australia—peak organizations with diverse membership bases— ultimately endorsed the deal, while more specialized industrial associations continued their opposition, including the ACCI and Minerals Council.[159]

Negotiations with the Liberal Party

With stakeholder support in hand, Prime Minister Rudd introduced the CPRS into the Australian Lower House. Rudd hoped for Liberal support to pass the policy through the Australian Senate, where Labor lacked a majority. Labor negotiators moved between two poles, engaging the SCCC on one end to gauge if concessions to Liberals would upset Labor constituencies and engaging Liberal leaders on the other end to see if common ground could be found.[160] After the 2007 election, Malcolm Turnbull, the environment minister in the late Howard years and a strong supporter of emissions trading, was elected Liberal leader. Yet, Turnbull's pro-ETS posture exposed persistent climate policy cleavages across the Australian right. Party leaders had been actively involved in the emissions trading debate during the late Howard years but sought political cover to bring wavering constituencies along.[161] However, other Liberal politicians continued to hold deeply skeptical and even denialist climate change views.[162] Meanwhile, the Nationals—the Liberal's perennial coalition partner—grew stridently anti-climate; Nationals leader Barnaby Joyce strongly opposed carbon pricing and publicly excoriated Liberals for even considering the policy.[163]

In early negotiation stages, Turnbull floated major CPRS changes.[164] For example, in his effort to leave an intellectual policy mark, he proposed an alternative structure for electricity pollution management that purported to weaken price impacts for consumers.[165] However, stakeholders were skeptical as to the realism and viability of his plan.[166] Labor negotiators refused to consider the idea, and it never gained traction.[167] More generally, Labor sensed Turnbull's internal difficulty in selling climate policy; Rudd's team did little to speed negotiations along, hoping to extract political advantage from the process by weakening their political rival.[168]

The Australian Senate voted the CPRS down in August 2010, without Liberal support. Even so, anti-climate Liberal factions grew restless with their leadership's policy dialogue with Labor. Responding to mounting internal criticism, Turnbull made a dramatic political statement in early October, proclaiming, "I will not lead a party that is not as committed to effective action on climate change as I am." He threatened to resign if the party would not follow his climate lead.[169] MacFarlane sent the Liberal caucus a copy of Howard's election platform, which had, of course, included an emissions trading scheme.[170] By late October, Turnbull received agreement during a protracted and divisive five-hour caucus meeting to negotiate a Labor ETS deal, provided the caucus could vote privately on whether to support it.

Broadly, pro-climate Liberals wanted to amend the CPRS to increase business compensation; they hoped these changes could mute the party's loudest internal critics and opposition from its carbon-intensive constituencies.[171] Turnbull's eventual deal with Labor reflected this agenda, shifting the CPRS even closer to Howard's original PMTGET architecture.[172] First, industry was given additional free permits.[173] Second, to appease the Nationals, agriculture was explicitly excluded from future emissions trading schemes.[174] Third, the package increased levels of direct industry assistance to offset the initial scheme costs. These included $1.5 billion AUD for the coal sector,[175] $150 million AUD to support food processing sectors, and $1.1 billion for the Transitional Electricity Cost Assistance Program.[176]

Unions supported the deal; however, environmentalists grew concerned that the ACF had endorsed the deal too early as part of the May 2009 compromise. They worried that increased carbon polluter compensation undermined the policy's ability to drive meaningful emissions reduction.[177] In the end, WWF stuck with the deal, believing that having an institutional framework in place was worth the policy's compromises. However, the ACF board put pressure on senior leadership to walk back their earlier endorsement and instead maintain a neutral position.[178] Reflecting the gradual compensation increases necessary to bridge political cleavages, the final ETS deal thus won the support of such industrial unions as the AWU, but not all of the policy's previous environmental backers.

Other groups, including Treasury bureaucrats, saw the deal in optimistic terms, believing the policy's compensation review process would soon intensify industry's cost burden. Under the CPRS, compensation review was delegated to the Productivity Commission, a neoliberal-oriented government body that bureaucrats believed would wind down free permit allocation much more quickly than industry expected.[179] These bureaucrats believed the Productivity Commission would be committed to the establishment of an undistorted market price, and the commission was deliberately chosen to facilitate this outcome. As one would later reflect: "we knew what [the Productivity Commission] would say—we could have written their report. They'd say no way. No business case. So the pressure was always going to be on dismantling [free permits]."[180]

However, Turnbull faced growing internal opposition. Emissions trading split the party and a leadership spill was called for December 1, 2009. In Australia, leadership spills allow elected representatives to vote for a new leader. By this time, three separate Liberal constituencies were mobilizing against Turnbull's ETS support: agricultural communities aligned with the Nationals, mining and resource constituencies, and coal-dependent electricity generators.[181] Broadly, the party remained divided between the urban, pro-environment Turnbull wing and a rural wing with closer ties to carbon-intensive sectors.[182]

The leadership spill's frontrunner, Joe Hockey, promised Liberal legislators could vote their conscience on the ETS. However, Tony Abbott also challenged Turnbull as a climate skeptic, promising to terminate Liberal climate policy engagement. Turnbull retained the pro-climate faction's support while Hockey's moderate position failed to attract anti-climate caucus members. Instead, by a razor thin vote of 42–41, Abbott seized party control. He immediately rejected the Liberal-Labor ETS deal from only weeks earlier and proclaimed the CPRS a "great big tax on everything." When the

Senate voted on the Labor–Liberal negotiated CPRS deal on December 2, Senate Liberals again rejected the policy.

Even with the Liberal Party's abrupt reversal, Labor had CPRS options. Labor could pivot to seek Green Party support, and then they would only need one or two Liberal Senate votes for a majority. At least a handful of pro-climate Liberals were believed to be willing to cross the floor and endorse the deal their party had just negotiated. Even if that failed, Labor could have called a "double dissolution" election—a full election of both houses of Parliament that the government can call when the upper house (the Senate) twice rejects a bill that received majority lower house (House of Representatives) support.

Yet, shaken by the failed Copenhagen climate conference in December 2009, Rudd became reticent.[183] Both the Copenhagen collapse and the Liberal Party reversal also tempered business belief that climate action was imminent, weakening the government's bargaining position.[184] For instance, the Australian Industry Group (AIG) leadership's consensus that carbon pricing was inevitable fractured.[185]

Furthermore, Labor remained hesitant to engage the Greens. Labor had deliberately avoided engaging Green Party senators on emissions trading during the CPRS process, believing Green involvement would compromise business stakeholder comfort with any eventual policy.[186] In Australia, Labor and Greens see their political fortunes as mutually exclusive, with significant enmity between senior Labor and Green staffers.[187] Some Labor officials believe they should not negotiate over any policy, even an environmental policy, unless strictly necessary.[188] Mainstream environmental advocates also have an uncertain relationship with Green politicians, keeping their distance to avoid being perceived as agents of the Green Party by other parties' politicians. Even environmental advocacy groups were in infrequent contact with the Greens over the course of developing the CPRS.[189] Thus, the Green Party found itself on the outside throughout this period, reliant on media and public debate to shape the emissions trading deal.[190]

Publicly, Green senators rejected the Liberal-Labor compromise, arguing it further "browned" the CPRS down.[191] Greens believed the CPRS risked locking in relatively ineffective carbon pricing policies for two decades with a weak target that would not mitigate climate risks but would foreclose other more ambitious policies.[192] Privately, Green Senate leader Christine Milne sent more constructive signals, suggesting the package could be acceptable with some small shifts in the compensation packages and long-term targets.[193] For instance, the Greens asked for a more compressed time-line over which EITE and energy sector compensation could be revisited.[194] Still, there was no real attempt to explore a potential deal.[195] Labor signaled to the Greens they would have to accept the deal as negotiated with the Liberals or vote the package down.[196]

The Greens did vote down Labor's plan in late 2009, anticipating the policy would come back for a new vote in January 2010. To support move-ment, they released a compromise in late January 2010, establishing a two-year transitional fixed-price period beginning on July 1, 2010, with pollution permits fixed at $20 per ton.[197] Yet, Labor refuse to bargain. Mean-while the Green base became upset with leadership's willingness to asym-metrically compromise with Labor. Labor politicians then grew incensed with what they saw as intransigence by Green Party legislators on the most significant climate reform ever proposed in Australia.[198]

Rudd reintroduced the CPRS a third time on February 2, 2010 including the Turnbull amendments but without Green Party support. By this time, carbon pricing reservations had emerged within the Labor Party too.[199] For example, within Rudd's inner Cabinet, Julia Gillard and Wayne Swan pushed Rudd to abandon the policy in the face of Abbott's criticisms.[200]

Rudd put the CPRS on indefinite hold in late April 2010, putatively until the emergence of a global climate consensus.[201] With this pivot, the labor community also tempered its public commitments, diluting public and web statements about the nature of the climate threat (Burgmann 2013). Meanwhile, industry was as split with Labor's turn away from emissions trading as it been with Labor's embrace of the policy. Carbon-intensive con-stituencies celebrated what they perceived as a tremendous victory. Other parts of the Australian business community felt betrayed by Rudd's decision to walk away from a deal they saw as the best they were ever likely to get; many business leaders had invested substantial personal capital in inducing support for climate policy among peak business lobby groups.[202]

Rudd's climate policy indecision damaged him politically. On June 24, 2010, Rudd was replaced by Gillard as Australia's prime minister in a Labor leadership spill. Just as Abbott used climate policy to undermine Turnbull, Gillard used Rudd's commitment to climate policy to consolidate power against him. Unlike Rudd, Gillard was affiliated with Labor's Left faction. Yet, indicative of climate policy's cross-cutting nature, Gillard held weaker climate policy preferences. Earlier in her career, Gillard had worked for the CFMEU; the environmental community did not view her as deeply committed to their cause.[203] Gillard was among senior ALP leaders who pressured Rudd to abandon the CPRS earlier in 2010, fearing economic and political climate policy costs.[204]

Gillard acted consistently with this perception. Despite entreaties from the Green Party, Gillard remained uninterested in revisiting climate issues; instead, she intensified attacks on the Greens.[205] In the 2010 federal election, held two months after Rudd's deposition, Gillard's Labor Party all but abandoned climate policy. Its central climate commitment was establishing a citizen's panel of 100 to 200 Australian volunteers to build a public consensus on future climate policy. This panel was widely panned by Australian environmental organizations.[206] Gillard also proposed that Labor establish emissions standards for new coal plants, but without flagging how aggressive these standards should be, or committing to retroactively applying these to the fifteen new coal-fired power plants then under construction or recently approved. While Labor remained generally committed to carbon pricing, Gillard categorically ruled out a carbon tax.

By contrast, Liberal leader Abbott spent the campaign outlining doomsday economic scenarios that would result from even modest carbon prices.[207] Abbott instead proposed a "Direct Action" plan. Announced in February 2010 as a CPRS alternative, the plan centered on an Emissions Reduction Fund (ERF) that would competitively buy emissions abatement from businesses. Abbott claimed the plan would allow Australia to meet Labor's 5 percent reduction target by "providing incentives rather than imposing massive balance sheet liabilities."[208] In design, the ERF was more similar to the AGO's failed GGAP program. It was designed with little business consultation except from some mining and resource interests who publicly opposed emissions trading.[209] Major Australian businesses, prepared to support carbon pricing, opposed the Direct Action plan. One organization even prepared a report criticizing the policy's economic and policy viability but left it unpublished for fear it might undermine future access to a Liberal government.[210]

The overall result was that, within six months, both federal Australian leaders—Rudd and Turnbull—who came from pro-climate factions of their respective parties were replaced by leaders holding tepid (in Gillard's case) or actively hostile (in Abbott's case) climate policy preferences. This completed a remarkable fall in the political fortunes of emissions trading. In December 2009, a bipartisan coalition broadly supported by unions, businesses, and some environmental groups had coalesced behind the CPRS. By the following August, emissions trading was off the short-term Australian political agenda altogether.

The Clean Energy Futures Package

The 2010 federal election resulted in an unexpected 72–72 seat tie between Labor and the coalition. In the lower house, the Green Party and rural political independents held the balance of power. The Greens also held the balance of power in the Australian Senate. Both Abbott and Gillard jockeyed for the political independents' support, who found themselves in the position to choose the next government. Climate policy commitments proved consequential to this decision. For instance, the Climate Institute organized a private meeting for two of the independents, Rob Oakeshott and Tony Wilson, with Nicholas Stern and Ross Garnaut in the election's aftermath; they would later describe these conversations as pivotal to their choice of Gillard as prime minister.[211]

While Green support for Labor was never in doubt, the Greens nonetheless demanded a multi-party carbon pricing dialogue as a condition for their support.[212] To the Green Party's surprise, Gillard agreed and set up the Multi-Party Climate Change Committee (MPCCC) as a Cabinet-level subcommittee that also included Oakeshott and Wilson.[213] The Liberals declined to participate in the MPCCC.[214] The MPCCC also included four expert members—a scientist, an economist, a social service provider and an energy stakeholder—who lacked voting rights but were otherwise committee participants.[215] The MPCCC met monthly, supported by the same Treasury bureaucrats who had managed the CPRS process and who remained committed to emissions trading.[216] The MPCCC largely focused on negotiations between political actors rather than social and economic stakeholder consultation.[217] Unlike CPRS negotiations, Labor did not seek policy validation from the SCCC or other constituencies.[218]

However, unions, if anything, enjoyed greater policy design influence through their closer connections to senior Labor policymakers under Gillard.[219] Still split over climate policy, the labor movement itself also tilted more toward its industrial unions' interests.[220] Service unions championed aggressive targets but deferred to the ACTU for public advocacy. In turn, ACTU only lobbied for policy costs acceptable to industrial unions.[221] By contrast, blue-collar unions aggressively mobilized on their own. For instance, responding to the MPCCC, AWU head Paul Howes called a "crisis meeting" of union officials in April 2011, and warned the government that AWU support for carbon pricing would disappear if even "one union job is gone" as the policy's result.[222] When the Minerals Council distanced itself from the negotiations, the coal mining unions, according to a MPCCC stakeholder, "stepped into the industry's shoes."[223]

By contrast, businesses were more cautious about the MPCCC process. Some businesses were concerned by Green Party influence, viewing the party as anti-business.[224] Others demanded that the CPRS as amended by Turnbull be the policy starting point.[225]

Against the MPCCC backdrop, the environmental community launched its "Say Yes" campaign. A joint initiative of nine labor and environment groups, the campaign sought to pressure ALP members and a skeptical green base to back emissions trading.[226] The campaign proved controversial since it directed environmentalist resources toward papering over factional cleavages in the Labor Party's base that had repeatedly stalled Australian climate policymaking, rather than pressuring legislators to increase the policy's ambition.[227] Green Party politicians felt the Say Yes campaign undermined their negotiating position within the MPCCC by prioritizing common ground over climate mitigation.[228]

Inside the MPCCC, politicians grappled with the same cross-cutting tensions. Labor wanted to reintroduce a weakened CPRS.[229] Yet, the Greens had voted down the CPRS for being too unambitious, and demanded a substantially stronger package. The MPCCC's two rural independents also brought specific agendas to the table. For instance, Oakeshott was concerned with land carbon-use issues, proposing to offset electricity sector emissions through land carbon sinks.

The eventual MPCCC deal announced in July 2011 navigated these tensions by expanding the scope of its policy bargain. Unable to agree on a narrow emissions trading policy, the MPCCC proposed a sweeping Clean Energy Futures Package (CEFP) that coupled carbon pricing with major new subsidies for clean energy and industrial support. The package's emissions trading scheme was the Carbon Pricing Mechanism (CPM), a proposal that largely reintroduced the CPRS architecture and sectoral coverage.[230] One major difference was the inclusion of a price floor. Labor initially resisted to the idea but conceded after the investment community threw their support behind the Green proposal to establish more price certainty.[231] Few Labor stakeholders believed the floor would ever come into play, so it was not seen as a major concession.[232] However, the MPCCC was unable to agree on how to calibrate the CPM. Green representatives advocated carbon-negative policy strategies;[233] they refused anything less than a 25 percent reduction against 2000 base levels. By contrast, Labor wanted to calibrate the scheme at a 5 percent reduction below 2000 levels.[234]

Unable to bridge this gap, the MPCCC punted its target decision to a new independent body, the Climate Change Authority (CCA), which would study the issue and report back to parliament with its recommendation. The CCA was loosely modeled after the UK Committee on Climate Change and was an attempt to depoliticize the target-setting process.[235] However, in the absence of parliamentary endorsement of the CCA-proposed target, the CPM would by default adopt Labor's 5 percent reduction goal preference.

Deferring a target decision meant the CPM needed a transitional carbon price before trading underneath the policy cap could begin. The parties agreed to establish an interim fixed price but struggled to agree on its level. Green representatives wanted a price of $40 AUD per ton of carbon over as long a period as six years. Labor wanted a shorter period and a carbon price between $10 and $20 AUD. The committee compromised on $23 per ton over three years.[236]

While the CPM was a clear CPRS successor, the remainder of the Clean Energy Futures Package constituted a break from the previous legislation. The CPRS had involved some compensation and crediting, but the CEFP was a full-scale industrial transformation initiative. Rather than simply shifting carbon pollution costs, the CEFP also offered the largest industrial policy package in Australian history.[237] This industrial policy focus helped bring blue-collar labor interests into alignment with white-collar climate policy preferences.[238] The package also included generous subsidies to compensate industry for policy costs. For instance, industrial unions pressed the ALP aggressively for 100 percent compensation to steel after the AWU threatened to withhold support for the package.[239] The government also established the Steel Transformation Package (STP), a $300 million AUD package viewed as cash subsidy rather than a serious effort to transform the industry's carbon intensity.[240] Yet, even with the substantial subsidies that largely defined the non-CPM parts of the CEFP, the Australian policy still distinguished itself from efforts in such countries as Norway with its willingness to impose some costs on individual sectors or industries. For instance, under the "Contracts for Closure" program, the government would help pay for the closure of about 2,000 megawatts of carbon-intensive coal-fired power plants.

For their part, the Greens demanded substantial new programs and subsidies to compensate for their compromise on the low interim fixed price, including a new Clean Energy Finance Corporation (CEFC) and an Australian Renewable Energy Agency.[241] The CEFC received enthusiastic backing from the investment community.[242] By contrast, Treasury officials opposed the CEFC because they were uncomfortable with establishing a $10 billion AUD fund to shape investment decisions.[243] Finally, the CEFC also included more substantial social benefits than the CPRS, including changes in the tax-free threshold for low-income Australians.[244] This social agenda was driven by Labor.[245]

The Australian parliament passed the CEFP in November 2011.[246] Abbott's Liberals offered sustained opposition and framed the policy as ruinous taxation.[247] Senior Liberals seized on Gillard's admission that the CPM's fixed price could be viewed as a tax to accuse her of breaking her campaign promise to not introduce an Australian carbon tax.[248] Yet, as always, these views were not homogenous. For instance, Turnbull broke with his party to vote for Gillard's plan.[249]

While the CEFP bridged cross-cutting policy preferences within the Labor coalition, it did not receive the same backing from business stakeholders as had the CPRS. In part, businesses felt a weaker need to strategically accommodate government policymakers.[250] Liberal opposition to carbon pricing, including promises to repeal climate policies under future Liberal governments, undermined the previous sense of policy inevitability. Even pro-climate business voices hesitated to enter the debate. For example, members of Australia Industry Group—who had overcome internal tensions to endorse the CPRS in late 2009—worried about the optics associated with endorsing emissions trading when the Liberals so vociferously opposed the policy. In an attempt to shape the debate, they hired a multinational consulting firm to evaluate the CEFP's merits versus Direct Action in an attempt to outsource their views.[251] By contrast, policy opponents, notably the Minerals Council, unreservedly expressed their opposition.[252]

The result was asymmetric mobilization. Pro-climate advocates tried to paper over cross-cutting cleavages on the left, while anti-climate advocates on the right mobilized aggressively against any deal.

Green Party involvement in carbon pricing policy dropped off after the MPCCC's work finished. Policy implementation and communication campaigns were managed exclusively by Labor,[253] which soon walked back some of its compromises with the Greens. The Contracts for Closure program first morphed into a program called the Energy Security Package, and then was jettisoned entirely after no generating facilities made reasonable bids.[254] Labor also made cuts to the Australian Renewable Energy agency and biodiversity components of the package, again key Green demands.[255] Next, in August 2012, Labor dropped the CPM's price floor.[256]

Instead, Labor moved toward developing a two-way trading linkage with the EU ETS.[257] The EU linkage and scrapped price floor had been key business demands, and part of the business sector's justification for not endorsing the CEFP.[258] Yet, many business leaders remained uncomfortable declaring victory and supporting the changes for risk of appearing to endorse an unpopular policy being amended by an increasing unpopular Labor government.[259]

With an election looming, Labor attempted to resuscitate its political fortunes by switching leaders again. In June 2013, Rudd challenged Gillard for the Labor Party leadership and returned as prime minister. Rudd immediately announced changes to the emissions trading system, including an attempt to moderate Abbott's damaging carbon tax attacks by moving the scheme to a floating price one year earlier, beginning in July 2014.[260] These changes did little to shift the contours of Australian climate policy conflict.

Repealing Australia's Carbon Price

Tony Abbott's Liberals won the 2013 election and began unwinding the country's climate policy infrastructure. A first act in office was disbanding the Climate Commission, a group of public experts, which the Gillard government established to engage the public on climate science and policy.[261] Abbott also instructed the CEFC to stop making new renewable energy investments.[262]

The Liberals then began efforts to repeal Australia's carbon price. Despite holding a majority in the lower house, Abbott did not initially not have an upper house majority.[263] A first CPM repeal was thus rejected in March 2014, where the Green Party retained the balance of power. This created an awkward transition period during which carbon pricing implementation continued despite the government's efforts to undermine its success. The apex of this tension occurred when, in the midst of efforts to wind down the CPM altogether, the Climate Change Authority delivered a mandated report on recommended emissions reduction targets. Using a carbon budget approach, the CCA suggested a target increase to at least 15 percent. The CCA cited recent actions by China and the United States and the affordability of this target.[264] These targets took some business stakeholders by surprise, who believed that the terms of the CCA report were biased against recommendations for aggressive action.[265] Whether the symbolic CCA report would have been sufficient to move Labor away from its default 5 percent reduction baseline is not clear; with the Liberals in government, the CCA report was simply ignored.

Carbon price repeal negotiations moved to the Australian Senate, whose composition changed in July to give the newly formed Palmer United Party (PUP) the balance of power, alongside an eclectic set of political independents.[266] The leader of PUP was the eccentric Australian industrialist Clive Palmer, a mining magnate who at the time of his election was in a legal dispute with the Australian government over carbon pricing. Palmer maintained a principled objection to pay any carbon price on carbon pollution his companies released within a CPM framework he viewed as unconstitutional.

Surprising political observers, Palmer initially refused to endorse the full repeal of Australia's domestic climate policies. In a June 2014 press conference, he was joined by Al Gore to trumpet his plan for a zero-price emissions trading scheme that would be legislated and implemented but would remain dormant until triggered by other countries' actions. Palmer also refused to support the abolishment of the Climate Change Authority, the country's renewable energy target, and the Clean Energy Finance Corporation.[267] Thus, Palmer rejected repeal legislation in early July because the government was unwilling to provide concessions to the PUP. After Abbott agreed to spare the CEFC and the CCA, Palmer let the repeal pass one week later.

Having repealed Labor's carbon price, Abbott began work on his Emissions Reduction Fund (ERF). An expenditure-focused program, Direct Action swung the climate policy pendulum back to the Department of the Environment from its decade-long home in Treasury.[268] Yet, ERF support among non-Liberal senators was tepid. A former government official involved in the climate policymaking process lamented that Direct Action is "tonne for tonne, the most expensive way of trying to reduce carbon emission."[269]

To broker a deal, the government began backroom talks with independent senators and, representing the environmental community, ACF chair Don Henry.[270] These talks delivered a "safeguards" mechanism—to limit emissions increases from the power sector and manufacturing so that emissions growth in one part of the economy would not counteract purchased reductions in another.[271]

The government released an ERF White Paper in October 2014, which included these safeguards mechanisms. If companies increased their emissions beyond committed levels at future points in time, the White Paper suggested they might have to purchase offsets from other emissions reduction projects. Additionally, companies that made emissions reduction commitments they could not deliver could make up for their shortfall by purchasing credits.[272] These efforts created a potential, implicit carbon market at the heart of the ERF proposal.

Anti-climate Liberals objected to even this unambitious ERF.[273] While some pro-climate voices pushed for Australia to meet its targets with international offsets—supported by much of business sector and some environmentalists—Abbott vetoed any forms of international linkages that might indirectly nurture emissions trading.[274]

The ERF was approved as a $2.5 billion AUD program over four years. To win votes from PUP in the Senate, the final agreement included an awkward commitment by the Liberals to study the possibility of introducing an emissions trading scheme, even as the government simultaneously denied that a coalition government would ever support a carbon price.[275] Palmer subsequently took public credit for "substantial progress towards delivering an ETS," despite having supported the repeal of Australia's emissions trading policy months before.[276] The review's terms of reference told the CCA to decide whether Australia should have an ETS and under what conditions, with a draft by November 2015 and a final report by June 2016.

More broadly, few stakeholders believed that the ERF would meet the 5 percent reduction target that Liberals claimed. For instance, commissioned research by the Climate Institute found Direct Action inefficient and would not meets its intended targets without another $4 billion.[277] At the same time, the repeal of Australia's carbon price and its replacement with a purchased abatement program received strong buy-in from some of the largest carbon polluters in the mining and resource sectors with strong ties to senior Liberals. However, the shift lacked buy-in from many other business community members, who remained skeptical. Actors who had supported the CPRS and CPM worried that emissions trading would return to the Australian political agenda later in the decade, but in a political context that would include less generous compensation for Australian industry.[278] Despite these reservations, these businesses were more concerned with securing policy concessions in other domains than lobbying for climate policy to an antagonistic Liberal leadership.[279]

Even during this period, significant support for emissions trading remained within the Liberal caucus. But these politicians lacked the political space to express their preferences in the context of strong party discipline imposed by Abbott.[280] Senior Cabinet ministers—from Julie Bishop to Joe Hockey—were on the record supporting emissions trading. Greg Hunt, the minister responsible for developing the Direct Action proposal coauthored a master's thesis arguing for introduction of market-based pollution taxes in the state of Victoria.[281] However, these pro-climate factions remained unempowered within the party.

These pro-climate factions' influence revived in September 2015 after Malcolm Turnbull won a leadership spill against Abbott to claim the prime ministership. Abbott's popularity had plummeted over the preceding year as a result of non-environment-related political missteps. Turnbull immediately broke with Abbott's antagonism toward renewable energy. Within days of replacing Abbott, Turnbull transferred authority of the Clean Energy Finance Corporation and the Australian Renewable Energy Agency (ARENA) to Hunt's Ministry of the Environment from finance and industry ministries, respectively; he simultaneously disavowed Abbott's efforts to repeal pro-renewable policy instruments. Finally, Turnbull also initiated discussions around how the Liberals' Direct Action policy might be strengthened through its "safeguard" mechanism to ensure that industrial growth would not undermine ERF-purchased emissions reductions.[282] However, Turnbull also won leadership by promising to respect existing policy directions and disavowing his belief that carbon pricing was necessary. His government's post-2015 record remained consistent with this position: Turnbull and his pro-climate Liberals were able to protect and sometimes extend subsidies for low-carbon technologies, but deferred to anti-climate Liberals in side-stepping any new costs on Australian carbon polluters.

The Trajectory of Australian Climate Policymaking

For three decades, Australian climate policy preferences have cross-cut existing party divides. Labor party politicians struggled to reconcile tensions between industrial unions and environmentalist demands. Liberal party leaders oscillated between embracing and rejecting climate reforms. From the emergence of the climate threat in the 1980s through the mid-2000s, the result was climate policy inaction. Under Labor Prime Minister Robert Hawke, industry-allied Labor politicians circumscribed domestic climate policies so that they would not disrupt the economic status quo. This "no regrets" approach to climate policymaking intensified under Labor's Keating, who worked to marginalize climate policy proponents within his own party. Liberal Prime Minister John Howard then entrenched this perspective, limiting Australian climate policy to weak voluntary measures throughout the late 1990s and early 2000s. While Howard would defensively establish the world's first integrated climate change agency, the Australian Greenhouse Office, this institution was overseen by Liberal cabinet ministers with ties to Australian industry.

As the climate threat increased in salience during the mid-2000s, opponents saw their grip on Australian climate policymaking weaken. During this period, cross-cutting preferences within Australian business and labor groups became increasingly exposed. Vocal, pro-climate policy stakeholders emerged, particularly within the financial and investment communities. Consequently, peak Australian business associations struggled to manage divided internal preferences around such issues as ratification of the Kyoto Protocol and the desirability of domestic carbon pricing. Similarly, the Australian labor movement, despite broad support for some climate policy action, found itself divided on what types of domestic climate policies to support.

During this period, pro-climate factions within both the Australian left and right seized control of the Australian policymaking agenda. Within the governing Liberals, a pro-carbon pricing coalition emerged that brought pro-environment Liberals together with Treasury officials. At the same time, opposition Labor politicians at the state level began to design an emissions trading architecture in an effort to split their right-leaning opponents. Both parties would commit to establishing an ETS as part of the 2007 election.

While Labor won this election, its efforts to design an ETS involved a continuation of preexisting Liberal plans, rather than a major departure. Just as Howard had given major carbon polluters a key seat at the table as part of his Prime Ministerial Task Group on Emissions Trading, Labor's Carbon Pollution Reduction Scheme (CPRS) accommodated carbon-dependent economic interests that were critical Labor constituencies. Struggling to build a coalition in support of the proposed policy, Rudd's Labor Party brought industry, unions, and environmental groups into a joint consultation to modify the policy in light of stakeholders' economic concerns.

Despite this accommodation, the CPRS was still a policy that, in comparative perspective, imposed some costs on producers. Rudd did not have a close relationship with his party's unions. Further, climate policy design details were delegated to Treasury bureaucrats who were comfortable imposing sectoral costs in ways that reshaped the types of profitable economic activities in Australia. Yet, a broad willingness to impose these costs was still constrained by a political need to accommodate the Labor party's carbon-intensive constituencies. The result was a proposed policy design that balanced producer and consumer costs. Ultimately, policy designers were comfortable imposing costs within a more adversarial, pluralist policymaking process, even as political actors simultaneously worked to compensate climate reforms' double-represented economic losers.

The CPRS initially drew support from the Liberal Party, then under pro-climate policy leadership. But, cross-cutting support for climate policy action within both Labor and Liberal parties remained constrained by the distributed presence of carbon-intensive constituencies within each party. These within-party cleavages would soon upend the politics of Australian climate reforms.

Within the Liberals, anti-climate factions mobilized to oust Liberal party leader Malcolm Turnbull over his support for emissions trading. The party then replaced him with an avowed climate policy opponent in Tony Abbott. Abbott immediately launched a campaign, supported by carbon-intensive businesses and allied National Party members, to increase the salience of carbon pricing on consumers. The intent was to mobilize climate policy conflict into the public domain to undermine electoral and legislative incentives to pass reforms. Without Liberal support, the CPRS stalled in the Australian Senate.

The Labor prime minister, Kevin Rudd, would also be replaced, in part over his poor handling of the climate file. The new prime minister, Julia Gillard, held weaker preferences for climate reforms and went into a 2011 election without a strong commitment to domestic climate policy action. However, Labor won an unusual minority in this election, with the Australian Greens holding the balance of power in both houses. This launched carbon pricing back onto the policymaking agenda. Ultimately, a coalition of Labor and Green party politicians passed a comprehensive climate reform bill in 2011 as the Clean Energy Futures Package. This package fused an emissions trading scheme similar to the CPRS with extensive consumer and producer support, and compensation packages. By balancing producer and consumer needs, the CEFP positioned Australia in comparative perspective as a late actor that imposed a mixed distribution of costs. However, Australian climate policy action did not reflect a fundamental disruption of carbon polluters' double representation.

Within this Labor–Green coalition, the economic needs of Labor's industrial wing, particularly the vocal Australian Workers Union, continued to constrain climate policy negotiations. At the same time, many parts of the environmental community focused on papering over divergent policy preferences between labor and environment groups, rather than pressuring negotiators to increase the climate reform's ambitions. Ultimately, the final agreement catered more to the "red" wing of the red–green coalition. For example, on the central issue of targets, Labor and Green actors could not come to an agreement. Instead, they deferred a decision on long-term policy costs to an independent commission. Yet, in the absence of a later parliamentary agreement to endorse these targets, Labor's less ambitious reduction target would remain the default. Analogously, Labor unilaterally abandoned a number of the Green Party's hard-won concessions during policy implementation.

As in the United States and Norway, climate policy opponents responded to the policy threat by mobilizing to increase the salience of policy-linked consumer costs. After a change in government in 2013, these anti-climate forces succeeded in killing the country's carbon price, making Australia the first advanced economy to repeal a major climate policy reform. Even pro-climate factions within the Liberal Party fell in line behind this effort, arguing that they had to accommodate a Liberal electoral mandate that included the repeal promise. The Liberal Party instead replaced the country's carbon price with a producer-friendly system that paid major carbon polluters to voluntarily provide carbon pollution abatement targets. While pro-climate Liberals later regained control of party leadership in 2015, they did not reverse their party's carbon pricing approach. On the left, the voice of carbon polluters within the Labor Party also continued. The party remained committed to climate policy, but Labor Party leader Bill Shorten was a former leader of the ACTU. The voice of industrial labor continued to exercise a privileged voice in constraining climate policy content within the left-leaning political coalition.

Australian climate policy volatility thus obscures substantial continuity in the dynamics of the country's climate policy conflict. What can appear at first glance to be simple ideological conflict over climate policy instead reflects complex distributive conflict within and across dominant political coalitions. Both the Liberal and Labor parties struggled to manage cross-cutting climate policy preferences, splits that appeared among both business and labor communities. Despite periods of support for climate reforms that spanned the Australian left and right, carbon-dependent economic interests used their cross-cutting representation across governments to circumscribe efforts to impose costs on national carbon pollution.

In this way, periods of Australian climate policy action and inaction were both structured by changes in the factional balance of power within and across political coalitions. Early climate policy inaction reflected not an absence of pro-climate factions within any party, but instead climate policy opponents' structural influence on both sides of the ideological spectrum. These opponents were able to limit efforts to impose domestic costs on carbon pollution. However, within an adversarial policymaking system, pro-climate factions sometimes took control of each political party and used their control of the political agenda to push domestic reforms.

However, while such reforms as the 2011 Carbon Pricing Mechanisms did impose some costs on producers, advocates' efforts were still constrained by the political need to accommodate important political constituencies with institutionalized ties to both left- and right-leaning political coalitions. Unlike the Obama administration's US climate regulations, Australian climate policy design was more constrained by ties between left-leaning political actors and carbon polluters, including carbon-intensive industrial unions. The result was enhanced access for carbon-dependent economic actors on climate policy design and policy sensitivity to both consumer and producer burdens. Even so, the distributive bargains struck during climate policy negotiations among business, labor, and environmental groups were insufficient to guarantee policy stability over time. Anti-climate factions on the country's right succeeded in raising the salience of consumer and producer policy costs and achieved electoral gains; these opponents then repealed the country's climate reforms and replaced them with an industry-friendly alternative. Yet, just as the enactment of carbon pricing before it, this repeal was not a simple function of ideological conflict between political parties but instead emerged as political allies attempted to navigate complex distributive conflict.

The Australian case thus complicates several common distributive politics theories of climate policy enactment. It points to the importance of thinking not just about sectoral balance of power explanations, but also the ways in which different interests interact with national policymaking institutions. It also complicates theories over the role environmental interests play in driving policy content. Australia's proportional representation electoral system allowed for the emergence of a Green Party that served in a coalition government with Labor between 2010 and 2013. However, the Australian Green Party was not a major player in climate debates prior to 2010 and, while the party did a play a significant role in shaping the climate policymaking agenda through the MPCCC, its ability to force costs onto domestic carbon polluters was limited. Much as we saw in Norway, the green partner in the red–green coalition remained beholden to carbon-intensive economic interests represented by the industrial wing of the coalition's "red" party.

Instead, to make sense of the trajectory of Australian climate politics it is vital to consider how the climate threat's emergence surfaced latent cleavages within existing political coalitions. Carbon-dependent economic interests gained a persistent voice in policymaking through the labor movement on the left and through business associations on the right. And this double representation of carbon polluters interacted with domestic political and policymaking institutions to shape climate reform content and timing. Australian climate politics thus echoes the dynamics of the Norwegian (chapter 3) and American (chapters 4 and 5) cases. Are these dynamics universal? I now turn to this question in chapter 7, which explores whether this book's account of climate policy conflict extends beyond its primary cases to other important advanced economies.

7 Exploring the Theory's Generalizability

To this point, my account of the double representation has drawn from within-case analysis of policymaking in Australia, the United States, and Norway. Do the causal processes that shape climate reforms in these three cases generalize to other advanced economies? In this chapter, I explore the theory's generalizability using four additional advanced economies as shadow cases.

Critically, I find evidence to support the theoretical scope conditions outlined in chapter 2. That chapter's propositions drew from cross-case comparisons of Australia, the United States, and Norway. Cross-case comparisons in small-n research should not be viewed as causal inferences; instead, they provide conjectures about the conditions under which within-case identified causal mechanisms may be present. Do the book's propositions correctly predict the timing and content of climate policymaking outside of the book's core cases? This chapter provides encouragement: climate reform timing and content in Canada, the United Kingdom, Germany, and Japan are largely consistent with theoretical predictions.

The chapter is structured as follows. First, I outline a principled method to select shadow cases. Second, I detail the empirical record of climate reforms in each shadow case. Finally, I reflect on the fit between shadow case policymaking dynamics and the book's theoretical claims.

My theory suggests that climate policy outcomes are structured by three sources of institutional variation: the relative corporatism or pluralism of a given society, the presence of formal linkages between labor and left-wing parties, and the presence of formal linkages between capital and right-wing parties. These conditions can be represented as a 2 × 2 × 2 cube, as in figure 7.1. However, in practice, intimate relationships exist between business interests and right-wing political parties in virtually every relevant case (and

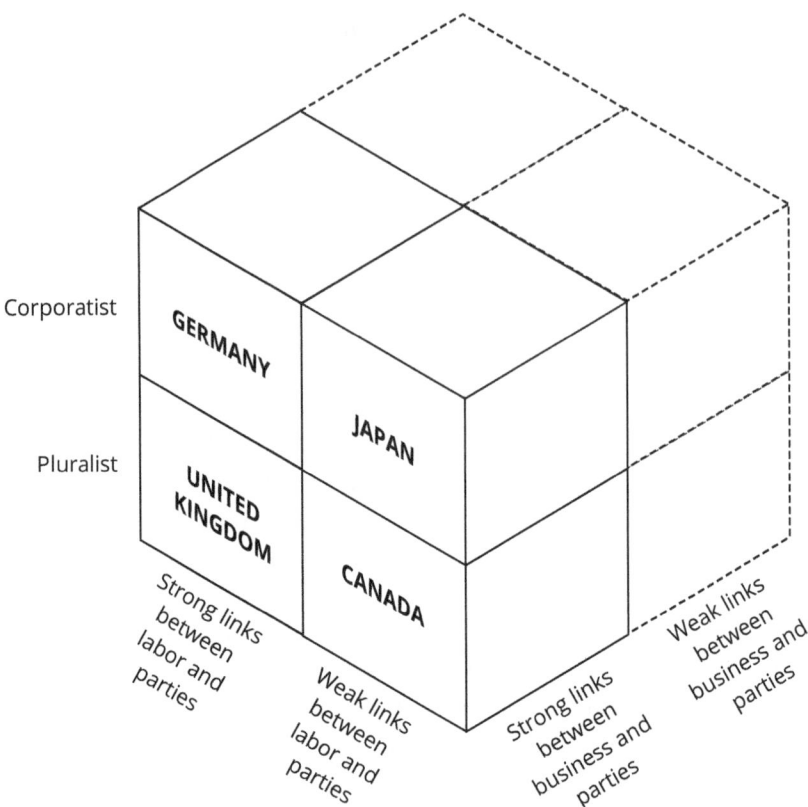

Figure 7.1
Shadow case selection strategy

between business and left-leaning parties in most cases); in the absence of variation on this dimension, predictive tests are not possible. Instead, shadow cases can be selected on two dimensions: policymaking institutions and formal links between labor and left-wing political coalitions.

A principled strategy for out-of-sample case selection is also required. In each quadrant of figure 7.1, I choose the advanced economy with the most significant level of carbon pollution. These countries are the relatively largest climate change contributors, conditional on institutional arrangements. They thus serve as the substantively most important cases for any theory of policy action. The corporatist shadow cases are Germany (with labor linkages) and Japan (without labor linkages). The pluralist shadow cases are the United Kingdom (with labour linkages) and Canada (without labor linkages). All shadow casework was undertaken after the theory-building stage of the overall project was completed to provide the most rigorous possible tests. Shadow casework relied on secondary and gray literatures.

In the pages that follow, I review critical policymaking episodes in all four cases. I show that my theory of climate policy conflict helps explain climate policymaking dynamics in each case. Consistent with book's primary casework, I limit analysis to climate policymaking before the 2015 Paris Agreement. Collectively, these shadow cases should increase confidence that this book's account of climate policy conflict helps explain the most important advanced industrial economies. Because I select shadow cases based on carbon pollution levels, the book's theory may apply less well to countries with lower carbon pollution levels or small industrial sectors. This limit should not be a major concern; theories of cross-national climate politics are most relevant when they explain the domestic politics of the countries most responsible for solving the climate threat. This chapter's efforts to assess generalizability prioritize this.

Germany

German policymaking involves a unique form of legal corporatism that differs in key respects from the Nordic social-democratic model. In Norway, peak industry groups conduct both industrial coordination and policy lobbying; by contrast, in Germany, these roles are differentiated (Martin and Swank 2012). By law, most German businesses must belong to a Chamber of Commerce and Industry, in turn grouped under the peak Association of German Chambers of Industry and Commerce (*Deutscher Industrie- und Handelskammertag*, DIHK). DIHK and its local chambers coordinate business training and services, manage industrial relations with labor, and consult with the German government over policies that impact particular sectors. Sectoral economic interests are also represented within policymaking debates by more traditional business associations, including the influential Federation of German Industry (*Bundesverband der Deutschen Industrie*, BDI). On the labor side, the most important peak labor union is the Confederation of German Trade Unions (*Deutscher Gewerkschaftsbund*, DGB), which represents more than six million workers. Within DGB, the largest single union is IG Metall, which represents more than two million workers in heavy industries.

Overall, German climate policymaking through 2015 maps onto my theory of double representation. Carbon polluters in Germany were embedded within both left-leaning and right-leaning political coalitions through links to labor and business organizations. In turn, major carbon polluters—most prominently the coal industry—were shielded from policy costs by ideologically dispersed political allies.

The earliest German climate policy debates were intertwined with debates over nuclear power's future. The climate threat first emerged on the political agenda in the early 1980s, raised by a coalition of economists and political actors hoping to frame nuclear power as pro-environment in the face of significant public opposition (Hatch 1995). These early efforts were contested by the coal industry, including representatives from the German coalminer's union (*Gesamtverbanddes Deutschen Steinkohlenbergbaus*).[1] During this initial period, German political parties struggled to manage internal tensions around climate. The Social Democratic Party (*Sozialdemokratische Partei Deutschlands*, SPD) could not reconcile anti-nuclear and pro-coal impulses, rooted in relationships with coal unions and German environmental activists respectively. The Christian Democratic Union (*Christlich Demokratische Union Deutschlands*, CDU) faced similar struggles (Hatch 1995).[2] Just as in other advanced economies, climate policy preferences cross-cut existing political coalitions.

Also paralleling other advanced economies, climate attention spiked in the late 1980s. This led the German Bundestag to establish a cross-party Inquiry Commission (*Enquete-Kommission Vorsorge zum Schutz der Erdatmosphare*) tasked with studying the climate threat and proposing an appropriate policy response. The commission's 1988 climate science report received unanimous multiparty support (Schreurs 2002); however, this initial report simply emphasized the serious risks associated with global climate change. Such abstract commitment to climate policy provided different political benefits to different parties without posing serious economic threats to any party's economic constituencies. The center-right coalition supported nuclear energy. It hoped climate change could weaken anti-nuclear interests and drive a wedge between SPD and its coal-dependent supporters. By contrast, the SPD wanted to position itself as a green force within German politics (Schreurs 2002).

The environment ministry coordinated the initial response to the Inquiry Commission's report, releasing a draft strategy in June 1990. This strategy led the German Cabinet to endorse a target of 25 percent carbon pollution reduction from 1987 levels by 2005 (Schreurs 2002). Yet, this target divided the center-right government of Chancellor Helmut Kohl. The right-wing CDU environment minister supported the 25 percent target, but this was resisted by the economics minister from the center-right Free Democratic Party (*Freie Demokratische Partei*, FDP) (Hatch 1995). Simultaneously, the Inquiry Commission chairman and CDU environment minister competed to position themselves as climate policy leaders, leading the Inquiry Commission to revise upward its target proposal to a 30 percent reduction below 1987 levels (Schreurs 2002). More generally, the ability to support nuclear power through climate policy bridged tensions between the government's economic and environment factions (Schreurs 2002).

A carbon tax commitment was subsequently included in the December 1990 CDU-FDP coalition agreement, even though there was no agreement on whether revenues should be used to lower income taxes or for climate mitigation (Hatch 1995). Yet, the economics ministry and its FDP minister continued to resist the proposal, joined by BDI. These opponents initially opposed any carbon fee but, as policy momentum increased, they instead demanded that Germany should only implement a carbon tax in the context of an EU-wide policy (Hatch 1995). Carbon taxation was also opposed by the SPD, the German Greens, and various environmental interests who all feared it would privilege nuclear power (Harrison 2010a). Reflecting its ties to the coal industry, the SPD also wanted coal exempted from any energy tax.

As the German economy slowed, Chancellor Kohl weakened his alignment with pro-climate factions. By the end of 1991, he pivoted to accommodate policy opponents' interests. This led to a strategic conditioning of any domestic climate response on Europe-wide action even as the German Cabinet committed itself, in principle, to a 25–30 percent carbon pollution reduction target. The European Union's efforts would, in turn, collapse, leaving Germany without an early carbon price.

During the 1990s, subsequent attempts to push environmental tax reforms found support from factions within all German parties but were systematically frustrated by cross-cutting policy preferences and German industrial opposition (Beuermann and Santarius 2006; Harrison 2010a). Correspondingly, Hatch (1995) reads corporatist institutions as critical to Germany's early carbon pollution reduction targets through the Inquiry

Commission's influence; yet, persistent distributive conflict stymied policies to meet these targets. Collaborative German policymaking helped to frame a common policy goal, but could not reconcile stark differences in policy preferences among domestic economic stakeholders.

As in Norway, Germany instead turned to voluntary industry agreements to reduce carbon pollution. A first agreement, signed in 1995, was immediately criticized for imposing few costs on industry. Critics contended the agreement could be met entirely through marginal energy efficiency gains. This controversy led to a 1996 commitment to strengthen the agreement by changing its baseline year from 1990 to 1987. Still, the agreement's ambition was questioned since German industry faced few policy costs (Skjærseth and Wettestad 2008). Meanwhile, German coal subsidies persisted, with the government paying coal producers the difference between the international market price for coal and domestic production costs (Cass 2006).

During this period, weak domestic action remained disconnected from Germany's international climate posturing. For instance, Germany pushed for binding commitments during Kyoto Protocol debates, and agreed to an ambitious 21 percent below 1990 reduction commitment under the EU burden-sharing agreement (Skjærseth and Wettestad 2008).

A renewed environmental taxation debate began in 1998 after an SPD-Green Party coalition replaced the long-standing CDU government. The German Greens leveraged participation in government to achieve several environmental reforms. However, this red–green coalition would remain constrained by the SPD's industrial base. Even modest climate proposals divided the coalition's environment factions from industry factions and their labor allies, including IG Metall (Schreurs 2002). The SPD and Green Party instead explored common ground through environmentally friendly economic growth strategies. These included ecological tax reforms and new investments in clean energy (Schreurs 2002).

First, in 1999, the red–green coalition enacted "ecological tax reform" to increase the tax rate on oil, transport fuels, and electricity while decreasing pension contributions. Yet, the law contained sweeping exemptions for SPD-aligned industrial actors. For instance, energy-intensive industries received up to 80 percent tax rate reductions and further rebates if tax increases outpaced reductions in corporate pension contributions (Beuermann and Santarius 2006; Harrison 2010a).[3] Further, while the tax included nuclear it exempted coal, reflecting close relationships between the SPD and German coal unions (Schreurs 2002; Harrison 2010a). Rather than bringing industry fully into the reform, the government instead negotiated a new voluntary agreement to reduce carbon pollution, signed in 2000 (*Erklrung der deutschen Wirtschaft zur Klimavorsorge II*); however, this agreement lacked enforcement and audit mechanisms (Hughes and Urpelainen 2015). Shielded from cost imposition by left-leaning political allies, coal was only brought into the German environmental tax system in 2006 after new EU rules forced its inclusion (Harrison 2010a).

Second, the SPD-Green coalition pushed renewable energy, efforts that have been profiled globally as the country's "*Energiewende.*" The 2000 Renewable Energy Act (*Erneuerbare-Energien-Gesetz*) set a 12 percent renewable energy target by 2010. It also expanded renewable energy deployment subsidies through feed-in-tariffs (FITs).[4] BDI opposed these reforms despite energy-intensive producers being exempted from contributions to implementation costs. Again, climate policy's distributive burden fell toward consumers and away from producers (Hughes and Urpelainen 2015).[5] Over time, these German renewable energy subsidies led to substantial wind and solar deployment, reaching over 30 percent of the country's mix by 2014. Yet, these reforms were acceptable within the German political system precisely because they bridged intra-coalitional climate policy tensions through generous subsidies for new industrial growth and employment. Initially, the Energiewende did not involve cost imposition on carbon polluters. Nonetheless, these subsidies later threatened incumbent utility interests. German energy debates now focus on capacity pricing and other policies to stabilize incumbent utilities' position in the face of shifting energy price structures.

Since the mid-2000s, the EU Emissions Trading System shaped German climate policy. German decision makers were initially skeptical of emissions trading, signaling their preference for voluntary domestic measures (Skjærseth and Wettestad 2008). Senior officials only engaged with the policy instrument after the threat of EU action forced domestic opponents to the table in early 2000. Still, emissions trading remained divisive, and Germany emerged as a significant obstacle to the establishment of a European carbon price (Skjærseth and Wettestad 2008).

Within EU-level debates, German interventions emphasized voluntary participation in emissions trading, at least during the pilot phase;

this position reflected business divisions over whether Germany should endorse emissions trading (Skjærseth and Wettestad 2008). After the EU released a draft emissions trading proposal in October 2001, both BDI and the German Chemicals Association (VCI) mobilized conflict into the public domain, including a media campaign to warn the public about the EU proposal's economic costs (Skjærseth and Wettestad 2008). Many German elites sympathized; negotiators worked to advance producer interests and weaken the EU directive. However, their efforts mostly failed and Germany was forced to accept more stringent EU emissions trading rules than some officials wanted (Skjærseth and Wettestad 2008). In this way, the costs imposed on German carbon polluters by the EU ETS reflected EU institutions, not domestic corporatist institutions. In fact, at the time of its passage, EU emissions trading directive was apparently opposed by both the left-leaning SPD under Chancellor Schröder as well as the conservative opposition under the CDU and FDP (Skjærseth and Wettestad 2008).[6] The SPD would later concede to Green demands to endorse the EU directive during a second SPD-Green coalition in 2002. However, the SPD would circumscribe this support to not interfere with existing voluntary carbon pollution agreements with domestic industry (Skjærseth and Wettestad 2008).

Subsequent German government efforts to weaken the EU ETS were largely unsuccessful, with EU negotiation momentum constraining the influence of German economic interests from above. Germany was only able to guarantee that emissions reductions commitments could be pooled, in an attempt to maintain an institutional fit between its sectoral voluntary approach and a binding continental cap (Skjærseth and Wettestad 2008).

German implementation of the ETS continued to signal reluctance. Germany's first National Allocation Plan (NAP) was rejected by the European Commission in 2004 for being too generous. The commission criticized Germany's proposals for ex-post adjustments in allocation decision rules to benefit specific economic actors.[7] Government officials also championed energy-intensive interests. For instance, during 2008 reform debates, Germany secured free allowances for such industries as steel, cement, chemicals, and coke (Skjærseth and Wettestad 2010).

While Germany reluctantly participated in European-wide carbon pricing, domestic German climate debates continued to be structured by carbon polluters' double representation. Like Norway, Germany proposed two CO_2 compensation schemes for energy-intensive industries. The European

Commission approved a policy to compensate trade-exposed industries their electricity costs' carbon content. However the Commission rejected a €40 million scheme to reimburse ETS implementation costs.

Similarly, coal production and consumption was championed by actors on both sides of the political spectrum, including coal-affiliated politicians within the left-leaning SPD party. For example, the December 2013 coalition agreement between Merkel's CDU and the SPD emphasized that "conventional power stations (lignite, anthracite, gas) will remain an essential part of the country's energy mix for the foreseeable future."[8] This provision was pushed by SPD parliamentarian Ulrich Freese, a former official with the German industrial and mining union (*Industriegewerkschaft Bergbau, Chemie, Energie*), in partnership with regional SPD leaders.[9] After two decades of climate reform debates, hard coal still enjoyed subsidies of about 2.4 c/kwH through domestic production support, while climate-unfriendly lignite coal was exempted from groundwater-usage fees worth 1.3 cents per kilowatt hour.[10] As a result, coal's share of the German electricity mix remained stable even as renewable energy production spiked.

As Germany seemed likely to miss its 2020 emissions reduction targets, environmental advocates criticized Germany's persistent support for coal production and coal-fired electricity generation. Germany's government had previously denied it would implement any policy requiring the closure of domestic coal-fired power plants.[11] Yet, the government faced controversy after a leaked policy proposal suggested capping coal-fired power plants emissions; the policy was opposed by some bureaucrats, regional left-leaning SPD governments, and industry.[12] Notably, these costs proposed, for virtually the first time, on German coal industry came from the German right, not left.

Potential coal regulations catalyzed dramatic conflict. German industrial unions, supported by the European Confederation of Trade Unions, threatened massive strikes, claiming the proposal could cost the country more than one hundred thousand jobs. German utilities and regional governments argued the plan would be economically catastrophic. These claims were echoed by opponents within both the CDU and SPD, emphasizing the persistent cross-cutting nature of German climate policy preferences.[13] Seeking to quell opposition, the German government promised to review job impacts before making a final decision.[14]

Overall, German climate policymaking through 2015 is consistent with the theory of double representation. Allies of German industry on both the left and right repeatedly worked to advance carbon polluter interests within German policy design, and blocked climate reformers' efforts to impose threatening costs. Where costs were imposed, these tended to come from Brussels, not domestic corporatist institutions. While Germany deployed substantial renewable energy, this outcome stemmed from subsidies that bridged intra-coalition divisions and appeared nonthreatening to incumbent carbon polluters. Further, subsidies were disproportionately funded through consumer costs while shielding large producers from contributions, again consistent with theoretical expectations. This book's theory thus sheds new light on the surprising trajectory of German climate policymaking: leadership in the clean energy domain despite persistent support for coal. This paradoxical outcome resulted from a political system seeking to act on climate policy, including during a period of red–green coalition government, while constrained by the institutionalized double representation of domestic carbon polluters.

More optimistically, some German subsidy-based programs generated substantial costs for incumbent carbon polluters through indirect price shifts. Incumbent German utilities did not fully anticipate the speed with which renewable energy would become cost competitive, and now face unexpected financial threats. Similar dynamics in the United States have been extensively problematized by Stokes (2020) who shows how utility incumbents misjudged long-term renewable energy threats during a "fog of enactment." In these ways, the German case does not offer an exception to the logic of double representation but instead suggests a possible mechanism through which the power of entrenched incumbents can be indirectly weakened, even where double representation is strongly institutionalized.

Japan

Japanese policymaking institutions have been described as a system of "corporatism without labor" (cf. Pempel and Tsunekewa 1979; Wilensky 2002). In Japan, business organizations enjoy intimate access to government to coordinate economic policies, constituting one-third of the country's "ruling triad" of business, the state, and the bureaucracy (Broadbent 1999). Businesses' interests are organized into cross-sector peak associations, notably the Japan Federation of Economic Organizations (*Keidanren*). Analogous to the Norwegian case, carbon-intensive polluters enjoy a privileged voice within these peak associations. For example, the Japan Iron and Steel Federation, the Japan Automobile Manufacturers' Association, and the Federation of Electric Power Companies effectively control Keidanren as a function of their historic importance to both the association and national economy (Oshitani 2006).

The right-leaning Liberal Democratic Party of Japan (LDP) dominated Japanese politics for most of the country's rapid post-World War II industrialization (Schreurs 2002). Among the LDP's distinctive organizational features were party *zoku*: small groups of like-minded legislators organized around specific policy issues. Japanese zoku surfaced factional conflicts within the LDP, bringing together Diet members with strong connections to or interest in specific policy domains; in turn, zoku developed strong influence over domain-specific ministerial decisions (Schoppa 1991).[15] However, over the past two decades, the influence and coherence of zoku have weakened.

Compared to other advanced economies, the Japanese bureaucracy enjoys strong policy autonomy (Wallace 1995).[16] The Japanese Diet often passes broad and open-ended policy frameworks ("Basic Laws") and delegates substantial implementation authority to bureaucrats. However, the absence of delegated enforcement authority incentivizes bureaucrats to build consensus with regulated private-sector entities (Oshitani 2006). This includes a network of policy committees that bring firms and researchers into consensus-based conversation with bureaucrats over policy design (Hughes 2012).

In Japan, three agencies play an outsized role in climate debates. First, the Ministry of Economy Trade and Industry (METI) coordinates most economic and energy policymaking.[17] The Ministry of Finance coordinates fiscal policy and has engaged with carbon pricing policies in this role. Finally, the Ministry of the Environment (MOE) has broad authority over environmental policymaking; however, unlike METI and Finance, the MOE did not enjoy full ministerial status until 2001. (It was previously called the Japanese Environment Agency). Over the past two decades, these environmental bureaucrats played a largely subordinate role, lacking resources and staff to compete with central economic ministries (Wallace 1995; Van Asselt, Kanie, and Iguchi 2009); one exception has been the ministry's control over environmental impact assessments, which has been important to power plant planning. Further, industry enjoys close institutional ties with central economic ministries, particularly METI.[18]

By contrast, labor and environmental groups have a weaker position within Japanese corporatist policymaking, forcing early environmental advocates to embrace mass politics and legal strategies to contest the severe air and water pollution that emerged during early Japanese industrialization (Schreurs 2002; Wilensky 2002; Oshitani 2006). Yet, while industry and its LDP allies ignored early civil society demands for environmental reforms, the party was not monolithic. Some early reform proposals exposed cross-cutting economic cleavages, particularly early nuclear debates that split the party into pro-coal and pro-nuclear factions (Lesbirel 1990). Nonetheless, the tight linkages between industrial interests and the LDP kept climate change off the policymaking agenda until the late 1980s (Schreurs 1997). Instead, early climate attention was limited to conversations among marginalized Environment Agency (EA) bureaucrats (Takao 2012).

Looking in from outside Japanese corporatist institutions and the LDP party, the labor movement affiliated with Japanese opposition parties, including the Democratic Socialist Party, the Clean Government Party, and the Japanese Communist Party (JCP) (McKean 1981). From an early stage, the movement experienced the same cross-cutting cleavages that shaped labor politics elsewhere. For example, some labor unions, in tandem with the JCP, supported citizen anti-pollution protests during early environmental debates; yet, others adopted positions of passive support or active opposition, concerned that anti-pollution reforms would harm Japanese jobs (McKean 1981). This generated splits between rural unions and urban unions aligned with high-polluting companies.

The climate threat's growing profile brought cross-cutting climate preferences into sharper relief. Inside the LDP, a pro-climate faction emerged led by Prime Minister Noburu Takeshita (Schreurs 1997, 2002; Oshitani 2006). Such forums as the 1989 Ministerial Council on Global Environmental Protection and the LDP's Special Committee on Global Environment, subsequently exposed political tensions (Oshitani 2006). In the private sector, nuclear and pollution control companies advocated for climate action, while more carbon-intensive sectors opposed policy action (Schreurs 1997). Within the bureaucracy, the EA and MITI (METI's former name) clashed over a 1990 effort to set a carbon pollution stabilization goal; the eventual agreement included both a per capita target (that MITI felt was the core position) and an absolute reduction target (championed by the EA but viewed as nonbinding by MITI). MITI viewed these goals as formalizing the trajectory of Japanese efficiency gains; it did not anticipate policies that would require cost imposition on Japan's carbon-intensive actors (Oshitani 2006).

Climate reforms splintered the dominant Japanese political and economic coalition. As in other advanced economies, carbon taxation emerged on the policymaking agenda during the early 1990s. The first serious proposal was a 1991 joint effort by the EA and the Finance Ministry, viewed as a direct response to Northern European efforts (Schreurs 2002; Oshitani 2006). By 1992, a sizeable green zoku had emerged within the LDP, buttressed by the LDP Investigatory Committee on Basic Environmental Problems (Oshitani 2006). While the tax found support from Takeshita as well as Finance Minister Hashimoto, the proposal was scuttled—just as in other advanced economies—by industrial opposition and LDP opponents (Oshitani 2006). Subsequent efforts to reconsider a carbon tax, including as part of a 1992 EA report, faced opposition from both business and union actors including, for example, both the Japanese Iron and Steel Federation and the Japanese Federation of Iron and Steel Workers Unions (Schreurs 2002; Oshitani 2006). Peak business associations coordinated to undermine the position of pro-carbon tax LDP factions and the government backed off (Oshitani 2006). For example, Keidanren proposed a consumer-facing consumption tax rather than a broad-based producer-facing carbon tax (Schreurs 2002).

Then, during 1992 debate over Japan's Basic Environment Law, EA bureaucrats pushed for authority to deploy economic environmental policy instruments (Wallace 1995). MITI officials worked to cut off this possibility, winning legislative language that ensured environmental taxes would have to be equitable, receive "positive consent" from the public, and take international policies into consideration (Oshitani 2006). Remarkably, Schreurs (2002) describes how EA officials had to sign a formal memo agreeing to limit their ability to propose carbon tax reforms as a condition of MITI and Ministry of Construction support for a broader 1993 environmental reform package.

Takeshita's green LDP faction weakened as the prime minister's influence declined during a 1993 corruption scandal (Schreurs 2002). Instead, Japanese climate policy in the late 1990s shifted toward voluntary efforts that comprised substantial concessions to carbon-intensive interests and their bureaucratic allies. The most prominent was Keidanren's 1991 voluntary pollution reduction program under its "Global Environmental Charter." In 1996, Keidanren expanded this charter into a coordinated effort to improve efficiency and reduce industrial emissions. Industry viewed these efforts as a strategic attempt to preempt more costly climate reforms (Oshitani 2006).

Japan's decision to host the Kyoto Protocol meeting brought climate issues back onto the Japanese policymaking agenda but did not bridge tensions.[19] Again, Japanese interests splintered over the development of a Kyoto carbon pollution-reduction commitment. The EA pushed for a binding target, supported by nuclear-dominated utilities. MITI, alongside such economic actors as the automobile and manufacturing industries, adopted a more skeptical posture (Schreurs 2002). MITI eventually conceded the adoption of an overall target, pressured not only by EA but also by pro-climate voices in a range of other Japanese ministries and sectors, including foreign affairs, transportation, agriculture, and forestry and fisheries. However, MITI ensured that the target did not pose any serious economic threat to Japanese industry (Oshitani 2006).

MITI similarly dominated post-Kyoto policymaking. Japan's 1998 Guideline for Measures to Prevent Global Warming empowered MITI while the Law Concerning the Promotion of Measures to Cope with Global Warming, including Japan's 6 percent Kyoto reduction target, lay largely within EA's remit. However, the MITI guidelines had more teeth and, in either event, simply reflected a codification of existing MITI energy reforms; the carbon pollution reduction target, managed by the EA, remained aspirational (Oshitani 2006).

The 2001 US withdrawal from Kyoto again exposed cross-cutting tensions in Japan (Tiberghien and Schreurs 2007). On one side, nuclear, pollution control, and insurance industries and such retail-heavy unions as the Japanese Consumer Cooperative Union argued for Japanese ratification. Yet, other industry and bureaucratic interests who stood to be distributive losers—including Keidanren, MITI, and some LDP interests—opposed action. LDP members also fractured (Tiberghien and Schreurs 2007). Eventually, Prime Minister Junichiro Koizumi sided with the LDP's pro-Kyoto faction and Japan ratified the protocol (Schreurs 2002). This episode thus offers a case where cross-cutting policy preferences boosted environmental reforms; in this case, distributed advocates across the Japanese system tempered the business community's ability to strongly oppose ratification (Tiberghien and Schreurs 2007).

Yet, this symbolic victory did not reshape the basic dynamics of national climate conflict. Carbon-intensive economic actors continued to leverage their access to policy design venues to limit climate reform costs. For example, MITI proposed a modest 2003 environmental tax package that included new taxes on coal, gas, and oil with revenues directed toward energy conservation.[20] This proposal split the MOE, since it was viewed as an effort to preempt environmental bureaucrats' efforts to propose future climate reforms. The two agencies eventually agreed that the MITI energy tax should not be understood as a carbon tax (even though it included carbon logic in its justification), so that the MOE still had the technical right to propose its own carbon tax policy (Oshitani 2006). However, industry argued that any new carbon taxes would be "double regulation" (Oshitani 2006); for example, carbon-intensive business interests successfully opposed a 2004 Cabinet proposal for a $6.55 USD per ton CO_2 tax to support Kyoto Protocol implementation (Kuramochi 2015).

In this way, METI used a modest energy tax to capture policymaking control over carbon taxation and to position that policy within an institutional structure where distributional losers held substantial influence. During this same period, the structural power of carbon-intensive interests also increased within the business community after the more pro-climate Nikkeiren merged with Keidanren in 2002. Much as with peak business associations in Norway and Australia, even pro-reform Nikkeiren officials found themselves constrained by the strong institutionalized voice of anti-reform industries within Keidanren (Oshitani 2006)

Japanese climate reforms from the mid-2000s onward continued this pattern of occasional lofty goals whose implementation was compromised by the embedded influence of distributional losers. For instance, in 2006, Prime Minister Asō proposed a 15 percent below 2005 levels by 2020 target. He described this as a *mamizu* or clear water target because it excluded reliance on emissions or carbon credits.[21] Yet, the Asō government enacted few policies that could credibly be understood as delivering on this commitment.

Instead, carbon pricing reforms were incremental and slow-moving. In 2005, the country established the JVETS (Japan Voluntary Emissions Trading Scheme) for firms not part of Keidanren's Voluntary Action agreements. Viewed as an opportunity to learn about the policy instrument, JVETS included subsidies for impacted industries although it only applied to only a small fraction of the country's total emissions base (less than 0.3 percent) (Kuramochi 2015).[22] In 2008, as part of the "Trial Integrated Market," the government then attempted to link JVETS to the Keidanren voluntary system but with provisions that guaranteed industry flexibility. Over the next five years, efforts to broaden this carbon pricing system stalled.[23] In 2009, a national ETS proposal faced sustained industry objections and was abandoned by December 2010. The 2010 proposal for a Basic Act on Global Warming Countermeasures, passed during a brief interregnum in which the Democratic Party of Japan (DPJ) controlled government, provided for the establishment of an ETS or carbon tax system. However, the proposal, which included a 25 percent reduction target for 2020 and 80 percent target by 2050, never passed the Diet (Kuramochi 2015).

After regaining power, the LDP introduced a Tax for Measures to Cope with Global Warming in October 2012. This tax applied to all fossil fuels on their CO_2 basis. The tax started at 0.95 USD per ton CO_2 in 2012 and reached 2.89 USD per ton CO_2 in 2016. Yet, even this tax reform extended exemptions for fossil fuel usage—including fuel for petrochemical production and coal used in iron, steel, cement, and coke (Kuramochi 2015). Utilities were allowed to pass through costs explicitly to consumers. Moreover, tax revenues were directed to a range of technological subsidies and transitionary support to impacted industries (Kuramochi 2015).

Under the 2014 Basic Energy Plan (BEP), Japan again embraced restarting its nuclear plans and moved away from the previous DPJ government's 2012 decision to phase out nuclear power by the 2030s. This plan weakened policy specifics, including the timeline for CCS requirements at Japanese fossil fuel power plants (Kuramochi 2015). The BEP also promoted the construction of coal-fired power plants in Japan without emissions controls, leading to a rash of new coal-fired power plant proposals by a range of Japanese regional utilities (Kuramochi 2015). As part of a negotiated agreement with the government, utilities agreed to a ceiling on the carbon intensity of power generation in return for the right to build these coal-fired power plants. This has resulted in expansion of Japan's coal fleet since 2014. By contrast, MOE efforts to develop a more meaningful national carbon price have yet to yield results.

In short, Japanese climate policies have been incremental, at best, and have systematically shielded privileged carbon-intensive economic actors from serious economic costs. This has been true even in the face of bureaucratic and political pressures to generate more serious reforms, either from within the LDP, from temporarily empowered voices within the EA, or from DPJ politicians during the LDP's brief period outside of government. Even where incremental costs were advanced, these policies distributed costs on consumers and largely shielded producers. While such efforts may reshape consumer carbon pollution trajectories, they do not reshape the distribution of political power in ways that might facilitate more stringent future reforms. More serious efforts to push costly reforms were systematically blocked by distributed economic interests across the Japanese state, industry, and bureaucracy. In these ways, the trajectory of Japanese climate policymaking is largely consistent with predictions offered by the logic of double representation.

Canada

Like Australia and the United States, Canada has a pluralist policymaking system. While the federal parliament is bicameral, the appointed upper Senate chamber is advisory in practice. With strong party discipline, a prime minister who controls the majority of lower house seats faces few policymaking obstacles. However, constitutional responsibility for environmental and resource policies largely resides with Canadian provinces, creating complex federalist debates over environmental reforms.[24]

The major Canadian political parties are the center-left Liberal Party and the center-right Conservative Party (formerly the Progressive Conservatives). The left-leaning, labor-aligned New Democratic Party (NDP) also wins both rural and urban seats, but has never been in government. These parties have divergent links to carbon-dependent economic interest groups. The Conservative Party has roots in the Canadian West, including in the oil-rich province of Alberta. Consequently, the party has strong links to fossil fuel-intensive constituencies. The NDP has organizational links to the Canadian farming community and such industrial unions as the Canadian Autoworkers. However, organizational ties between the NDP and Canadian labor communities are weaker than in countries with formal Labor parties.

Climate change emerged on the Canadian policymaking agenda in the late 1980s. As in Australia, Norway, and the United States, the earliest climate policy debates revolved around carbon pollution reduction targets.

In 1988, a Progressive Conservative government under Brian Mulroney endorsed the Toronto climate conference target of a 20 percent reduction below 1988 levels by 2005. However, a subsequent 1990 Green Plan offered a weaker commitment of stabilization at 1990 levels by 2000. After the Liberals took power in 1993, they restored the 20 percent below 1990 by 2005 target in an effort to position their party as greener than their rivals (Harrison 2010b). Still, early climate policy debates were characterized by cross-party support for domestic carbon pollution reduction targets.

Yet, in designing domestic policies to meet these targets over the following two decades, these same parties navigated significant cross-cutting policy preferences. A federal carbon tax was briefly debated during consultations for the 1990 Green Plan; however, the proposal faced opposition from carbon-dependent economic interests, particularly within Alberta, and was seen as politically unviable among Ministry of Environment bureaucrats (Hoberg and Harrison 1994). When the Liberals won the 1993 election, carbon tax discussions stalled entirely (Harrison 2012).

Under the Liberals throughout the 1990s, Canada played a largely constructive role within climate negotiations, agreeing to a relatively ambitious 6 percent reduction below the 1990 level target at Kyoto. Yet, federalism created challenges for Canadian climate policy implementation, since constitutional responsibility for the environment and energy reside at the provincial level. In Kyoto's aftermath, the federal government consulted with provinces to ensure no region would shoulder a disproportionate policy burden (Harrison 2012). Few policies resulted from this process.

Domestic policymaking became increasingly contentious after Bush abandoned Kyoto in 2001. The ensuing debate over Canadian Kyoto ratification split the Liberal Party. Most Canadian business associations opposed ratification for fear that Canadian action without corresponding US efforts would be economically disadvantageous, from the Canadian Council of Chief Executives to the Canadian Association of Petroleum Producers (Harrison 2010b). The decision also divided the Liberal Cabinet; Prime Minister Jean Chrétien had to force Cabinet colleagues to support ratification (Harrison 2010b). As part of this debate, pro-climate advocates within government informally agreed that policy costs would not surpass $15 per ton and that emissions reductions would be capped at 15 percent below business-as-usual projections through 2010 (Harrison 2010b). Again, cross-cutting preferences for climate policy action constrained the types and levels of costs that could be imposed on major carbon polluters. Further, the government's November 2002 federal implementation plan was vague and imposed few concrete policy instruments or costs on carbon polluters across the economy (Harrison 2010b).

Pro-climate factions gained more ground in a revised climate policy plan offered by a new Liberal minority government in 2004. This plan included a modest sectoral emissions trading proposal; however, the ambition of overall reductions was shaved from 15 percent down to 12 percent to secure support for increased producer obligations (Harrison 2010b). Even this carbon pricing proposal was limited in its coverage, with the bulk of the proposed reductions still coming from subsidies, financial transfers to provinces, and offsets. Then, after the Conservatives won a minority government in 2006, they canceled climate policy expenditures, backed away from Canada's Kyoto commitments, and initiated a new round of consultations on domestic climate reforms. In March 2008, the government formally proposed a baseline credit scheme that used intensity-based targets to promise 20 percent reductions below 2006 levels by 2020, and a promise to transition to binding caps from intensity-based targets between 2020 and 2025.

In 2006, former Environment Minister Stéphane Dion captured control of the Liberal leadership in a tightly contested party primary to become opposition leader. Just a year earlier, Dion had proposed green tax reforms while serving as the environment minister, but could not bring the divided Liberal Cabinet on board with his plan (Harrison 2012). Dion's ascension represented an abrupt switch of Liberal Party control to the party's pro-climate faction. Dion faced huge resistance from inside the party after, in March 2008, he cued his interest in taking a major carbon tax to the next federal election as the party's central platform. However, strong party discipline gave Dion the capacity to push his preferred policy forward despite internal opposition (Harrison 2012). He announced his Green Shift in June 2008, with an initial $10 CAD tax that would increase annually in increments of $10 until reaching $40 per ton of CO_2 in 2012. The proposal initially exempted gasoline, structuring the tax to privilege consumers and households. The tax was to be revenue-neutral, and revenues were to be allocated toward a wide range of social spending from child poverty to income inequality. The policy's emphasis on reducing consumer cost salience matches predictions regarding relative cost emphasis in pluralist countries when pro-climate factions seize control of the climate policymaking agenda.

Posing a major policy threat to carbon-dependent economic interests across the country, the Green Shift faced attacks from both sides of the ideological spectrum. On the Liberal Party's right, the Harper government excoriated the proposal as a "green shaft" on the Canadian Public.[25] From the left, the New Democratic Party criticized the carbon tax proposal for its burden on low-income communities. Party leader Jack Layton promised to make sure a "carbon tax never sees the light of day."[26] The NDP's vocal rejection of carbon taxation divided the party, with environmentalist members upset at the party's political positioning. The Liberal Party suffered historical electoral losses during the 2008 election, leading to widespread political perceptions that carbon taxation was politically unviable at the federal level in Canada (Harrison 2012). Policy opponents dispersed across the political system had succeeded in undermining any electoral or legislative incentives for party leaders to push carbon pricing policies over subsequent years.

The Conservative government would not immediately implement the baseline credit scheme that its leadership had promised in 2007–2008. Instead, after the Conservatives' election victory, the government stepped away from prior climate policy commitments altogether. With the federal government's climate policymaking agenda controlled by political actors closely allied with the oil industry, federal Canadian climate policymaking came to a standstill. First, the Conservative government stalled on its promise to publish a draft emissions trading scheme, suggesting it was premature to design a US plan until the details of the developing US domestic regime became clear (Harrison 2010b).

While the Conservative government continued to introduce some climate policies, these accommodated the interests of carbon producers. For instance, in 2012, the Conservative government introduced coal-fired power plant regulations in Canada, which were enacted with only minimal public controversy. In contrast to the US Clean Power Plan, these Canadian coal regulations only applied to new plants built after 2014, rather than existing plants. The regulations were not expected to lead to significant reductions in national carbon pollution in the short or medium term.[27] Thus, while Canada moved to enact coal regulations in advance of the United States, their design protected utility interests in Canada through 2030.

In this way, the federal Canadian policymaking in the decade before Prime Minister Justin Trudeau's election in October 2015, just before the Paris climate conference, was dominated by regional interests aligned with the fossil fuel industry. This gave carbon polluters significant access to climate policy design during a period in which many other advanced economies engaged in more consequential reform efforts. However, even the Canadian right experienced cross-cutting policy preferences. For example, one of the most vocal Canadian carbon tax proponents was former Reform Party leader Preston Manning. The Reform Party was a large Alberta-based conservative populist party that merged with the Progressive Conservatives to form the current Conservative Party. Manning continued to be a vocal champion of carbon pricing and carbon taxation in Canada even as Conservative leaders resisted reform enactment.[28]

Over two decades, federal Canadian climate policy has been limited, with climate policy opponents effectively stalling efforts to enact ambitious domestic climate reforms under both Liberal and Conservative governments. Until 2015, the most consequential effort to impose costs on carbon pollution in the country emerged in the late 2000s when green factions of the Liberal Party seized control of the centrist party's political leadership. Opposition leader Stéphane Dion then pushed forward with an ambitious carbon tax reform despite cross-cutting policy preferences within his own party. Consistent with expectations for climate policy proposals within pluralist policymaking contexts, the proposed carbon tax reform was sensitive to consumer costs, and attempted to increase consumer support through social program spending. Further emphasizing the ways in which domestic climate reforms cross-cut ideological divides within advanced economies, the proposal was attacked by both industry-aligned Conservative politicians, and labor and consumer-aligned New Democrats. These parties raised the salience of consumer costs during a high-profile election campaign. This mobilization subsequently undermined the short-term political incentives to enact federal carbon pricing policies in Canada.

United Kingdom

The United Kingdom is mostly pluralist; however, in contrast to Canada, its dominant left-leaning political coalition has formal ties to the country's labor movement. The UK Labor Party was founded in 1900 as a vehicle for trade unions to have direct representation in the British parliament. British unions formally affiliate with the Labor Party through a membership fee; this fee allows unions' representation in the party's National Executive Committee and at its annual party conference. Under the Trade Union & Labor Party Liaison Organization, union officials and Labor Party leaders meet regularly to discuss political and policy decisions. Members of affiliated Labor unions must opt out of party membership; otherwise, members are automatically registered to vote in Labor leadership elections.

The sectoral composition of Labor Party-affiliated unions has shifted over the past three decades. Historically, British coal unions were a major political force, particularly the National Union of Mineworkers (NUM). However, reduced profitability and coordinated attacks by the Conservative Thatcher government during the 1980s dealt the industry a fatal blow. These Conservative politicians were willing to impose economic costs on a labor base that Thatcher famously described as "the enemy within." Moreover, British coal mines themselves were government-owned through the National Coal Board (NCB), creating a relative absence of coal business representation within right-leaning political coalitions. In this book's terms, there was effectively "single representation" of British coal. Arguably, this made British coal uniquely vulnerable from a political perspective. NUM membership thus plummeted from nearly one million in 1920 to slightly over 170,000 in the early 1980s to fewer than 100 active members by 2015.[29] Labor unions are also represented politically in the United Kingdom by the Trades Unions Congress (TUC), a peak federation that claims to represent the collective interests of British workers.

Business representation in British politics parallels other advanced economies. The largest British business lobby is the Confederation of British Industries (CBI), which formed as a merger of smaller manufacturing and industrial associations in 1965. The CBI includes a diverse sectoral membership of 190,000-plus businesses. Other prominent associations are the British Chamber of Commerce, a national organization that integrates the voices of more than seventy-five thousand British businesses, and the British Federation of Small Businesses. British fossil fuel interests are represented by such lobbies as Oil & Gas UK. As in other pluralist economies, these business groups do not have formal institutional ties to policy design. However, they are frequently consulted on policy debates that affect sectoral material interests (Oshitani 2006). Some scholars have described historic patterns of British environmental policymaking as "courteous negotiation" between polluters and regulators, rather than being structured by conflictual relations between government and the private sector (Jordan et al. 2003). Notwithstanding these norms, voluntary approaches to pollution management have been rare and, where present, localized and flexible (Jordan et al. 2003).

Climate change emerged on the British policymaking agenda in the late 1980s during Margaret Thatcher's Conservative government. Thatcher herself surprised observers by personally advocating for climate action and supporting such global institutions as the IPCCC.[30] Thatcher's climate motives have been the subject of substantial debate; some suggest her science degree predisposed her to trust climate scientists while others suggest she embraced the climate file to boost her international profile (Oshitani 2006). Whatever the reason, her efforts legitimized climate change on the British political agenda, and led to funding for new climate science research. Under Thatcher, the United Kingdom also set its first carbon pollution reduction target: carbon pollution stabilization at 1990 levels by 2005. While some Conservative voices, including Environment Secretary Chris Patten, argued the UK should adopt the EU's more ambitious stabilization timeline, Thatcher resisted (Oshitani 2006).

Prime Minister John Major's Conservative government following Thatcher's resignation in 1990 reflected continuity with Thatcher's term, despite some party members' interest in stronger action (Carter 2009). During Major's tenure, his environment minister again proposed increasing the country's stabilization goal, but this continued to face resistance from such figures as then Energy Secretary John Wakeham (Oshitani 2006). However, secular changes in the British economy, particularly fuel switching from coal to gas, led to naturally declining emissions levels. In 1995, under Major, Environment Minister John Gummer eventually got Cabinet approval for the more ambitious target of 5 to 10 percent reduction below 1990 levels by 2010; Cabinet target support was apparently a function of promises that the targets would be met without policy interventions (Oshitani 2006).

Despite these debates, Conservative governments enacted few deliberate climate policies. While open to climate science, Thatcher remained wary of climate mitigation measures; post-government, she would take a skeptical tone and criticize climate policy instruments as costly agents of socialism (Thatcher 2002). Nonetheless, as in other advanced economies, carbon taxation emerged on the British policymaking agenda during the early 1990s, pushed forward by Patten and his senior advisor David Pearce. Reform efforts were backed from outside government by the environmental lobby, including Friends of the Earth. However, carbon-dependent economic actors inside and outside government stymied serious consideration of the policy instrument. Within government, officials from the Treasury, the Department of Trade and Industry, and the Department of Energy all expressed reservations. Similarly, British officials opposed EU-level consideration of a carbon tax, a function of both sectoral interests and skepticism of EU-level policymaking on the issue (Oshitani 2006; Skjærseth and Wettestad 2008). During this time, business communities through the CBI and the labor community through the TUC opposed any carbon tax as a function of its potential economic and job impacts (Oshitani 2006).

Instead, Conservative-era climate reforms were, at best, indirect. For instance, an increase in the fuel and power value-added tax (VAT) was introduced in 1993. While motivated by revenue needs, some actors apparently viewed the policy as linked to the country's carbon stabilization goals (Oshitani 2006).[31] Even these modest measures were viewed by some interests as defensive maneuvers to block carbon taxation (Oshitani 2006). Notably, in this case, Labor opposed the quasi-environmental VAT tax from the left on a social justice basis (Darkin 2006). As in other advanced economies, the British government also promoted voluntary action, which involved loose coordination with business associations to undertake voluntary environmental compliance actions (Oshitani 2006).

Labor took power in 1997 under Prime Minister Tony Blair. Blair immediately offered strong rhetorical commitment to climate policymaking, particularly within British foreign policy. Blair was quick to criticize American counterparts for being climate laggards, both during his first US visit for the Denver G7 meeting and in his July 2003 address to Congress. Under Blair, the UK also upped its carbon pollution target to 20 percent below 1990 levels by 2010. However, Labor domestic reforms were generally more modest and centered on the Blair government's 2000 British Climate Change Programme (BCP). The programme itself was partly inspired by the government-commissioned Marshall report, coordinated by a former head of the CBI, which explored prospects for market-oriented instruments in environmental policymaking (Darkin 2006).

Blair's Labor government embraced market-based policy instruments in a way preceding Conservative administrations had not (Jordan et al. 2003). This BCP thus set the stage for such climate policymaking instruments as the UK Renewables Obligation in 2002 (a hybrid policy that combined a renewable mandate with tradeable compliance certificates) and an Energy Efficiency Commitment (a regulation directed at energy suppliers' home energy provisions). However, the most contentious policy measure was Labor's Climate Change Levy (CCL), first introduced as part of the 2000 Finance Act. Coming into effect in April 2001, the CCL taxed the energy intensity of different fuel sources.[32] It passed alongside a 0.3 percent reduction in employer National Insurance Contributions and new renewable energy-oriented R&D funds.

The CCL was not a pure carbon tax. While it did exempt most forms of renewable energy, it still included carbon-pollution free nuclear energy. In rejecting a pure carbon tax, Labor rejected the Marshall report's recommendations out of concern they would unduly burden the coal industry (Darkin 2006). Moreover, it offered substantial producer flexibility through industry-level Climate Change Agreements (CCAs). CCAs exempted businesses from up to 80 percent of the levy if they agreed on voluntary carbon pollution-reduction benchmarks.[33] By 2002, forty-four sectoral associations had signed CCAs, including aluminum and steel (Bailey and Rupp 2005). Further producer-oriented flexibility was introduced with the April 2002 UK Emissions Trading Scheme (ETS), a voluntary program that allowed participants to trade emissions reduction permits relative to an absolute target baseline (the average of a participants' 1998–2000 emissions); CCA signatories could then buy and trade these permits as insurance against failure to meet CCL carbon pollution-reduction benchmarks.

Even with these flexible provisions, industry mobilized against the CCL. For instance, business lobbies and environmental groups clashed over the policy's projected impacts. A report by the Engineering Employers' Federation, the UK Steel Association, and the Chemical Industries Association suggested the policy would kill 95,000 manufacturing jobs.[34] By contrast, environmental groups forecasted net employment gains of 12,000 by 2002. The intensity of this business opposition apparently tempered Gordon

Brown's (then chancellor) enthusiasm for climate policymaking over the following decade, including during his subsequent term as prime minister (Carter 2014).[35] Moreover, partly responding to dramatic fuel price protests in September 2000, Labor leaders were also acutely concerned about the consumer costs of their policies; the real yield of consumer fuel taxes including the CCL actually decreased by 4 percent from 2000 to 2007 (McLean 2008).

In fact, the UK ETS itself emerged out of British industry's effort beginning early in Blair's first term to unsuccessfully preempt the CCL. In 1998, business actors in consultation with government agreed to explore emissions trading through the Advisory Committee on Business and the Environment (Skjærseth and Wettestad 2008). British business associations then formed an emissions trading group in 1999 to design the architecture for a CCL alternative. This effort was backed by companies like BP that already had carbon pricing experience and by financial interests in London looking to exploit new market opportunities associated with Kyoto Protocol carbon markets (Jordan et al. 2003). The emergence of explicit business community splits on climate policy continued to surface throughout the early 2000s. For instance, a number of major British corporations founded the Corporate Leaders Group on Climate Change to advance business interests related to climate risk mitigation, from Lloyds to Shell to Tesco (Carter 2014). The group's communications appear to have increased Blair's interest in undertaking climate reforms, partly because the emergence of pro-climate business interests genuinely surprised him (Carter 2008). Splits in British business interests were also reflected within more established lobby groups. For instance, CBI set up a task force in 2005 to work through cross-cutting climate policy preferences within the British business community (Lockwood 2013).

Domestic climate policymaking under Blair proceeded in parallel to EU-level efforts to negotiate a common climate policy. Unlike Germany, the British government was one of the strongest backers of emissions trading within the EU beginning in the late 1990s when other major actors were skeptical of EU-level action. British positioning on this issue was particularly salient given the country's previous hard opposition to EU-level carbon taxation (Skjærseth and Wettestad 2008). British preferences may have been driven by a desire to preempt tax instruments and by growing interest in emissions trading by business actors hoping to preempt carbon taxation domestically (Skjærseth and Wettestad 2008). However, despite this leadership, British political officials imagined a far less ambitious EU proposal than would eventually emerge; they viewed EU-level carbon pricing as best organized through coordination of domestic systems rather than centralized policy. Accordingly, the UK lobbied the EU to design the EU ETS using a weak and voluntary architecture similar to its domestic scheme, not the substantially more ambitious scheme that would eventually emerge (Jordan et al. 2003; Skjærseth and Wettestad 2008). In this way, the simple fact of British leadership in pushing EU emissions trading confounds the unambitious content of its preferred policy. Like Germany, the UK was forced into the EU ETS. By 2009, the EU ETS covered almost half of British carbon emissions (Bowen and Rydge 2011). At the same time, partly because of British experiences during its domestic emissions trading experiment, the UK was a European leader in managing EU ETS implementation (Skjærseth and Wettestad 2008).

During 2005–2006, climate change shot to the top of the UK political agenda.[36] Several converging developments vastly increased media coverage of climate change, from the government-commissioned Stern Review on climate change economics to high-profile speeches by climate scientists (Carter 2014). As climate change reached the top of the agenda, within-party divisions over climate policy became more prominent. On the Labor side, David Miliband became Secretary of State for Environment, Food and Rural Affairs in 2006 and began using climate advocacy as an issue to raise his own party profile (Carter 2014). On the Conservative side, David Cameron brought the party's green faction into power on the opposition benches (Carter 2008). After Blair's 1997 election, the Conservatives had rotated through a series of leaders who took varied levels of passive or active anti-environment positions. For example, during the 2001 election, Conservatives campaigned on removing the CCL.[37] Cameron used climate change to distinguish himself from these earlier leaders. Early on, he visited a Norwegian glacier as a symbolic effort to increase his party's green brand and proclaimed that "tackling climate change is a key part of my ambition for the Conservative Party to lead a new green revolution" (Carter 2009). The party began to campaign by urging voters to "Vote Blue, Go Green."

Sensing a policymaking window of opportunity, Friends of the Earth began a "Big Ask" campaign in May 2005 to enshrine carbon pollution reduction budgets in a piece of national legislation. Other environmental groups organized as part of the Stop Climate Chaos lobbying umbrella, which brought in student union movements as well as the UNISON service sector union, the country's second largest. In September 2006, David Cameron embraced the campaign and pledged his endorsement, joining a growing group of MPs from across party lines who had offered support for the effort. Quickly, Labor leaders also embraced the proposal, reluctant to cede strategic "green" political ground (Carter 2014). This multiparty signaling effort led to passage of the 2008 UK Climate Change Act.

Under the Climate Change Act, the British government endorsed a 34 percent reduction relative to 1990 levels by 2020 into law. The act also included an 80 percent reduction target for 2050. This legislation was first of any country to enshrine targets into law (Lockwood 2013). The Climate Change Act also created the Committee on Climate Change as an independent body to manage national carbon budgeting, evaluate national progress toward the target, and advise the government on mitigation and adaptation measures (Bowen and Rydge 2011). The committee was tasked with proposing five-year carbon budgets for the country; the government would then develop policies to meet these carbon budget plans. By 2009, national carbon budgets through 2022 had been established and a new Department for Energy and Climate Change had been tasked with developing policy measures to meet them.

Notably, this ambitious policy framework was not driven predominantly by the left-leaning government then in power. Instead, green groups built a relationship with pro-climate factions across the ideological spectrum that subsequently empowered pro-green voices within the Labor government to pass the act (Carter 2014). Similar dynamics played out with such other policies as feed-in-tariff subsidy support for renewable energy (Carter 2014). On a number of occasions, Cameron opposed Labor's efforts as insufficiently green; for instance, he suggested that Labor's efforts to increase the cost of air travel in 2006 was insufficient given the scale of the climate risk (McLean 2008). In turn, Prime Minister Gordon Brown, who replaced Blair in 2007, preemptively enacted green policies to ensure green voters would not migrate to Cameron's Conservatives (Carter 2008). Similarly, Cameron's 2008 Blue/Green Charter appears to have motivated Labor officials to accelerate environmental policymaking in an effort to regain an electoral advantage (Carter 2009). And the Conservatives and Liberal Democrats criticized a proposed coal-fired power plant in Kent that Labor's business secretary wanted to approve, forcing Labor to propose a CCS requirement on new plants (Carter 2014).

Of course, Cameron's climate positioning did not eliminate intra-party cleavages on the right. Prominent Conservative backbenchers resisted climate science and criticized the Cameron effort to green the party, including support for the Climate Change Act, and also the role of former Environment Secretary John Gummer who intervened in support of Cameron's shift (Carter 2009; Lockwood 2013). This criticism of "bunny-hugging" was particularly pronounced around green taxes (Carter 2009).

Cross-cutting preferences within the Conservative party surfaced again after Cameron ascended to the prime ministership in 2010. A minority government in coalition with the Liberal Democrats, Cameron promised to lead the "greenest government ever." The centrist Liberal Democrats tended to have the strongest green platforms among major parties (Clements 2014; Carter 2015). Conservative green factions were empowered by the Liberal Democratic presence. Yet, brown, polluter-aligned factions of the Conservative party also began to assert themselves more prominently. For instance, discussions of post-2022 carbon budgeting faced resistance from such figures as Business Secretary Vince Cable and Chancellor George Osborne who pushing for weaker policies to sidestep purportedly growing policy costs. This faction won the right to re-review Cameron's 2014 carbon budget, passed subsidies for oil and gas exploration and shale gas, and blocked decarbonization targets in a Conservative energy bill that had the support of cross-party MPs, relevant parliamentary committees, and some business voices (Carter 2014). In other words, despite Liberal Democratic support for ambitious climate policies, fossil fuel–aligned wings within the Conservative party blocked efforts to increase carbon polluter costs, even in the presence of a persistently divided business community. The prominent exception here was the Conservative introduction of a carbon price floor in 2013, which topped EU ETS carbon prices up to a net price of €18 per ton of carbon, in effect amplifying the domestic carbon price signal. The price floor was intended to continue until coal power was completely phased out of the UK electricity market.

Notwithstanding the carbon price floor, the Conservative Party's green voices continued to lose ground and even Cameron himself purportedly lost interest in environmental issues.[38] At the same time, Conservative policymakers became increasingly responsive to producer interests (Carter and Clements 2015). At the extreme, Conservatives removed the renewable energy exemptions entirely from the country's Climate Change Levy in 2015, effectively ending its carbon logic. As before, neither Conservative Party members nor business unanimously embraced this retrenchment. Instead, some Conservatives along with energy sector interests and investors fought against the change, worried it would undermine investment in the energy transition (Lockwood 2013). Yet, these voices were not strong enough to protect the policy under attacks from anti-climate party factions.

Discussion

Each shadow case offers a probe of the theory of double representation's generalizability. On balance, the dynamics of climate policymaking in Germany, Japan, Canada, and the United Kingdom all underscore the logic of double representation's considerable explanatory power. In each case, climate policy conflict closely shadows the dynamics detailed in previous analysis of Norwegian, US, and Australian policymaking.

First, climate policy preferences cross-cut preexisting political cleavages. Climate reform proposals split business organizations and labor movements in all four shadow cases. Likewise, new climate change cleavages emerged within both left-leaning and right-leaning political coalitions. Just as in Norway, Australia, and the United States, these splits surfaced in the early 1990s and persisted through the book's entire study period up until the 2015 Paris climate conference.

In turn, carbon-intensive economic interests in all four shadow cases enjoyed a double representation within national climate debates. At different moments, political actors with ties to carbon-dependent economic actors on both sides of the ideological spectrum blocked climate reforms. In the UK, both the Confederation of British Industries and the Trade Unions Congress mobilized to oppose carbon pricing in the early 1990s. In Canada, the Liberal's Green Shift proposal was opposed by both the right-leaning Conservative Party with ties to Albertan fossil fuel producers and the left-leaning New Democratic Party with ties to the national labor movement. In Japan, fossil fuel voices dominated a divided business community and blocked efforts by pro-climate factions to impose costs on carbon pollution. In Germany, both industrial unions and major industrial producers mobilized in opposition to government plans to limit coal-fired power plant emissions, amplified by opposition from within both the right-leaning Christian Democratic Union and left-leaning Social Democratic Party (SPD) benches.

These shadow cases also offer evidence in favor of a key simplifying assumption in this book's theory: that there has been broad convergence across advanced economies in the timing of climate policy proposals. This assumption finds empirical validation across all four shadow cases. In each country, carbon taxes were considered in the early 1990s—contemporaneously with debates in the United States, Australia, and Norway. Such events as the Kyoto climate meeting and the 2007 IPCC reports led to parallel increases in climate change salience's on each of the four countries' domestic policymaking agendas. Moreover, while carbon taxes were debated in each country in the early 1990s, emissions trading schemes all became the subject of substantial policymaking debate during the 2000s. This empirical regularity, of course, is central to this book's subsequent efforts to explain variation in climate policy enactment (compared to variation in the timing of climate policy proposals).

Further, these shadow cases highlight the same limitations to existing explanations that this book's empirical case studies elaborated. While sectoral balance of power accounts correctly highlight distributive politics in these cases, changes in the presence or size of either green or brown industries did not clearly precede moments of climate policymaking in any of the shadow cases. Neither the British Climate Change Levy or the German Energiewende passed because of balance of power shifts; for instance, coal interests in Germany continued to prosper during and after these climate-oriented reforms, buoyed by strong representation within the SPD party. This is not to say that sectoral conflict is immaterial. Climate splits in the British business community appear to have pushed Prime Minister Tony Blair into taking climate reforms seriously. Sectoral conflict in Japan appears to have facilitated Prime Minister Junichiro Koizumi's decision to stick with the Kyoto Protocol. However, theories of distributive conflict must examine not just the distribution of preferences in a given country at a given moment in time, but also the interaction of different preference coalitions with political and policymaking institutions.

These cases also highlight the surprisingly weak importance of environmental lobbies to climate policymaking outcomes. Early reforms in all four cases developed in the absence of systematic pressure from green lobbies. In Germany, environmental groups were more preoccupied with nuclear policy than climate policy during early debates; in fact, CDU factions promoted the climate issue as a wedge to undermine environmental opposition to nuclear development. Japanese environmental groups barely appear in most scholarly accounts of national climate policy; environmental activism tends to focus instead on legal and public activism outside of formal policy-making venues. Environmental NGOs have occasionally played oversized roles, such as Friends of the Earth efforts to push British climate reforms by skillfully exploiting cross-cutting climate policy preferences across the British left and right in 2007–2008. However, these moments of substantial green group influence tend to be the exceptions.

Third, in all shadow cases, climate policies drew support from both left-leaning and right-leaning political actors. The cross-cutting nature of climate preferences also distributed climate policy proponents on both sides of the ideological spectrum. In these shadow cases, some of the strongest pro-climate voices emerged not from the left but from center-left or even right-leaning parties. In Canada, it was the centrist Liberal Party that pushed the costliest climate policies during the party's 2006 election campaign. During that campaign, the Liberals were attacked from the right by the oil sands–linked Conservative government and from the left by the labor-affiliated New Democratic Party. In the United Kingdom, Labor leaders between 2006 and 2010 passed climate reforms with cross-party consensus; during this period, Cameron's Conservatives strategically positioned themselves to the green side of Gordon Brown's Labor. Overall, while there is an ideological gradient in support for climate reforms, shifts in carbon pricing regimes have not been a clear function of governing party ideology.

Fourth, the shadow cases provide support for the book's down-weighting of proportional representation electoral institutions as an explanation for the presence or absence of carbon pricing policies. The German proportional representation system provides for significant Green Party representation in government. However, in Germany, the formal red–green coalition between the German SPD and the Greens produced the same type of constrained policy output previously described in Australia and Norway. Green voices could champion environmental reforms broadly and even won major subsidy-based policy fights (e.g., the Energiewende package);

however, they remained unable to impose any costs on carbon polluters because of industrial representation within the carbon-intensive wing of the "red" coalition party. While the UK Liberal Democrats are not a green party, they nonetheless have a relatively ambitious green policymaking agenda. Yet, even in coalition with the British Conservatives led by a prime minister who promised the country's "greenest" government ever, pro-climate factions could not push reforms over the objections of persistent climate skeptics and industry-minded politicians within Conservative ranks.

The empirical record of climate policymaking across all shadow cases also revealed substantial variation in the timing and content of domestic climate reforms. Is this variation consistent with the propositions introduced in chapter 2? Each proposition can be probed using the shadow cases to unpack the theory's explanatory power in more detail.

First, the theory of double representation predicted that climate policy should emerge earlier in systems where the double representation of carbon polluters was more strongly reinforced by domestic institutions. The institutionalization of double representation across shadow cases was selected to vary according to the presence of formal institutional links between labor actors and the dominant left-leaning coalition, and according to the policymaking system. We can consider the labor dimension first, making cross-wise comparisons between the two pluralist and two corporatist shadow cases. Here, we do find evidence consistent with the theoretical proposition. German climate policy emerged earlier than in Japan, and British climate policy emerged earlier than in Canada. Additionally, we might also expect that, relative to pluralist countries, corporatist countries more strongly institutionalize the voice of double-represented polluters by guaranteeing access to policy design venues. For instance, the theory would also have predicted earlier action in Germany in comparison to the UK. In this respect, the shadow cases offer ambiguous guidance on the theory's scope conditions. Germany and the UK first enacted serious climate policies around 2000, though Germany's policies were arguably more far-reaching, given the perceived weaknesses of the British Climate Change Levy. However, direct comparisons of British versus German policy timing are also confounded by both countries' EU membership and their joint participation in the EU emissions trading system beginning in the early 2000s. Similarly, we might expect Japan to act in advance of Canada. Again, we see this is only true to a limited extent as both countries largely resisted substantial climate policies throughout the study period. At the same time, after emissions trading emerged on the policymaking agenda in both countries, a national voluntary program sensitive to producer interests was enacted in Japan even as climate proposals stalled in Canada. Generally, this suggests that the book's theory may overpredict the ability of corporatist systems to deliver consistently earlier policy (even though this is definitely true in many other corporate systems such as the Netherlands, Sweden, and Denmark) relative to the most prominent pluralist cases of Canada, Australia, and the United States.

Of course, descriptive patterns consistent with the theory of double representation's predictions offers a cursory form of empirical support. Instead, we should also interrogate whether differentiated policy timing across countries is rooted in the policymaking pathways that different countries take in pursuing climate reforms. Where double representation is strongly reinforced, the theory predicts early action precisely because carbon-intensive interests have a strong voice in climate policy design and can ensure that proposed policies do not threaten serious costs. We do see this in Germany where early policies eschewed threatening costs on producers. A broad reluctance to impose costs on institutionally privileged carbon polluters pushed policymakers to subsidize renewable energy policy as a central climate policy plank rather than impose costs on coal-intensive economic interests.

By contrast, where carbon polluters have less access to policy design, more threatening policies were introduced onto the political agenda. This was clearly the case with the Canadian Liberal's Green Shift proposal which precipitated efforts by carbon-dependent economic interests to mobilize against these policies in the public domain, attempting to undermine legislative and electoral incentives associated with climate policy enactment. Subsequently, a new climate policymaking window only emerged in Canada after the period covered by this book, as part of the Justin Trudeau government's national carbon pricing effort.

Second, the theory predicted that the logic of double representation skews domestic climate reforms relatively toward carrots and away from sticks. Given that we find evidence of double representation in all four shadow cases, we should also expect an absolute reluctance to embrace policymaking sticks in all four countries. This is indeed the empirical pattern observed. While the logic of double representation privileges the influence of distributional losers within policymaking debates, supporters of climate policy are also distributed across political coalitions. For instance, left-leaning political coalitions have struggled to reconcile an ideological commitment to climate policy with resistance to the idea of imposing costs on carbon-intensive labor constituencies. These coalitions have often turned toward technological subsidies or international investments as a way to reconcile these intra-coalition tensions. This is clear in Germany and in the internal debates of the Canadian NDP. Split beliefs in the United Kingdom on both the right and left also mirror this dynamic.

We should also expect that, while all countries continue to prioritize carrots over sticks, the tendency is less pronounced in settings where the logic of double representation is more weakly reinforced by domestic institutions; in these settings, policy proponents can sometimes partially exclude carbon polluters from positions of policy influence, even carbon polluters that are aligned with their political coalition. Subsequently, there is a relatively stronger tendency to impose costs on carbon polluters in pluralist settings, even as there is a global tendency to prioritize carrots. Here we find the pluralist systems are more likely to propose such sticks even if they are not always enacted. The UK's Climate Change Levy did involve some direct costs, unlike Germany's Energiewende policies. Canada's Green Shift was among the most aggressive proposals put forward by any political party during a national election during this book's time period. Japanese policies, by constrast, remained extremely producer-focused.

Third, the final proposition predicted that where the double representation of carbon polluters is strongly institutionalized, climate policies are more producer oriented; where it is more weakly institutionalized, policies are often more consumer oriented. Where distributive conflict is characterized by the logic of double representation, the focus of climate policymaking continues to be the accommodation of distributional losers, particularly industrial producers. In some contexts, producers are accommodated through weak policies that impose few overall costs on any actors, for instance voluntary climate policy measures. In cases where the overall costs of a climate policy are higher, these costs are imposed on diffuse, unorganized consumers. Similarly, while absolute cost levels are sometimes higher in these producer-oriented policies, they are differentiated at a sectoral level to avoid the imposition of threats to the economic status quo.

Climate policy in all four countries takes this form, with perhaps the strongest producer costs in Canada and the weakest in Germany and Japan. The United Kingdom stands as an ambiguous case in this theory—as producer interests were substantially accommodated at every stage of the policymaking process. However, the British government was similarly attentive to the salience of consumer costs.

Overall, these shadow cases offer policymaking trajectories consistent with the logic of double representation and the institutional conditions that shape this logic's effects on policy outcomes. The rough outlines of each proposition appear to be consistent with primary casework in Australia, Norway, and the United States. This chapter thus offers credible evidence that the logic of double representation characterizes the dynamics of climate policy conflict across advanced economies, at least during this book's pre-2015 study period.

Of course, we should not expect any parsimonious theory of distributive conflict to fully capture political dynamics across advanced economies. However, these generalizability probes suggest that the logic of double representation captures an essential part of the policymaking dynamic in many important cases. Given the importance of these cases as sources of global carbon pollution, it becomes even more important to understand how double representation may be disrupted in the post-Paris Agreement world of decentralized, domestic climate policymaking. The post–Paris Agreement world has also been marred by the uncertainty of President Trump's withdrawal from the multilateral process in June 2017. Chapter 8 thus pivots to consider the lessons that the logic of double representation offers for comparative climate politics over the coming decade.

8 Disrupting the Logic of Double Representation

Our governments have failed us. Once a distant threat to future genera-tions, the climate crisis is now a problem for the here and now. You and I—absent extraordinary policy interventions—will experience economic and social hardship from climate change over our lifetimes. And our chil-dren will face profound suffering on an unrecognizable planet. The ques-tion is no longer *whether* we will experience collective loss, but just how severe these losses will be.

Worse, the problem of climate change has become harder to solve over time, not easier. Thirty years ago, a different future was available. Gradual climate policies could have slowly steered our economies toward gently declining carbon pollution levels. The costs to most citizens would have been imperceptible. That future is no longer available. Scientists estimate that our world can release 2,620 gigatons of carbon pollution into the atmosphere, relative to pre-industrial levels, before the climate warms to dangerous levels.[1] Our society has already consumed 84 percent of this available carbon budget. And, with every passing year of policy inaction, more of our remaining budget is squandered. We must now decarbonize at a punishing pace.

At the same time, various ecological tipping points threaten to upend even the narrow mitigation pathways still available to us. For instance, warming soils could release as much carbon into the atmosphere each year as an industrialized economy the size of the United States (Crowther et al. 2016). These biological feedbacks threaten to create runaway climate change that will overwhelm all but the most dramatic policy interventions. For human society to flourish in the twenty-first and twenty-second centu-ries, immediate, strong, and fast action is necessary.

This book provides a new diagnosis for an unfolding political and environmental catastrophe. It helps explain how carbon-intensive business and labor groups captured the climate policymaking process, endangering each of our lives. It describes how some policymaking efforts still succeeded despite entrenched opposition by carbon polluters. And it encourages us to rethink how political systems should confront the existential threat of climate change.

In this chapter, I revisit the received wisdom about climate politics in light of my empirical findings. Conventional accounts typically involve two claims. First, at the domestic level, carbon pricing policies are the best strategy to mitigate climate change. Second, at the international level, we should prioritize a binding global climate treaty that undermines free-riding incentives. I argue that both claims are misguided. Neither carbon pricing nor a binding climate treaty are viable strategies to stabilize the planet's climate. Both proposals aim to reshape incentives to pollute. They assume that, once economic incentives change, the economic and political power of carbon polluters will decline. But both put the cart before the horse. We need to *first* disrupt the political power of carbon polluters before we can effectively reshape economic incentive structures.

In other words, carbon pricing and a binding climate treaty offer an improved equilibrium state of the world. But they do not offer a clear political pathway to reach this new policy equilibrium. Both policies are well suited for late stages in society's decarbonization efforts. They may help lock in policy gains. They may help optimize carbon pollution reduction at the margins. But they are flawed strategies for the short term.

Instead of economic theory, we must center politics in climate policymaking. Economic considerations matter, but they cannot be the dominant lens through which climate policies are evaluated. In particular, we must move away from a focus on economic efficiency toward a focus on the distribution of political power. For example, from a political economy perspective, carbon pricing is flawed. Carbon pricing makes consumer and producer costs transparent without making economic or environmental benefits salient. This allows opponents to politicize public costs even when policies are designed to benefit the public. Consequently, ambitious carbon pricing policies will not pass until carbon polluter influence is weakened. Advocates may instead benefit from concentrating their short-term efforts on policies that weaken the political influence of powerful carbon polluters, not carbon pricing.

Three decades of global efforts to negotiate a global climate treaty suffered from a similar misunderstanding about policy sequencing. These efforts presumed that domestic policymaking would follow once the problem of global free-riding was solved. But this logic is also backwards. International negotiations will not reshape the distribution of political power domestically. Instead, effective climate negotiations require that domestic actors have already won difficult political conflicts at home. A binding global climate treaty can lock-in domestic policy wins that have already occurred. It cannot solve the distributive conflict that has stymied most climate reforms over the past thirty years.

As carbon pricing and global negotiations falter, the loudest voices now begin to suggest we *depoliticize* climate change. Many scientists and policymakers argue that a threat as large as climate change requires society to transcend partisan and ideological conflict. This appeal is also misplaced. Carbon polluters will not voluntarily relinquish their power. Climate reforms must pass in the presence of contentious politics. And the intensity of this political conflict will increase as faster and more disruptive decarbonization trajectories become necessary.

Above all, climate change mitigation is a political act. We cannot move forward without a deep understanding of climate policy conflict. Unpacking the logic of that conflict has been this book's goal.

Centering Politics in Our Analysis of Climate Policy Design

For the past three decades, economists have dominated climate policy debates. Consequently, most climate policies have been evaluated on the basis of their economic efficiency: the best policy instruments should reduce carbon pollution at the lowest marginal costs. Carbon pricing performs well by this metric, and so policy conversations have centered on the optimal means to set up a carbon price. Should countries directly tax carbon pollution? Or should the price emerge from emissions trading?

This book joins a different dialogue. It focuses not on the theoretical merits of carbon pricing, but on the political economy of carbon pricing reforms (Jenkins 2014; Rabe 2018). Putting a price on carbon pollution makes policy costs very salient. At the same time, the policy keeps environmental and economic benefits hidden. This approach thus fore-grounds the most politically difficult dimensions of climate reform while backgrounding the dimensions necessary to nurture political support coalitions. This is a grim recipe for success.

Worse, by raising the salience of policy costs, carbon pricing gives political ammunition to the very economic actors who profit from releasing pollution into the atmosphere. These carbon polluters can easily mobilize opposition by isolating concrete policy costs from abstract benefits, reducing electoral and political incentives to support climate reforms. Hoping to forestall such opposition, most existing carbon prices shield powerful producers from substantial policy costs, either through sectoral exemptions from carbon taxation or free allowance distributions. The result: today's carbon prices don't really undermine the political power of climate reform opponents. But without disruption of these opponents' policymaking influence, more ambitious carbon pricing policies remain unlikely.

Recognizing that carbon pricing foregrounds costs over benefits, more attention has been given to the redistribution of carbon pricing revenues back to the public. The most prominent proposed approach are "cap and dividend" policies. These policies redistribute carbon tax revenues back to the public through annual rebates. Advocates hope that by creating direct financial benefits associated with carbon pricing reforms, interest groups' ability to mobilize against consumer costs will be weakened. Thus, Switzerland's 2008 carbon tax pays out health insurance subsidies each year. And in some parts of Canada, a federal carbon tax now provides annual income tax rebates.

We still don't know if linking carbon pricing to consumer financial benefits will be sufficient to remake the political economy of carbon pricing. There are reasons to be skeptical. The public has a weak baseline understanding of even large government benefits (Mettler 2011). And climate policy opponents have not shied away from aggressive attacks against carbon prices even when rebates are present. For example, in Ontario, the Conservative provincial government has attacked the federal carbon tax by aggressively highlighting policy costs without mentioning any of the policy's individual benefits. This includes a proposal to put stickers on gas pumps that exclusively highlight the carbon tax's costs. More broadly, while carbon price rebates create short-term financial benefits for some members of the public, they do nothing to make salient the longer-term economic and social benefits associated with a stable climate.

In contrast to carbon pricing, large-scale industrial policy, such as a Green New Deal, may offer more political benefits. These packages foreground benefits without making individual costs as salient. They may also help bridge divisions between labor and environmental constituencies, which this book highlights as a major obstacle to ambitious climate policymaking on the left. For example, Australia's Carbon Pricing Mechanism, which included substantial industrial transformation, passed with labor support while the Carbon Pollution Reduction Scheme, a standalone carbon price, failed.

A focus on political power in climate policy design also emphasizes the potential for economically inefficient policies like renewable energy mandates and subsidies (Stokes 2020). Some economists argue that once a pollution cap is established through an emissions trading scheme, other climate and energy reforms become superfluous (Montgomery 1972; Stavins 2003). Politicians allied with carbon polluters echo this claim. In the United States, such conservative Democrats as Jim Matheson opposed the inclusion of a renewable portfolio standard in the Waxman–Markey bill because—once a cap had been established—a mandate to produce renewable energy was irrelevant.[2] Similar attacks were levied in Australia against such entities as the Clean Energy Finance Corporation. The intuition is that renewable energy deployment will not reduce net emissions; instead they will just create additional slack under the cap for carbon-polluting industries.

Yet, these perspectives ignore the political implications of different policy designs. Renewable energy mandates actively promote the development of new actors that can counterbalance the power of carbon polluters. These policies may be inefficient. However, they reshape the distribution of political power and, in doing so, can increase the ambition of future climate reforms (Stokes 2020). Analogously, we should engage skeptically with claims that carbon pollution is cheaper to address abroad than at home. Norwegian climate policy in the 2000s focused on deforestation overseas rather than imposing costs on domestic industrial actors. Underlying this perspective is the assumption that a ton of carbon pollution abated in one part of the world is equivalent to a ton of carbon pollution abated elsewhere. Since climate change is a global problem, policymakers should therefore minimize the costs associated with climate policy by targeting pollution abatement opportunities without regard for national borders.

However, one ton of CO_2 reduced in one sector is *not* the same as a ton of CO_2 reduced in a different sector or in a different part of the world. This is because the right to continue releasing carbon pollution into the atmosphere shapes the distribution of economic actors in that country or sector. This, in turn, dramatically constrains the politics of climate change. Allowing domestic actors to continue to release carbon pollution enhances their profitability and increases the odds they will exist to contest future rounds of climate policymaking. In other words, policy losers won't disappear on their own without active political conflict (Breetz, Mildenberger, and Stokes 2018). If a country relies on lowest-cost international carbon pollution reduction opportunities, they allow the status quo distribution of domestic carbon-intensive actors to persist. This can stymie future efforts to ratchet up climate policymaking ambition over time.

Once we accept that there is no *political* equivalency in which tons of carbon pollution get reduced, this changes our approach to climate policy choice. Effective policies must reduce carbon pollution while also disrupting opponents' political power. In this way, a policy that imposes high costs on select consumers while shielding producers from any costs may shape consumer carbon pollution while remaining *politically unambitious* when evaluated according to the policy's ability to catalyze efficient and effective long-term policy outcomes. Policies must unlock the capacity for society to decarbonize ever more quickly in the future.

Correspondingly, stringent environmental standards across advanced economies are not a legitimate reason to expand carbon pollution in these countries. For instance, in the early 1990s, Australian officials justified expansion of carbon pollution on the basis that Australia had higher environmental standards and therefore it was globally more efficient for Australian emissions to grow.[3] Analogous arguments have been advanced by the Norwegian oil industry. Because Norwegian oil has some of the world's most stringent production standards, the theory goes that expansion of Norwegian oil and gas industries can substitute foreign production to decrease global net emissions (Hovden and Lindseth 2002).[4] Similar arguments have been raised in favor of the Albertan tar sands and Californian oil development.

But what these arguments ignore is that continued fossil fuel production entrenches carbon polluters as long-term actors within these countries' climate debates. Climate policymaking is not a one-shot game. It will require iterative reforms that increase in ambition over time. The political decisions and policies passed today shape the distribution of actors who will survive to contest future policymaking debates. And so, short-term support for fossil fuel extraction reinforces the profitability and presence of carbon-intensive actors in policymaking debates. Putting off difficult economic conflict will not make that conflict easier. Instead, it will become *harder* to act when faster and more disruptive rates of decarbonization are necessary.

In short, decisions on where and how to transform an economy matter not only in terms of their short-term economic implications, but also their long-term political effects. Inefficient policies can disrupt the distribution of political power by reshaping the resources and influence of economic stakeholders. In doing so, they can unlock more ambitious long-term carbon pollution reduction opportunities.

Rethinking the Role of Global Climate Institutions

At the international level, many climate proponents have focused their efforts on establishing international institutions to facilitate climate policy cooperation. These actors assume the most important barrier to policy action is the incentive to free-ride. Rooted in a collective action interpretation of the climate threat, unilateral action by any country is viewed as irrational. Why? Theoretically, it would involve the imposing costs domestically while countries only received a small fraction of globally distributed benefits. Correspondingly, global climate negotiators have assumed that the key to unlocking action and establishing a global climate agreement is to coordinate pollution reduction across countries. A binding global climate treaty will, according to this perspective, override disincentives to act.

However, this perspective also rests on flawed assumptions about political sequencing. As this book has highlighted, climate policy inaction is not primarily shaped by concerns over free-riding but by the persistent power of carbon-intensive economic interests that stymie reform efforts. In every advanced economy, serious efforts to enact major climate reforms have occurred over the past two decades, even in the absence of a binding global climate treaty; yet, pro-climate actors have been routinely frustrated by the power of double-represented carbon polluters.

A global climate agreement will not erase national-level distributive conflict. International negotiations will not reshape the distribution of domestic policymaking interests. Instead, the outcomes of domestic climate policy conflict determine the success of global negotiations.[5] Global negotiations are a *function* of domestic inputs. For example, the stringency of the Obama administration's climate policies opened the door for the United States to make credible international commitments. Thus, in November 2014, President Obama announced a joint climate accord with Chinese President Xi Jinping. Under the accord, Obama agreed to increase the ambition of the US carbon pollution reduction target to 26–28 percent below 2005 level by 2025. China agreed to its first-ever carbon pollution constraint, promising to stabilize emissions by 2030.[6] It was not the existence of the US-China accord that empowered US pro-climate factions to impose these costs; instead, the ascendant power of US climate reformers made possible the bilateral agreement.

Paris negotiators understood this. Seeking to avoid the failures of Kyoto and Copenhagen, they embraced a new architecture. The Paris Agreement forgoes binding climate commitments. Instead, it coordinates voluntary carbon pollution reduction targets set independently by each country. The Paris pledge and review structure allows each country to define its own carbon pollution reduction schedule consistent with domestic conditions. These inputs can then be aggregated to estimate whether global commitments add up to addressing the full scale of the climate problem. The agreement's architects hoped this "trust but verify" process would dampen free-riding fears and generate momentum for mutually reinforcing domestic climate reforms.

Echoing this logic, climate policy observers feared the worst when President Trump said the United States would withdraw from the agreement in June 2017. Would the agreement collapse in the absence of one of the world's most significant carbon polluters? To some surprise, it did not. Such diverse countries as China, Brazil, Canada, Costa Rica, and Colombia immediately reiterated their unconditional support for the Paris Agreement irrespective of Trump's climate intransigence. If anything, Trump's rejection of the Paris accord appears to have strengthened the hand of climate policy proponents in some countries—enhancing the probability of climate policy action in these countries—even as it reduced the probability of action in others by providing rhetorical cover for policy opponents.

Some advocates assumed that, under the Paris Agreement's "trust but verify" architecture, climate policy's economic winners would continuously and synchronously control the domestic levers of power over time. On the contrary, climate conflict will occasionally empower economic losers who will repeal or retrench existing climate policy commitments, as we saw in Australia in 2014, in the United States in 2017, and in Ontario, Canada, in 2018. Elections will bring climate losers into power and these opponents will seek to retrench or repeal climate reforms. Paris will be successful not through a trust but verify process, but instead to the degree that it helps climate reformers across the world entrench and extend their political influence.

Embracing Political Conflict in Climate Policymaking

Ultimately, there are no short-cuts to circumvent the fundamentally distributive nature of climate policymaking. This means that scholars need to understand the political logic of this conflict, and identify the conditions under which climate policy advocates can disrupt entrenched opponents.

When will carbon polluters lose control over the policymaking process? As this book suggests, some political actors traditionally viewed as green— left-leaning political parties, European countries—may not usher in the dramatic climate reforms that advocates hope for. These limits may become particularly apparent as global climate stability comes to depend on rapid decarbonization. For example, a sudden wave of left-wing political victories will not automatically generate climate progress. Instead, climate reforms have historically come from both the left and right sides of the ideological spectrum, since cross-cutting climate policy preferences distributed pro-climate actors across the political system. Yet, climate advocates on both the left and right also faced off with policy opponents inside their parties and political coalitions.

Instead, some of the most aggressive efforts to impose costs on carbon polluters have come from centrist parties that lacked historical ties to carbon-intensive capital or labor. In Norway the most stringent climate policy proposals emerged during the center-right Christian Democratic government. It was this first Bondevik Cabinet, in coalition with the center-right Venstre and the rural-agrarian Center Party, that made the strongest push to impose costs on carbon-intensive producers. This same Bondevik Cabinet also became the first global government to fall on the issue of climate change in 2000, when it refused to allow *carbonization* of the Norwegian electricity system. In other countries, centrist parties with an urban or service-sector constituency can also be influential. For instance, the Canadian Liberals went into the 2007 election proposing a carbon tax that was attacked from the right by the resource-linked Conservative Party and from the left by the union-allied NDP party.

For ambitious climate reforms to emerge from the left, progressive political coalitions will have to bridge labor and environmental community divides. Labor unions continue to be a major force resisting non-incremental climate reforms around the world. Some US unions have forcefully opposed the Green New Deal. The head of the Hyundai worker's union in South Korea described electric cars as "evil" disasters.[7] North America's Building Trades Unions railed against "hyper climate activism" and "misinformation campaigns by environmental extremists" that worked to block oil pipeline developments.[8]

To date, labor–environmental coalitions have succeeded only when conversations focused on common ground. For example, in Australia, green interests partnered with labor actors during carbon pricing negotiations, believing that Labor Party efforts would require a consensus between the union and environmental wings of the party. In doing so, they agreed to avoid conversations about policy stringency and other topics where industrial labor's preferences conflicted with environmental groups' goals. As a result, industrial labor needs moderated the environmental movement's advocacy.

However, the time for incrementalism has passed. Any realistic decarbonization trajectory will require the left to impose serious costs on both carbon-intensive labor and carbon-intensive capital. Jobs will be lost. Some communities will face economic hardship. And so, progressive political coalitions must engage in immediate and thorough debates over compensation strategies for impacted workers and communities. Advocacy coalitions must be grounded in the targets and policies demanded by climate science, not the incremental policies that satisfy the most carbon-intensive labor unions.

Like some centrist parties, Green parties also lack ties to carbon-intensive interests. Yet, this book emphasizes their limited power to weaken entrenched opponents, at least to date. Unlike centrist forces, Green parties have not seized sufficient control of the political agenda in any country to control the policy design process. Instead, they have had to partner with left-leaning parties through red–green coalitions. However, in both Australia and Norway, climate reforms under red–green governments were limited. Red–green coalitions were consistently stymied by the industrial "red" wing of the red–green coalition—typically Labor Party members with ties to industrial unions and capital. Because Green parties' presence in government did not fundamentally reshape the cross-cutting distribution of climate policy preferences on the left, it did not catalyze ambitious climate reforms. However, if Green parties grow in strength over the coming years, they may eventually disrupt the double representation of carbon polluters.

The environmental community will also need to play a larger role in counterbalancing the voice of big carbon polluters. To date, the environmental community has played a weak role in shaping climate politics across advanced economies. The first decade of climate reforms were driven almost exclusively by political elites. During this period, environmentalists were, at best, followers that mobilized after the fact in support of elite-driven policy agendas. In the 2000s, the community mobilized more consistently to push climate reforms; however, these efforts were largely ineffective. Environmental groups in the United States and Australia that continued to oppose policies like the Waxman–Markey bill and the Carbon Pollution Reduction Scheme were criticized by center-left political actors. These actors assumed that any comprehensive climate reform was a victory for the environmental movement and should have been met with robust support.

Instead, climate reforms have repeatedly splintered environmentalist groups. Some groups partner with labor and mainstream political actors to promote compromise policies, even if these policies do not meet climate science targets. Other groups are reluctant to endorse policies that they view as insufficiently ambitious. This has left several prominent policy proposals in an awkward position. Climate reforms emerge onto the agenda in the context of carbon polluter double representation. Thus, even ambitious policies are slanted toward the needs of carbon-intensive labor or capital, dampening environmentalist enthusiasm for these policies. But this does not stop carbon polluters from vocally opposing the same policies that they have already weakened during policy design. This leaves climate policies such as the US Waxman–Markey bill with few public champions during political debates.

Recognizing their strategic missteps, climate advocates have begun to invest in climate movement-building. Leaders including Swedish climate activist Greta Thunberg and initiatives including the UK's Extinction Rebellion and United States' Sunrise Movement may foreshadow a new wave of public organizing that could counterbalance the entrenched power of carbon polluters. The study of these movements must be a central part of political science scholarship moving forward. Scholars have an important role to play in supporting these efforts by studying how the public can build "collective power" (McAdam 2017; Han and Barnett-Loro 2018).

At the same time, we cannot focus on disrupting the power of entrenched opponents without recognizing the institutional contexts that reinforce this power. This book emphasizes how different countries may also have comparative advantages for climate reforms as a function of their national political institutions. Rather than pushing similar reforms across every country, climate advocates will need to tailor their domestic demands to align with each country's political context.

Pluralist countries often pursue the least ambitious climate reforms. This occurs when opponents control the political agenda, as under the George W. Bush and Trump administrations in the United States. During these periods, only abatement efforts that provide net economic benefits to individual carbon polluters are pursued. Voluntary climate policies pursued by the US and Australian governments in the late 1990s also fall into this camp.

Corporatist societies, by contrast, can pass more costly policies. These policies rarely pose existential threats to the economic status quo. However, while carbon polluters still enjoy access to climate policy design, they must still compromise with a more diverse set of social and economic stakeholders during the policy design process. Often, policies are calibrated to a level that is acceptable to producers, rather than those producers' cost levels. However, because incremental reforms do not threaten status quo, the distribution of political power remains unchanged through time. Carbon polluters stick around, ready to contest future efforts to non-incrementally ratchet up policymaking ambition.

The most disruptive policies can occur in pluralist settings when reform efforts lead to policy enactment *despite* producer mobilization. Unlike corporatist contexts, carbon polluters can be shut out of the policy design process when climate advocates hold power. In these cases, pro-climate coalitions that do not depend on carbon-intensive constituencies can impose serious costs on carbon polluters. If these reforms are successfully implemented, they may drive some polluters out of business and reshape the distribution of future political lobbies. However, in the face of threatening costs, producers mobilize and often successfully kill these reforms by mobilizing conflict into the public sphere.

These institutional dynamics suggest that different tactics are necessary to drive climate reforms in different countries. In corporatist systems, indirectly imposing costs on powerful polluters may hold more promise. For instance, Germany's substantial investments in renewable energy imposed no direct costs on the country's coal industry, which maintained its institutional presence even as renewable energy generation accelerated. However, by reshaping the relative cost of different energy sources, these subsidies have since eroded the profitability of some German utilities. The key to this success was that renewable energy policies were not understood as threatening at the moment of enactment (Stokes 2020). In this way, corporatist carrots can generate pressure for decarbonization if they shift technological cost curves in unexpected ways (Breetz, Mildenberger, and Stokes 2018). Unfortunately, it is not always clear a priori which subsidies will successfully disrupt political power, and which will instead serve as ineffective spending that papers over intra-coalitional political tensions without providing long-term climate benefits. For example, we can contrast the relative success of the German Energiewende with the expensive failure of Norwegian investment in CCS technology. Norwegian investments bridged tensions on the Norwegian left in the short term but ultimately had no effect on Norwegian carbon pollution trajectories.

Indirect pathways to cost imposition can also disrupt the climate policy opponents' long-term profitability in pluralist settings; however, contentious climate politics can stymie effort to pass subsidies at the scale necessary to disrupt existing economic forces. In these settings, more direct forms of cost imposition may be politically feasible. For instance, advocates may find the most success using regulatory measures to directly constrain the economic activity of carbon polluters.

Climate policies in pluralist countries also need to be more sensitive to consumer costs, since opponents will attempt to profile costs in their attempts to undermine reform efforts. As such, Skocpol (2013) criticizes efforts to enact climate policies in the United States through insider bargains between business and environmental elites. Instead, she argues that climate policy enactment necessitates broad public mobilization to counteract efforts by carbon-dependent economic actors to block reforms. This book echoes Skocpol's call by emphasizing the fundamental importance of distributive conflict to climate policymaking in pluralist countries. Efforts to impose costs on carbon polluters will necessarily entail dramatic mobilization against climate reforms by economic losers. Climate policy advocates will need vibrant, inclusive political coalitions that can push back against these efforts. They will need to establish persistent electoral and legislative incentives for climate policy enactment in the face of producer mobilization.

These conclusions about the relationship between institutions and climate policymaking cut against some previous work. A number of scholars have suggested that national political institutions—democracy and corporatism—might facilitate ambitious climate policy outcomes by facilitating collective action (Scruggs 2003; Bättig and Bernauer 2009). Yet, this book offers a theory of distributive climate policy conflict that troubles the existence of a simple relationship between institutions that facilitate collective action and those that deliver ambitious climate policy. This is because institutions that promote collective action also tend to accommodate distributional losers within policy design, and thus reinforce the privileged influence of carbon-dependent economic actors on climate policymaking.

For instance, when facing strong exogenous shocks, the Norwegian social welfare system is likely to have a comparative advantage; a strong social safety net will provide support to workers facing adjustment costs. This corporatist compensation capacity is seen as a key advantage that European countries have in managing the economic disruption that climate reforms will create (Finegan 2019). However, guaranteeing a safe and stable climate will require more: it will require a substantial disruption of the economic status quo. And it will require deep decarbonization. Corporatist systems may lack the institutional capacity to easily deliver such disruptive climate reforms. Instead, there may be a class of public goods that even social welfare states are poorly equipped to deliver to their citizens. The very collaborative aspects of cooperative policymaking, which facilitates the delivery of certain classes of social protections, may also exert bias against non-incremental transformation of carbon-intensive economies. This means that, ironically, the countries that are best able to support the economic displacement that results from deep decarbonization may be most hesitant to disrupt the economic status quo.

Comparative environmental politics scholars have previously framed Europe as a policymaking tortoise and the United States as a policymaking hare (Lundqvist 1980). Under this framework, incremental European environmental policymaking ultimately overperforms more punctuated policymaking efforts in the United States. Yet, it is unclear whether these broad insights will carry over into the realm of climate policymaking. In corporatist policymaking systems, incremental change might be easy, but is transformative change possible? By contrast, in pluralist policymaking systems, transformative change is possible. However, do producer efforts to mobilize political conflict into the public domain undermine the possibility of sustaining a broad-based public coalition in favor of climate reforms? In other words, can Europe decarbonize fast enough? And can the United States start decarbonizing soon enough?

Finally, climate reformers must also consider issues of scale. This book echoes a broader call to focus on "bottom-up" climate policymaking. At the same time, the growing literature on decentralized climate policymaking has tended to focus on regional, state, and city-level policies, rather than policy conflict at the national level. State and local climate policymaking have achieved some tangible gains (Hoffmann 2011). However, most analyses fail to note that the cities or regions that have proposed the most ambitious climate policies are often the jurisdictions in which climate policy is the least costly, and therefore easiest to implement. For instance, in the United States local efforts generally proceed where carbon-dependent economic actors have the weakest preexisting policy influence. In this way, California and the Northeastern states have enacted emissions trading schemes while the most carbon-intensive states in the South and Mountain West actively contest any efforts to impose costs on carbon pollution.

A narrow focus on local climate policymaking risks bifurcated policy outcomes, where climate policies are implemented in the least carbon-intensive contexts while the carbon-intensive areas that need the strongest reforms lag behind. The value of national climate reforms is that they include both distributive winners and losers in their scope. By encompassing the most significant carbon polluters, national policy may be able to impose costs on important sources of carbon pollution that are otherwise protected by regional and local political actors with carbon-dependent constituencies. In short, national policymaking efforts provide the political venue in which the broadest coalition of pro-climate actors can band together to impose the furthest-reaching costs on a policy's economic losers.

All Hands on Deck

In this book, I have advanced a new theory to explain cross-national variation in climate policy timing and content. I have shown how, in most advanced economies and across most time periods, carbon polluter double representation has given industrial capital and labor control over climate policymaking. While individual countries have occasionally implemented serious climate reforms, they remain collectively insufficient to mitigate the serious economic and social risks associated with the climate crisis.

Countries across the world must do better. Ultimately, political responses to the climate crisis need to be judged based on our collective capacity to respond to the risks as defined by climate science. Managing global climate change may necessitate not only quick stabilization of current atmospheric pollution levels, but also a reduction from current carbon pollution levels of about 415 parts per million (ppm) to concentrations as low as 350 ppm (Hansen et al. 2008). Yet, atmospheric greenhouse gas concentrations continue to increase. Moreover, one-third of global oil reserves, one-half of global gas reserves, and nearly four-fifths of global coal reserves must remain in the ground—including 90 percent of Australian and US coal and nearly all of Canadian oil (McGlade and Ekins 2015). Unfortunately, the development, extraction, and consumption of fossil fuel reserves continues unabated.

Can the global community change course to mitigate the most devastating projected climatic impacts? Can global economic institutions be reformed to reflect the true costs associated with releasing carbon pollution into the global atmosphere? Even in the presence of shifting cost structures, there is nothing automatic about the pace and structure of energy transitions (Breetz, Mildenberger, and Stokes 2018; Stokes 2020). Addressing climate change in a timely fashion will require sustained political efforts.

And it will require scholars to center politics in their analysis of climate change. We need to move beyond vague concepts like "political will." Instead, we need to take seriously the political economy of climate reforms. What incentives and institutions structure policymaking debates? What political coalitions can disrupt entrenched incumbents and under what conditions will such political coalitions arise? Policy designers must examine the ways in which short-term economic decisions can reshape the long-term political context in which climate policies are negotiated. Short-term policies should be designed and implemented in a way that opens new strategic opportunities for future, anticipated rounds of climate policymaking in which the ambition of a given policy regime must be increased.

Above all, we need to wrest control of our futures back from the carbon polluters whose actions threaten each of us. This will not be easy. As this book has documented in sobering detail, several generations of policy advocates have failed before us. We will have to succeed where these previous generations fell short. This will involve a wholesale renegotiation of our economic institutions. It will require us to create new economic losers with every incentive to resist reforms. It will stretch the capacity of our political institutions. But we have no choice. To protect our societies from economic and environmental ruin, we must disrupt the double representation of carbon polluters.

Glossary

ACCE	American Coalition for Clean Coal Electricity
ACCI	Australian Chamber of Commerce and Industry
ACES	American Clean Energy Security Act
ACF	Australian Conservation Foundation
ACOSS	Australian Council of Social Services
ACTU	Australian Council of Trade Unions
AFL-CIO	American Federation of Labor and Congress of Industrial Organizations
AFP	Americans for Prosperity
AGO	Australian Greenhouse Office
AIG	Australia Industry Group
AIGN	Australian Industry Greenhouse Network
ALP	Australian Labor Party
AMWU	Australian Manufacturing Workers' Union
API	American Petroleum Institute
ARENA	Australian Renewable Energy Agency
AUD	Australian dollar
AWU	Australian Workers Union
BCA	Business Council of Australia
BCP	British Climate Programme
BCSE	Business Council for a Sustainable Energy
BDI	Bundesverband der Deutschen Industrie (Federation of German Industry)
BEP	Basic Energy Plan
BGA	BlueGreen Alliance
BTU	British thermal unit

CAA	Clean Air Act
CAD	Canadian dollar
CBI	Confederation of British Industries
CCA	Climate Change Authority (Australia)
CCA	Climate Change Agreement (Great Britain)
CCL	Climate Change Levy
CCS	carbon capture and sequestration
CDU	Christlich Demokratische Union Deutschlands (Christian Democratic Union of Germany)
CEFC	Clean Energy Finance Corporation
CEFP	Clean Energy Futures Package
CERES	Coalition for Environmentally Responsible Economies
CFMEU	Construction, Forestry, Mining, and Energy Union
CIO	Council of Industrial Organizations
CO_2	carbon dioxide
COP	conference of parties
CPM	Carbon Pricing Mechanism
CPP	Clean Power Plan
CPRS	Carbon Pollution Reduction Scheme
CSE	Citizens for a Sound Economy
DGB	Deutscher Gewerkschaftsbund (Confederation of German Trade Unions)
DIHK	Deutscher Industrie-und Handelskammertag (Association of German Chambers of Industry and Commerce)
DPJ	Democratic Party of Japan
EA	Environmental Agency of Japan
EAQ	Energy and Air Quality Subcommittee
EDF	Environmental Defense Fund
EEI	Edison Electric Institute
EFTA	European Free Trade Association
EHM	Environment and Hazardous Materials Subcommittee
EITE	emissions-intensive trade-exposed
ENGO	environmental non-governmental organization
EPA	Environmental Protection Agency
EPW	Environment and Public Works
ERF	Emissions Reduction Fund
ETC	Miljøavgiftsutvalget (Environmental Tax Committee)

ETPA	Energy Tax Policy Alliance
ETS	emissions trading scheme
EU	European Union
EU ETS	European Union Emissions Trading System
FDP	Freie Demokratische Partei (Free Democratic Party)
FIT	feed-in tariff
GCC	Global Climate Coalition
GDP	gross domestic product
GGAP	Greenhouse Gas Abatement Program
GGAS	Greenhouse Gas Abatement Scheme
GHG	greenhouse gas
GNP	gross national product
GTC	Grenn Skattekommisjon (Green Tax Commission)
HSH	Handels-og Servicenæringens Hovedorganisasjon (Federation of Norwegian Enterprises)
IBB	International Brotherhood of Boilermakers
IBEW	International Brotherhood of Electrical Workers
IGCC	Investor's Group on Climate Change
IPCC	Intergovernmental Panel on Climate Change
IUC	Industrial Union Council
JCP	Japanese Communist Party
JVETS	Japan Voluntary Emissions Trading Scheme
LDP	Liberal Democratic Party of Japan
LIUNA	Laborers' International Union of North America
LO	Landsorganisasjonen i Norge (Norwegian Confederation of Trade Unions)
METI/MITI	Ministry of Economy Trade and Industry
MOE	Ministry of the Environment
MPCCC	Multi-Party Climate Change Committee
NAF	Norsk Arbeidsgiverforening (Norwegian Employer's Confederation)
NAM	National Association of Manufacturers
NAP	National Allocation Plan
NCEP	National Commission on Energy Policy
NDP	New Democratic Party
NETT	National Emissions Trading Taskforce
NFIB	National Federation of Independent Business

NGO	non-governmental organization
NGRS	National Greenhouse Response Strategy
NHO	Næringslivets Hovedorganisasjon (Confederation of Norwegian Enterprise)
NOK	Norwegian krone
NRDC	Natural Resource Defense Council
NSR	New Source Review
OECD	Organisation for Economic Co-operation and Development
OMB	Office of Management and Budget
PMTGET	Prime Ministerial Task Group on Emissions Trading
PPM	parts per million
PSD	prevention of significant deterioration
PUP	Palmer United Party
QDR	qualitative data repository
R&D	research and development
RGGI	Regional Greenhouse Gas Initiative
SCCC	Southern Cross Climate Coalition
SEIU	Service Employees International Union
SFT	Statens Forurensningstilsyn (Norwegian Pollution Control Authority)
SPD	Sozialdemokratische Partei Deutschlands (Social Democratic Party of Germany)
TRAX	transparency appendix
UAW	United Automobile Workers
UMWA	United Mineworkers of America
UNIO	Utdanningsgruppenes Hovedorganisasjon (Confederation of Unions for Professionals)
USCAP	United States Climate Action Partnership
USD	US dollar
USDA	US Department of Agriculture
VAT	value-added tax
WRI	World Resources Institute

Notes

1   The Puzzle of Climate Policy Action

This book makes its sources as transparent as possible by providing a Transparency Appendix (TRAX) that follows American Political Science Association guidelines. This TRAX is available online at https://doi.org/10.5064/F6GYLSON and on the author's personal website at http://www.mattomildenberger.com. The TRAX offers more detail on the book's methodological approach and offers additional details on the book's sources. This includes expansions of many of this text's endnotes. Where chapter endnotes include an asterisk (*), readers can consult the TRAX for additional data or commentary in support of my claims. I also make sources referenced in these notes public, where possible, by depositing archival, media, and web materials in a Syracuse University Qualitative Data Repository (QDR), also available online at https://doi .org/10.5064/F6GYLSON.

1. Sarah Tincher, "14 Union Representatives Arrested at UMWA Anti-Clean Power Plan Rally in Pittsburgh," *The State Journal*, July 31, 2014; Kris Maher, "EPA Emissions Plan Draws Protest in Pittsburgh," *Wall Street Journal*, July 31, 2014.

2. EPA, Transcript of Public Hearing on Proposed Clean Power Plan for Existing Power Plants, 9:00 am to 5:36 pm, Thursday July 31, 2014, Room B. William S. Moorhead Building, Pittsburgh, PA.

3. Ibid.

4. US National Oceanic and Atmospheric Administration, "Trends in Atmospheric Carbon Dioxide," updated daily, https://www.esrl.noaa.gov/gmd/ccgg/trends/ monthly.html, accessed December 5, 2018.

5. Evan Lehmann. "Slow Stirrings among Conservatives on Adaptation—Just Don't Mention Climate Change," *New York Times*, July 28, 2011.

6. *Throughout this text, costs are provided as historical prices.

7. Other scholars propose even more disaggregated carbon-price design typologies (e.g., Gulbrandsen et al. 2017).

8. This figure uses data from the World Bank's Carbon Pricing Dashboard, http://carbonpricingdashboard.worldbank.org/.

9. A range of definitions for "corporatism" exist. Here, I focus on corporatism as a form of interest group intermediation. Smith (1993) reviews other approaches.

10. *Chicago Policy Review*, "Policy and Politics: A Candid Conversation with Tim Phillips," April 21, 2015, http://chicagopolicyreview.org/2015/04/21/policy-and-politics-a-candid-conversation-with-tim-phillips/, accessed June 4, 2019.

11. This appendix can be found online at https://doi.org/10.5064/F6GYLSON.

2 The Logic of Double Representation

1. Of course, carbon pollution often co-occurred with more visible forms of air pollution, such as the nitrous and sulfur oxides. Efforts to regulate these other air pollutants indirectly structured patterns of domestic carbon pollution.

2. I present this theory chapter up front so that interested readers can evaluate the theory's merits themselves while reading the case studies that follow. However, chapters 3–6 are not empirical tests of this chapter's claims; instead, this chapter consolidates these later chapters' within-case qualitative causal inferences. By contrast, chapter 7 probes the theory's generalizability using empirical analysis of four additional advanced economies—Canada, Japan, the United Kingdom, and Germany.

3. Mildenberger et al. (2017) offer evidence for persistent pro-climate beliefs among many Republicans.

4. Interview 31 with senior policy advisor to political party, Oslo.

5. It is difficult to identify episodes when LO acted against its industrial membership's preferences (Reitan 1998; Kasa 2000); Interview 21 with elected official, Oslo.

6. *Interview 22 with senior union official, Oslo. LO member unions that support ambitious climate policies have often been marginalized.

7. *Peak associations occasionally enjoy direct control over the political agenda. Interview 10 with senior business lobby official, Oslo.

8. *Interview 22 with senior union official, Oslo.

9. *Interview 72 with senior union official, Sydney.

10. Interview 66 with senior business lobby official, Melbourne.

11. Eric Lipton, "Energy Firms in Secretive Alliance with Attorneys General," *New York Times*, December 6, 2014.

12. UNFCC, "GHG Emissions Profile for Australia," https://unfccc.int/ghg_data/ghg_data_unfccc/ghg_profiles/items/4625.php, accessed March 14, 2014.

3   Climate Policy Cooperation in Norway

1. US EIA, "Norway," December 28, 2016, http://www.eia.gov/countries/cab.cfm?fips=no, accessed January 8, 2018.

2. *Interview 13 with senior policy advisor to political party, Oslo.

3. *These include UNIO, Akademikerne, and YS.

4. Virke was previously the Handels-og Servicenæringens Hovedorganisasjon (HSH).

5. Interview 32 with senior policy advisor to political official, Oslo.

6. *The ETC deepened a Norwegian turn to market-based policymaking instruments and growing influence of neoliberal economics on Norwegian economic policymaking (Kasa 2000).

7. Interview 14 with senior bureaucrat, Oslo.

8. *For example, natural gas consumed offshore was taxable, but natural gas consumed onshore—including by onshore industries—was exempted. Interview 14 with senior bureaucrat, Oslo.

9. *Despite Brundtland's green credentials, climate issues do not factor strongly in her 2002 political memoirs.

10. Marginal tax rates in the offshore sector are as high as 78 percent, so the carbon tax was a minor cost that could be deducted from corporate profits. Interview 4 with senior business lobby official, Oslo; Interview 14 with senior bureaucrat, Oslo; Interview 40 with senior union official, Oslo.

11. *While the tax sparked some efficiency gains, it did not shift oil and gas extraction patterns. Interview 14 with senior bureaucrat, Oslo; Interview 40 with senior union official, Oslo.

12. *By 1995, the country's carbon tax regime brought in $930 million US in revenue (Godal and Holtsmark 2001).

13. "Dette er en blå skatt, Gro," *Dagbladet*, June 7, 1996. As cited by Reitan (1998).

14. Interview 79 with mid-level bureaucrat, Oslo.

15. *Norwegian Official Report (*Norges Offentlige Utredninger*) (1996), 9. *Grønne skatter—en politikk for bedre miljø og hoy sysselsetting.*

16. Interview 27 with mid-level bureaucrat, Oslo.

17. Ibid.

18. Bernt Lund, "Grønne skatter—et klimapolitisk virkemiddel." Commentary published online by Besteforeldre mot Global Oppvarming (Grandparents Against Global Warming), http://www.besteforeldreaksjonen.no/?p=1391, accessed November 30, 2014.

19. *Tax rates would increase over time as other countries took domestic policy actions. Interview 27 with mid-level bureaucrat, Oslo.

20. *Potential distributional losses were corroborated by subsequent economic analyses (Godal and Holtsmark 2001).

21. Lund, "Grønne skatter."

22. Bernt Lund, "Grønne h °ar i suppa," *Natur & Miljø Bulletin*, Naturvernforbundet, Oslo.

23. Bureaucratic committee members were listed on the final report as representatives of their hometowns, not of their government departments. In Norway, this signaled an intention that bureaucrats participate as independent citizens not government agents. Interview 19 with academic, Oslo; Interview 79 with mid-level bureaucrat, Oslo.

24. Interview 19 with academic, Oslo; Interview 27 with mid-level bureaucrat, Oslo.

25. *Brundtland and Stoltenberg also pressured Lund to delay the GTC report's release until after a critical parliamentary vote on gas-fired power plants, so that the report would not cloud estimates of those plants' long-term viability. Lund refused his copartisan's request as a matter of principle, and released the GTC report on schedule. Lund, "Grønne skatter."

26. Interview 27 with mid-level bureaucrat, Oslo.

27. *Interview 5 with elected official, Oslo; Interview 13 with senior policy advisor to political party, Oslo; Interview 16 with senior religious official, Oslo.

28. *This doubled the GTC's proposed tax rate.

29. For example, the policy's parliamentary proposal described the tax as compromising the profitability of future gas-fired power plants. Norwegian Ministry of Finance, Storting Proposition no. 54 (1997–98); Lund, "Grønne skatter."

30. Compensation was also offered to fishing fleets and domestic aviation.

31. Interview 36 with senior union official, Oslo.

32. Ibid.

33. Ibid.

34. Ibid.

35. Unlike the GTC, industrial labor and capital were only represented indirectly in the Quota Commission, through a secondary advisory group (Moe 2010).

36. Norwegian Official Report [*Norges Offentlige Utredninger*], "Et kvotesystem for klimagasser," *A Report of the Quota Commission* (2000): 1.

37. Commission economists wanted everyone to pay full permit costs while the Statoil representative wanted free allowances. Bureaucratic members again abstained because the decision was "political."

38. Naturkraft, "Om gasskraftverk," http://www.naturkraft.no/ om-gasskraftverk/, accessed September 25, 2014; United Nations Climate Change Secretariat, "Summary of GHG Emissions for Norway," archived at: https://web.archive.org/web/20101007141037/http://www.naturkraft.no/om-gasskraftverk.

39. No author. "1.000 vil bruke sivil ulydighet mot gasskraftverk," *Aftenposten*, May 4, 1997; Interview 12 with senior environmental leader, Oslo.

40. *No author. "Berntsen overrasket over tallene," *Aftenposten*, May 9, 1997; Rosaleen Østorm and Jan Gunnar, "Kirken med bønn mot gasskraftverk," *Aftenposten*, April 27, 1997; Interview 12 with senior environmental leader, Oslo; Interview 16 with senior religious official, Oslo.

41. Jan Gunnar and Ole Nygaard, "Ikke flertall for gasskraftverk," *Aftenposten*, May 9, 1997.

42. No author. "Skjerpet gasskraftstrid i LO," *Aftenposten*, April 30, 1997. Industrial union leaders emphasized the need for new electricity production to enhance worker welfare.

43. LO, Minutes from the 29th Congress [Referat fra den 29 ordinære kongress, Kongressens vedtak 1997], Oslo, May 10–16, 1997. In particular, see 79, 81, 82, 85, 98, 142, 149, 174, 175, 206, and 417.

44. Ibid., 68.

45. Gunnar Magnus, "Jagland: Gasskraftverket bygges uansett," *Aftenposten*, May 12, 1997; Gunnar Magnus, Roar Østyård Gjelta, Kjersti Løken, and Lars Sæthere, "Håper på ny teknologi," *Aftenposten*, May 11, 1997.

46. Norwegian Ministry of Oil and Energy, "Naturkraft askonsesjon for bygging og drift av gasskraftverk på Kårstø og kollsnes—handsaming av klagesaka," press release, June 5, 1997.

47. *The Pollution Control Act, passed in 1981, flexibly incorporates new pollutants. Interview 30 with mid-level business lobby official, Oslo.

48. It also, in turn, suggested that any regulatory decision to restrict carbon pollution was at the discretion of political officials. Halvor Tjønn and Sveinung Berg Bentzrød, "SFT kan ikke stoppe gasskraftverk," *Aftenposten*, June 6, 1997.

49. Ibid.

50. Jagland was keen to avoid upsetting the party's pro-climate youth wing (Tjernshaugen and Langhelle 2009). Interview 12 with senior environmental leader, Oslo.

51. The permit allowed emissions to be balanced with carbon credits in lieu of CCS. Environmental groups lobbied to exclude this credit option. Bondevik government officials may have agreed to strip this provision but the government fell before a formal decision was made. Interview 12 with senior environmental leader, Oslo.

52. The Conservatives held significantly more seats than the Christian Democrats; however, unusually, the Conservatives conceded the prime ministership to Bondevik's Christian Democrats during coalition bargaining.

53. *Christian Democrats also controlled the Ministry of Oil and Energy, what Bondevik saw as the "second Ministry of the Environment." Interview 12 with senior environmental leader, Oslo.

54. *The second Bondevik government's record on gas plant construction was mixed. Despite promising no additional concessions, Bondevik approved a gas plant at Snøvit to support offshore oil development.

55. Interview 32 with senior policy advisor to political official, Oslo.

56. Ibid.

57. Ibid.

58. Ibid.

59. *Ibid.; Interview 37 with senior business lobby official, Oslo.

60. Interview 32 with senior policy advisor to political official, Oslo.

61. Interview 36 with senior union official, Oslo.

62. As a senior business official lamented, "our enemies—industry's enemies, the economists—they established a commission to work out a framework for emissions trading and that committee was dominated by our enemies from the University of Oslo." Interview 37 with senior business lobby official, Oslo.

63. *Interview 32 with senior policy advisor to political official, Oslo.

64. Ibid.

65. Interview 30 with mid-level business lobby official, Oslo; Interview 32 with senior policy advisor to political official, Oslo.

66. Interview 41 with senior policy advisor to political party, Oslo.

67. Interview 22 with senior union official, Oslo.

68. Interview 4 with senior business lobby official, Oslo.

69. Interview 10 with senior business lobby official, Oslo.

70. Interview 19 with academic, Oslo; Interview 38 with senior policy advisor to political party, Oslo.

71. Norway negotiated its entry into the emissions trading scheme alongside Iceland and Liechtenstein through an amendment to the Agreement on the European Economic Area. Talks continued until October 2007.

72. Interview 37 with senior business lobby official, Oslo.

73. Norwegian firms could buy credits from Europe; however, the European scheme did not allow the reciprocal purchase of Norwegian credits. Interview 4 with senior business lobby official, Oslo. Interview 8 with two senior bureaucrats, Oslo.

74. Interview 14 with senior bureaucrat, Oslo; Interview 27 with mid-level bureaucrat, Oslo.

75. Interview 4 with senior business lobby official, Oslo.

76. This agreement proceeded quickly because the proposed obligations were easily met through adoption of a new technology. Interview 37 with senior business lobby official, Oslo.

77. Interview 37 with senior business lobby official, Oslo.

78. *Interview 4 with senior business lobby official, Oslo.

79. *Interview 41 with senior policy advisor to political party, Oslo.

80. Interview 4 with senior business lobby official, Oslo.

81. *As Norway shifted to an ETS system, its ability to use the Pollution Control Act to directly regulate carbon pollution was partly preempted.

82. Under revised EU ETS Phase II sectoral coverage rules, a small number of process industries remained exempt. Norsk Industri negotiated a second voluntary agreement in 2009 for these exempt industries. However, the agreement was unambitious and was viewed, even by participants, as symbolic cover for secular industry trends. Interview 4 with senior business lobby official, Oslo. Interview 41 with senior policy advisor to political party, Oslo.

83. *Government of Norway, "Norwegian National Allocation Plan for the Emissions Trading System in 2008–2012," Technical Paper, March 2008.

84. *Interview 10 with senior business lobby official. EFTA Surveillance Authority Decision of 16 July 2008, Case 62345, Event No 483630, Dec. No. 504/08/Col.

85. *Pro-climate bureaucrats attempted to use technical rule-making processes to stretch policy ambition.

86. Interview 8 with two senior bureaucrats, Oslo; Letter from Norwegian Ministry of the Environment to EFTA Surveillance Authority, May 30, 2008, "Response from the Norwegian Government on the EFTA Surveillance Authority's request for additional information on the Norwegian National Allocation Plan for greenhouse gas allowances in the 2008–12 trading period"; Interview 26 with mid-level bureaucrat, Oslo.

87. *Under Prop. 68 (2011–12), Norway changed its domestic policy to comply with EU Directive 2009/29/EC, 23 April 2009.

88. Norwegian Klima og-Miljødepartementet, "Endringer i klimakvoteloven." http://www.regjeringen.no/nb/dep/kld/dok/regpubl/prop/2011-2012/prop-68-l-20112012/3.html?id=674350, accessed November 25, 2014.

89. *Interview 23 with mid-level business lobby official, Oslo.

90. Interview 39 with senior policy advisor to political party, Oslo.

91. Interview 23 with mid-level business lobby official, Oslo.

92. Interview 41 with senior policy advisor to political party, Oslo.

93. Interview 37 with senior business lobby official, Oslo.

94. Interview 21 with elected official, Oslo.

95. *Interview 25 with two senior union officials, Oslo; Interview 41 with senior policy advisor to political party, Oslo.

96. Interview 21 with elected official, Oslo; Interview 39 with senior policy advisor to political party, Oslo; Interview 41 with senior policy advisor to political party, Oslo; Thomson Reuters, "Norway Backs Off Lofoten Drilling, Eyes Barents," March 11, 2011.

97. Mikael Holter, "Norway Labor Party Approves to Study Lofoten Oil Exploration," *Bloomberg News*, April 21, 2013.

98. *Interview 22 with senior union official, Oslo; Interview 25 with two senior union officials, Oslo.

99. Interview 23 with mid-level business lobby official, Oslo.

100. These ranged from generous depreciation regimes for offshore oil platforms to tax deferrals until companies profited in mature fields. Interview 23 with mid-level business lobby official, Oslo; Norwegian Ministry of Oil and Energy, *An Industry for*

the Future—Norway's Petroleum Activities, White Paper to the Norwegian Parliament, *Melding til Stortinget,* no. 28 (2010–2011), June 24, 2011.

101. *Interview 19 with academic, Oslo. Successive Norwegian governments have been unwilling to consider serious restrictions on offshore oil and gas extraction.

102. *A discussion paper from Statistics Norway (*Statistisk sentralbyrå*) suggesting that it would be efficient to substitute demand-side emissions reductions with supply-side reductions (e.g., avoided oil extractions) ignited controversy throughout the political system. It was rejected by all but the Norwegian Green and Socialist Left parties. *Climate Policies in a Fossil Fuel Producing Country: Demand Versus Supply Side Policies,* Discussion Paper 747, Oslo: Statistisk Sentralbyrå, June 2013; Interview 6 with elected official, Oslo.

103. As background to this review, the government invited Sweden to peer review Norwegian climate policy in Spring 2007. Part of a broader EU program, the review was largely favorable to Norway. Interview 35 with senior union official, Oslo; Interview 41 with senior policy advisor to political party, Oslo.

104. Interview 39 with senior policy advisor to political party, Oslo. There was also debate over how the government should measure progress against its reference pathway target.

105. Under the Kyoto Protocol, Norway was permitted an increase in net emissions intensity, because the country's carbon-free electricity sector provided fewer low-cost opportunities for substantial emissions reductions. It was believed that the carbon price necessary to stabilize Norwegian emissions would be too high relative to the marginal price of emissions reductions elsewhere in the OECD. Interview 35 with senior union official, Oslo.

106. *The government argued that "it is not possible at present to obtain reliable figures for the emission reductions that will be achieved through existing and new measures in the next 10–15 years. Nor do we know which emission reduction measures will be implemented in Norway in the period up to 2020." Report No. 34 to the Storting (2006–2007), *Norwegian Climate Policy,* June 22, 2007, 5.

107. Press release from Conservative, Christian Democrat, and Venstre parties, "Vedlagt klimapolitisk plattform," November 8, 2007.

108. In this proposal, 1990 levels are the baseline year, eliminating the government's reference curve ambiguity.

109. Interview 13 with senior policy advisor to political party, Oslo.

110. Interview 21 with elected official, Oslo. They were particularly concerned about ambiguous targets, believing that commitments measured against a reference curve baked in pollution expansion over time.

111. Interview 41 with senior policy advisor to political party, Oslo.

112. Interview 38 with senior policy advisor to political party, Oslo.

113. Interview 12 with senior environmental leader, Oslo; Interview 21 with elected official, Oslo.

114. Interview 38 with senior policy advisor to political party, Oslo.

115. Ibid.

116. Interview 39 with senior policy advisor to political party, Oslo; Interview 25 with two senior union officials, Oslo.

117. CO_2eq is a greenhouse gas accounting unit that transforms different greenhouse gases into CO_2 as a function of their relative contribution to global warming.

118. This worked out to approximately 12 percent below 1990 levels when converted into a less ambiguous absolute target.

119. Labor, Socialist Left, Conservatives, Christian Democrats, Liberals and Centers. *Avtale om klimameldingen* [Agreement on the Climate Report], January 17, 2008. https://www.regjeringen.no/contentassets/fbe5a5829a5d468fab6e4eec0a39512d/avtale_klimameldingen_2008_01_17.pdf, accessed June 16, 2019.

120. Interview 24 with energy consultant, Oslo; Interview 19 with academic, Oslo.

121. Interview 12 with senior environmental leader, Oslo. However, the Christian Democrats also take credit. Interview 13 with senior policy advisor to political party, Oslo; Interview 38 with senior policy advisor to political party, Oslo.

122. The size of the initiative was derived by calculating Norway's share of the *Stern Report*'s estimated cost of saving global rain forests. Interview 12 with senior environmental leader.

123. Norwegian Climate and Pollution Agency, *Climate Cure 2020: Measures and Instruments for Achieving Norwegian Climate Goals by 2020*, TA 2678/2000, June 2010.

124. The costs associated with these measures were variable. The report estimated that twelve million tons of CO_2eq by 2020 could be achieved at cost levels up to 1100 NOK ($149.61 USD) per ton, with forestry measures providing another three million tons reduction to meet the goal of fifteen million tons.

125. Interview 41 with senior policy advisor to political party, Oslo; Norwegian Climate and Pollution Agency, *Climate Cure 2020*.

126. Interview 4 with senior business lobby official, Oslo. Some actors accused the government of significantly understating the costs of compliance by hiding high marginal costs toward the end of the compliance period with the report's emphasis on average costs. Interview 35 with senior union official, Oslo.

127. Confederation of Norwegian Enterprise (NHO), *Industry's Climate Action Plan: Norwegian Climate Policy—Time for Action*, December 2009.

128. *For instance, industry recommended enhanced efforts to improve offshore energy efficiency but did not recommend electrification of offshore installations.

129. *Domestic investment was limited to suggestions that existing carbon tax revenues be channeled into a Climate Fund that subsidized and promoted sectoral energy and climate technologies. NHO, *Industry's Climate Action Plan*, 25.

130. Interview 10 with senior business lobby official, Oslo.

131. Ibid.; Interview 34 with mid-level bureaucrat, Oslo.

132. Office of the Prime Minister of Norway, "New Policy Platform for the Red–Green Coalition Government," Press Release No. 156/09, October 13, 2009.

133. This was delayed because the 2011 Oslo terror attacks significantly disrupted the government's policymaking agenda. Interview 39 with senior policy advisor to political party, Oslo.

134. Norwegian Government Press Release No. 53/2012, "Offensiv klimamelding," April 25, 2012.

135. Interview 41 with senior policy advisor to political party, Oslo.

136. Ibid. Senior Progress Party members had voiced skepticism about climate risks throughout this period.

137. *Interview 39 with senior policy advisor to political party, Oslo.

138. Interview 10 with senior business lobby official, Oslo; Interview 12 with senior environmental leader, Oslo; Interview 39 with senior policy advisor to political party, Oslo. In contrast to industry, Norwegian environmentalists were mostly unconcerned about this change.

139. Interview 35 with senior union official, Oslo.

140. Through the state-owned organization Transnova, Norway aggressively invested in alternative fuels and electric charging stations across the country. Interview 18 with mid-level bureaucrat, Oslo.

141. Interview 18 with mid-level bureaucrat, Oslo.

142. *Interview 28 with two senior environmental officials, Oslo; Interview 41 with senior policy advisor to political party, Oslo.

143. Interview 12 with senior environmental leader, Oslo.

144. LO and Arbeiderpartiet, *Ta naturgassen i bruk!* [Use the natural gas!]. Report of the Henriksen Committee, July 2001.

145. *These efforts had mixed outcomes. New gas-based industries have not gained traction. Interview 31 with senior policy advisor to political party, Oslo. Even the gas-fired power plants that Stoltenberg assertively pushed through government after Bondevik's controversial attempt to block them have proven uneconomical.

146. Interview 31 with senior policy advisor to political party, Oslo.

147. Ibid.

148. Prime Minister Jens Stoltenberg's New Year's Address, January 1, 2007.

149. *Riksrevisjonen, Riksrevisjonens undersøkelse av statens arbeid med CO_2-håndtering, Dokument 3, no. 14 (2012–2013), September 17, 2013.

150. Interview 35 with senior union official, Oslo.

151. Interview 24 with energy consultant, Oslo.

152. Interview 19 with academic, Oslo.

153. Thema Consulting Group, "Future at Peril: Power Intensive Industry & EU ETS," TCG Insight Nr 1–2012, Oslo. Norwegian prices will be particularly affected by shifts in the marginal prices of coal generation in Europe.

154. *Art. 10a para. 6 of Directive 2009/29/EC.

155. Interview 22 with senior union official, Oslo; Interview 4 with senior business lobby official, Oslo; Interview 35 with senior union official, Oslo.

156. Rolleiv Solheim, "Hydro Threatening to Flag Out," *Norway Post,* June 7, 2012.

157. Interview 15 with two-mid level union policy advisors, Oslo; Norwegian Klima og-Miljødepartementet, "Endringer i klimakvoteloven."

158. Interview 38 with senior policy advisor to political party, Oslo; Interview 41 with senior policy advisor to political party.

159. Interview 31 with senior policy advisor to political party, Oslo; Interview 26 with mid-level bureaucrat, Oslo.

160. Interview 38 with senior policy advisor to political party, Oslo.

161. Interview 41 with senior policy advisor to political party, Oslo.

162. Interview 13 with senior policy advisor to political party, Oslo.

163. Interview 28 with two senior environmental officials, Oslo; Interview 31 with senior policy advisor to political party, Oslo.

164. "We fear now that the financial ministry has got a foot into the compensation scheme." Interview 10 with senior business lobby official, Oslo.

165. Interview 22 with senior union official, Oslo; Interview 35 with senior official, Oslo.

166. Interview 40 with senior union official, Oslo.

167. Interview 41 with senior policy advisor to political party, Oslo; Interview 4 with senior business lobby official, Oslo; Interview 22 with senior union official, Oslo.

168. Interview 40 with senior union official, Oslo.

169. Government of Norway, "Mandate for a New Green Tax Commission," 2014, https://www.regjeringen.no/en/whatsnew/Ministries/fin/press-releases/2014/Ny-gronn-skattekommisjon/Mandate-for-a-new-green-tax-commission/id764701/, accessed June 15, 2019.

170. GHG Emissions Profile for Norway, https://unfccc.int/ghg_data/ghg_data_unfccc/ghg_profiles/items/4625.php, accessed March 14, 2014.

171. Interview 5 with elected official, Oslo; Interview 14 with senior bureaucrat, Oslo; Interview 15 with two mid-level union policy advisors, Oslo.

172. Interview 4 with senior business lobby official, Oslo.

173. Interview 24 with energy consultant, Oslo.

174. Interview 19 with academic, Oslo; Interview 35 with senior union official, Oslo.

4   US Climate Policy Inaction, 1988–2006

1. European Parliament Resolution RC-B5–0267/2001, "European Parliament Resolution on the Kyoto Conference Objectives," April 4, 2001.

2. Tim Dickinson, "Inside the Koch Brothers' Toxic Empire," *Rolling Stone*, September 24, 2014.

3. IUC member unions included UMWA, the UAW, the Boilermakers (IBB), and the Electrical Workers (IBEW).

4. Already in 1956, oceanographer Roger Revelle warned members of Congress about global warming during congressional testimony: Testimony of Roger Revelle, US Congress, House 84 H1526–5, Committee on Appropriations, Hearings on Second Supplemental Appropriation Bill (1956).

5. *Climate change received minimal attention under the Johnson, Nixon, and Carter administrations.

6. *Before Nixon, climate research had been spread across government branches.

7. *US Environmental Protection Agency, *Can We Delay a Greenhouse Warming?*, September 1983. The EPA report contained some of the first analyses of US climate policy options. Exploring the value of carbon taxes and command and control regulations, it concluded the United States had few policy levers to delay climate change's onset; EPA carbon taxes as high as 300 percent of market prices would only have a marginal effect on climate outcomes in the absence of a global climate agreement. Instead, "economically unfeasible" bans on the use of coal by 2000 or shale oil would be more effective.

8. Phillip Shabecoff, "Global Warming Has Begun, Expert Tells Senate," *New York Times*, June 24, 1988.

9. *These legislative proposals varied in their detail and policy content.

10. The Reagan administration agreed to the IPCC's creation because it believed the organization would mute the voices of the most radical climate scientists (Weart 2008).

11. George H. W. Bush, campaign speech delivered August 31, 1988 in Erie Metropark, Michigan.

12. Industrial efforts to preempt climate policy began as early as 1957, when a Shell International Chemical Company representative ridiculed the idea that "our furnaces and motor car engines will have any large effect on the CO_2 balance." As cited in Weart 2008.

13. Senior GCC leaders censored climate science recommendations from in-house scientists before distributing skeptical materials to political actors. Andrew Revkin, "Industry Ignored Its Scientists on Climate," *The New York Times*, April 23, 2009.

14. *Just three days after being sworn in, Secretary of State James Baker made climate action the subject of his first public address.

15. James Scheuer, "Bush's 'Whitewash Effect' on Warming," *New York Times*, March 3, 1990.

16. *Sununu purportedly believed that acid rain legislation would "shut down the US economy." Bush overruled his advice (Pooley 2010).

17. *Bush, campaign speech, August 31, 1988.

18. *The administration's pro-climate voices were marginalized. For example, pro-climate Secretary of State Baker was pressured to recuse himself from climate decision-making because, ironically, his oil stocks allegedly posed a conflict of interest. Leslie Gelb, "Sununu vs. Scientists," *New York Times*, February 10, 1991.

19. Philip Shabecoff, "US, in a Shift, Seeks Treaty on Global Warming," *New York Times*, May 12, 1989.

20. Paul Montgomery, "U.S., Japan and Soviets Prevent Accord to Limit Carbon Dioxide," *New York Times*, November 8, 1989.

21. Allan Gold, "Bush Administration Is Divided over Move to Halt Global Warming," *New York Times*, October 27, 1989.

22. *CQ Almanac, "Congress Pressures Bush to Attend Rio Summit," *Congressional Quarterly* 48 (1982): 285–286.

23. *Interview 95 with elected official, Washington, DC.

24. Bush-Quayle Campaign, "Bill Clinton Will Raise Your Taxes," Backgrounder: Issues Office, October 22, 1992.

25. Elizabeth Kolbert, "For the Most Negative Ads, Turn on the Nearest Radio," *New York Times*, October 30, 1992.

26. Interview 93 with environmental leader, by phone. For instance, through Gore green groups became involved in the Clinton administration's transition planning. Thomas Lippman, "Energy Tax Proposal Has 'Green' Tint," *Washington Post*, March 2, 1993.

27. S. 201 (102nd Congress), World Environmental Policy Act of 1991.

28. Co-sponsored by seventeen Democrats and Republican Jim Jeffords (VT), the bill also allowed citizen lawsuits against government officials who failed to comply with climate regulations. S. 2668 (102nd Congress).

29. *He also suggested revenues be recycled to low-income Americans and technological subsidies (Gore 1993).

30. Likewise, the labor movement was not yet seriously engaged; only the UMWA began tracking the issue in the early 1990s. Interview 84 with senior union official, Washington, DC.

31. Rabe describes reservations within the Clinton campaign about Gore's ambitious climate proposals (Rabe 2018, ch. 3, n17).

32. Keith Schneider, "Gore Meets Resistance in Effort for Steps on Global Warming," *New York Times*, April 19, 1993.

33. Ibid.

34. Interview 94 with senior policy advisor to elected official, by phone.

35. Clinton, "Remarks on Earth Day," April 21, 1993.

36. Gore, along with Representative Thomas Downey, Democrat, had lobbied for energy and environmental tax integration into 1991 tax reform efforts.

37. The initial proposal set a $0.342 per million BTUs as its base rate, with a supplemental rate of $0.599 per million BTUs for refined petroleum products. The tax was to be phased in over three years.

38. *The team was particularly concerned about the distributive impacts of these different policy instruments. Memorandum from Bill Pitts, Republican Leader Group re: Energy Taxes, January 25, 1993; Memorandum to Republican Members Re: Talking Points on Clinton Tax Proposals, Bill Archer, Ranking Republican Members, Committee on Ways and Means, May 5, 1993. Robert H. Michel Collection, Karen Buttaro Files Box 15 (hereafter KB15), Dirksen Congressional Research Center, Pekin, IL.

39. William Jefferson Clinton, "Address before a Joint Session of Congress on Administration Goals," February 17, 1993.

40. *Interview 94 with senior policy advisor to elected official, by phone.

41. White House Briefing Paper, "Broad-based Energy Tax," February 2, 1993, Robert H. Michel Collection, Karen Buttaro Files Box 2 (hereafter KB2).

42. This compromised the potential climate benefits of the tax. For example, taxing nuclear and hydroelectric sources increased (not decreased) the costs of low carbon energy.

43. Office of Tax Policy, Department of the Treasury, "The Administration's Modified BTU Energy Tax Proposal," April 1993, Robert H. Michel Collection, KB2.

44. Interview 94 with senior policy advisor to elected official, by phone.

45. Interview 93 with US environmental leader, by phone.

46. Interview 84 with senior union official, Washington, DC.

47. Steven Pearlstein and Thomas Lippman, "Industry Analysts See Broad-Based Energy Tax in Clinton's Future," *Washington Post*, January 1, 1993.

48. Note from Jeff White, Alliance Against a Carbon Tax on Friday 8 January [1993] Meeting; Memorandum from Jeff White, Alliance Against a Carbon Tax, February 11, 1993, Congressman Philip R. Sharp Papers, 1970–1994, Ball State University Archives, Muncie, IN.

49. *Thomas Hayes, "Bracing for Shock of Energy Tax," *New York Times*, February 26, 1993.

50. See, for instance, the testimony of Jerry Jasinowski, National Association of Manufacturers on the Comparative Merits of the Administration's Energy Tax Proposal and a Broad-Based National Consumption Tax before the Committee on Energy and Natural Resources, United States Senate, February 25, 1993, Robert H. Michel Collection, KB2.

51. Peter Stone, "The Cultivation of Grass Roots," *The Plain Dealer*, April 11, 1993; Martin Kasindorf, "Stirring Up the Grassroots for Industry," Newsday, March 8, 1993.

52. Joel Brinkley, "Cultivating the Grass Roots to Reap Legislative Benefits," New York Times, November 1, 1993.

53. Ibid.

54. Statement of the American Farm Bureau Federation, May 5, 1993, Robert H. Michel Collection, KB2.

55. TED Case Studies, "US BTU Tax," Trade and Environment Database. Archived at: https://web.archive.org/web/20090110191307/http://www.american.edu:80/TED/USBTUTAX.HTM.

56. *Affordable Energy Alliance, BTU Briefing Book, May 14, 1993, Robert H. Michel Collection, KB2.

57. Faxed memorandum from Illinois Farm Bureau to office of Congressman Robert Michel, May 24, 1993, Robert H. Michel Collection, KB15.

58. Memorandum from Meredith Kincaid to Indiana Congressional Delegation re: Energy Tax Impact on Farmers, May 24, 1993, Congressman Philip R. Sharp Papers, 1970–1994.

59. *Michael Duffy, "I Hear You, I Hear You," *TIME*, June 21, 1993. CSE's role in nurturing opposition to the BTU tax has more recently been interpreted as one of the organization's major political victories. Matthew Continetti, "The Paranoid Style in Liberal Politics," *The Weekly Standard*, April 4, 2011.

60. The Business Council's Views on the President's Economic Package, Remarks by Gerald Alderson before the Committee on Ways and Means, US House of Representatives, March 23, 1993, Robert H. Michel Collection, KB2.

61. *Rick Wartzman, "Treasury Plans a Campaign for Energy Tax," *Wall Street Journal*, March 15, 1993.

62. Alliance to Save Energy, "Savings from Low-Cost Energy Efficiency Measures Can Offset the BTU Tax for Average Households," Congressman Philip R. Sharp Papers, 1970–1994.

63. Memorandum from Jeff White, Alliance Against a Carbon Tax, February 11, 1993, Congressman Philip R. Sharp Papers, 1970–1994.

64. US Senate Republican Policy Committee, "Clinton's Energy Tax: What Every Taxpayer Should Know," May 14, 1993, Robert H. Michel Collection, KB2.

65. Ibid. They cited costs to the US GDP of over $38 billion and 600,000 jobs. This would cost the average family $450 annually; House Republican Policy Committee,

Statement of Republican Policy on the BTU Tax, April 29, 1993, Robert H. Michel Collection, KB2.

66. *Memorandum from Bill Archer to Ranking Republican Members on Committee on Ways and Means, Re: Talking Points on Clinton Tax Proposals, May 5, 1993, Robert H. Michel Collection, KB15.

67. Brian Cashell, *Macroeconomic Effects of Deficit Reduction: Are Spending Cuts Less Painful than Energy Taxes*, CRS Report 93–589 E, June 16, 1993. Advocates also argued the tax was small relative to oil price volatility. Testimony of Congressman Phil Sharp to Budget Committee of US House of Representatives, March 5, 1993. Congressman Philip R. Sharp Papers, 1970–1994.

68. Letter from Sen. J. Bennett Johnston to Sen. Patrick Moynihan, March 23, 1993, Robert H. Michel Collection, KB2; David Cloud, "Clinton Revises Energy Tax in Bid for Support," *Congressional Quarterly*, April 10, 1993.

69. *Steven Greenhouse, "The White House Struggles to Save Energy Plan," *New York Times*, May 10, 1993.

70. *Cloud, "Clinton Revises Energy Tax in Bid for Support."

71. The American Petroleum Institute had lobbied aggressively against this switch. API, "Reject Special Treatment for Ethanol," n.d., Robert H. Michel Collection, KB2.

72. The BTU tax exempted energy consumption in such industries as steel because the energy is converted into the final product, making it a feedstock. This discriminated against aluminum in the initial proposal.

73. While the Clinton administration had systematically resisted this provision, Brewster appeared to have sufficient committee votes to force the change. Jackie Calmes and David Wessel, "Clinton Changes Course on Part of Energy Tax," *Wall Street Journal*, May 11, 1993.

74. American Farm Bureau Federation, "Panels Approve Tax Hike, Agriculture Spending Cuts," *Farm Bureau News* 72 (20), May 17, 1993, Robert H. Michel Collection, KB2.

75. Steven Greenhouse, "Manufacturers and Farmers Oppose Clinton Energy Tax," *New York Times*, May 6, 1993.

76. *Republicans proposed a series of unsuccessful amendments to kill the tax.

77. Duffy, "I Hear You, I Hear You."

78. Boren worked tirelessly to kill the proposal in the Senate. Interview 94 with senior policy advisor to elected official, by phone.

79. Paul Horvitz, "Clinton Retreats on Energy Tax in Fight over Budget," *New York Times*, June 9, 1993. The only remnant of the proposal was a 4.3 percent increase in

the Transportation Fuels Tax on gas, diesel, and certain motor fuels. Even this policy had many sectoral exemptions.

80. Vice President Gore cast the deciding vote in favor of a plan that had abandoned any trace of his signature environmental tax proposal.

81. Interview 93 with US environmental leader, by phone.

82. *These included nearly $2 billion to support technology commercialization. Reflecting the policy's modest scale, other package elements included a proposal for workers to receive cash value for not using employer-paid parking and promotion of natural gas power production.

83. *Senior US officials recognized that the Berlin Mandate would be controversial domestically, but wanted to maintain international negotiating momentum (Royden 2002).

84. Memorandum for the President on Climate Change Recommendations from Gene Spirling, Kathleen McGinty, Daniel Tarullo, Jim Sternberg, and Todd Stern, October 18, 1997. Clinton Presidential Records, Speechwriting, Lowell Weiss, Folder: Climate Change 10/22/97 Decision Memos, OA/Box Number: 17200.

85. Clinton appears to have favored his environment team's advice, writing "this is better" in the memorandum's margins. Ibid.

86. Ibid.

87. Ibid.

88. Ibid.

89. Ibid.

90. William Jefferson Clinton, "Remarks at the National Geographic Society," October 22, 1997. This speech also included a new stabilization target of 1990 levels between 2008 and 2012.

91. In 1998, the Western Fuels Association founded the Greening Earth Society to defend coal-fired power plants by arguing that carbon dioxide would benefit welfare as an "amazingly effective aerial fertilizer." As quoted in Pooley 2010.

92. Interview 84 with senior union official, Washington, DC.

93. Industrial Union Division-AFL-CIO, "IUD Resolution on UN Climate Change Negotiations," November 13, 1996.

94. Interview 84 with senior union official, Washington, DC.

95. S. Res. 98 (105th Congress).

96. Hagel entered statements by various climate skeptics into congressional debates. In introducing the resolution, he cited both AFL-CIO fears that the treaty would cost the US 1.25 to 1.5 million jobs and an economic impacts warning from the National Association of Manufacturers.

97. Interview 96 with senior policy advisor to elected official, Washington, DC.

98. Interview 84 with senior union official, Washington, DC.

99. *Executive Order 13123 of June 3, 1999, "Greening the Government through Efficient Energy Management," Federal Register: Presidential Documents 64 (109), June 8, 1999.

100. *The two-page brief asserted that the Clean Air Act gave the EPA broad capacity to regulate nitrogen oxides, sulfur oxides, carbon dioxide, and mercury. Transcript of Congressional Hearing for Departments of Veterans Affairs and Housing and Urban Development, and Independent Agencies, March 10, 1998.

101. Memorandum from Jonathan Z. Cannon, EPA General Counsel, to Carol Browner, EPA Administrator, "EPA's Authority to Regulate Pollutants Emitted by Electric Power Generation Sources, April 10, 1998." By contrast, the memo said the EPA could only establish an emissions trading scheme if states were unwilling to take their own actions.

102. *In practice, climate had not featured heavily during the 2000 presidential campaign, despite Gore's candidacy.

103. *Speech by George W. Bush, "A Comprehensive National Energy Policy," September 29, 2000, Saginaw, Michigan. While Bush cited mandatory reductions he had signed in Texas as governor, and his government's support for renewable energy laws, his support for state-level environmental policies in Texas was more passive than active (Stokes 2020).

104. Interview 96 with senior policy advisor to elected official, Washington, DC.

105. Darren Samuelson, "Obama, Illinois File NSR Lawsuit against Midwest Generation," *E&E News*, August 29, 2009.

106. Ibid.

107. Ibid.

108. Ibid.

109. President George W. Bush, "Letter to Members of the Senate on the Kyoto Protocol and Climate Change," March 13, 2001.

110. Interview 96 with senior policy advisor to elected official, Washington, DC.

111. *Darren Samuelson, "Four-Pollutant Bill Edges Out of Senate EPW Panel," *E&E Daily*, June 28, 2002.

112. Speech by President George W. Bush at NOAA, "President Announced Clear Skies & Global Climate Change Initiatives," February 14, 2002.

113. Andrew Revkin, "Bush Offers Plan for Voluntary Measures to Limit Gas Emissions," *New York Times*, February 15, 2002.

114. Whitman beat back multiple efforts to weaken NSR standards during her tenure from copartisans, trying to argue that regulatory relief could be a carrot for other environmental reforms (Whitman 2005).

115. Most of the 2002 rule and the entirety of the 2003 rule were struck down by later court rulings.

116. Interview 99 with senior policy advisor to elected official, Washington, DC.

117. *White House Office of the Press Secretary, "President Bush Discusses Global Climate Change," June 11, 2001.

118. *As a requirement of the Global Change Research Act of 1990, the US government had to issue a climate impacts report every three years. However, the first major report took a decade to complete, and was only released in 2000 in the Clinton administration's final months.

119. *The Global Change Research Act mandated a single report. Instead, the Bush administration promised twenty-one briefs. Only some were completed and, even in their entirety, they would not have clearly met the Act's requirements.

120. Andrew Revkin, "Bush Aide Softened Greenhouse Gas Links to Global Warming," *New York Times*, June 8, 2005; Shulman 2008; Letter from Rick Piltz to US Climate Change Science Program agency principals; "On Issues of Concern about the Governance and Direction of the Climate Change Science Program," June 1, 2005.

121. After these edits became public, Cooney left the administration to work for ExxonMobil.

122. The vast majority of these bills were introduced by Democratic legislators without Republican support, gained little legislative traction, and were not discussed in committees.

123. Later, the Natural Resources Defense Council (NRDC) would also become involved.

124. *Initially, the bill required reduction of emissions to 1990 levels by 2016, but this was weakened in an effort to widen the bill's support base. Jennifer Lee, "Sponsors Ease Bill on Gases that Warm the Climate," *New York Times*, October 2, 2003.

125. *Allowance allocation decisions were delegated to the Commerce Secretary according to bill's principles.

126. Jennifer Lee, "Critics Say E.P.A. Won't Analyze Clean Air Proposals Conflicting with President's Policies," *New York Times*, July 14, 2003.

127. Republican Majority Leader Frist agreed to move the bill directly to floor debate in exchange for dropping the amendment (Layzer 2012).

128. Jennifer Lee, "Two Senators Aim to Put Others on Record on Emissions Cap," *New York Times*, October 2, 2003.

129. Interview 87 with senior policy advisor to elected official, Washington, DC.

130. "New Players on Global Warming," *New York Times Opinion*, January 15, 2003.

131. Even after the bill failed, the Farm Bureau, National Association of Manufacturers, and the US Chamber of Commerce sent a joint fax to senators urging legislators to reject any reconsideration: "Fax from American Farm Bureau, National Association of Manufacturers, and U.S. Chamber of Commerce to James Inhofe, June 1, 2004, re: Climate Stewardship Act, S.139," Voinovich Collections, Ohio University.

132. Maine Republican Senator Olympia Snowe cosponsored the McCain–Lieberman bill. Democrats Byrd (WV), Conrad (ND), Dorgan (ND), Baucus (MT), Levin (MI), Breaux (LA), Miller (GA), Lincoln (AR), Pryor (AR), and Landrieu (LA) voted against the bill. In addition to McCain and Snowe, Republicans Collins (ME), Chafee (RI), Lugar (IN), and Gregg (NH) voted for the bill.

133. Andrew Revkin, "Election over, McCain criticizes Bush on Climate Change," *New York Times*, November 16, 2004.

134. *The 2003 version of the bill only allocated revenue toward low-income Americans and dislocated workers. The 2005 bill also directed revenues toward renewables, CCS technology for coal-fired power plants, biofuels, and nuclear energy.

135. In a creative attempt to undermine climate policymaking, Senator Inhofe worked to disincentivize utility support for emissions trading (Layzer 2007). In June 2005, he introduced the Ratepayers Protection Act that would have required the CBO to study the impact of utility pollution reductions on "disadvantaged consumers." If impacts existed, the bill would have prohibited utilities from passing these costs to consumers.

136. S.Amdt.866 to H.R.6. [Energy Policy Act of 2005]. The resolution included the caveats that policy action should not "significantly harm the US economy" and should "encourage" action by other major economies. Republican votes came from Senators Chafee (RI), Collins (ME), DeWine (OH), Domenici (NM), Graham (SC), Gregg (NH), Lugar (IN), McCain (AZ), Snowe (ME), Specter (PA), and Warner (VA). Among Democrats, only Nelson (NE) voted against the resolution.

137. US EPA, *Inventory of U.S. Greenhouse Gas Emissions and Sinks: 1990–2005* (2007).

5 US Climate Policy Action, 2007–2015

1. The BCSE would become one of the Kyoto Protocol's highest-profile champions (Meckling 2011; Leggett 2001).

2. *This shift accelerated in the late 1990s and early 2000s (Meckling 2011).

3. *ICCP was founded in 1991; it became more prominent over the 1990s.

4. By 2000 these included Ford, Shell, Chrysler, GM, Dupont, and Texaco. The GCC disbanded altogether during the early Bush administration.

5. RGGI began allowance auctions in September 2008 for a January 2009 first compliance period.

6. The targets promised to stabilize Californian carbon pollution at 2000 levels by 2010 and reach 80 percent below 1990 levels by 2050. Office of the Governor of California, "Executive order S-3-05." June 1, 2005.

7. By 2011, only California remained in the initiative, alongside Quebec as a full member and some Canadian provinces as observers.

8. This strategic posture was also reflected in opponents' advocacy statements. For example, an American Petroleum Institute letter to Congress emphasized a strong preference for federal climate policy "in lieu of ill-suited federal and state regulatory programs," even as API continued to oppose federal policy overall. API, Letter from President Jack Gerard to Members of Congress, June 23, 2009.

9. They built on Pew's previous Business Environmental Leadership Council.

10. *Participants had diverse motives, but many believed it was in their economic interests to shape the anticipated climate deal (Meckling 2011).

11. *Centre for Global Development, Carbon Monitoring for Action (CARMA) Dataset; "Carbon Dioxide Emissions from Power Plants Worldwide," *ScienceDaily*, November 15, 2007. Rogers was simultaneously the chair of the Edison Electrical Institute.

12. USCAP, "A Call for Action: Consensus Principles and Recommendations from the US Climate Action Partnership: A Business and NGO Partnership," January 2007.

13. USCAP advocated for 100–105 percent of 2007 levels by 2012, 90–100 percent by 2017, 70–90 percent by 2022, and 20–40 percent by 2050. The plan recommended free allowances for trade-exposed industries and offsets from noncovered sectors and international markets as a cost containment mechanism. It included transitionary assistance for impacted workers, and government investments in low-carbon technologies, particularly CCS. Ibid.

14. Dupont, BP, and PG&E testified in support of climate policy action, earning Inhofe's dismissal as "climate profiteers." S. Hrg. 110–992 (110th Congress), "The US Climate Action Report," hearing before the Committee on Environment and Public Works, February 13, 2007.

15. Ibid.

16. *Other USCAP participants also faced pressure.

17. Rogers had pushed EEI to revisit its climate policy posture before he revealed USCAP's existence to fellow utility executives (Pooley 2010).

18. Matthew Wald, "Utility and Sierra Club Deal Aims to Cut Carbon Dioxide," *New York Times*, March 20, 2007.

19. For instance, the Bali COP in December 2007 was the first set of climate negotiations that the AFL-CIO engaged intensively. Interview 11 with two senior union officials, Washington, DC.

20. *See: United Steelworkers, *Securing Our Children's World*, June 1, 2006.

21. Interview 90 with mid-level environmental official, Washington, DC.

22. Ibid.

23. The Sierra Club and Steelworkers had a close relationship dating back to the Sierra Club's support for labor interests during NAFTA negotiations. Ibid.

24. Ibid.

25. International Centre for Technology Assessment and others, "Petition for Rule Making and Collateral Relief Seeking Regulations Greenhouse Gas Emissions from New Motor Vehicles under Section 202(a) of the Clean Air Act," October 20, 1999. Clean energy advocates included the California Solar Energy Industries Association and an energy efficiency company, Applied Power Technologies.

26. *Petitioners argued the administration had already implicitly conceded this determination.

27. *EPA press release, "EPA Denies Petition to Regulate Greenhouse Gas Emissions from Motor Vehicles," August 28, 2003.

28. *Massachusetts v. EPA*, 5.

29. Interview 1 with senior business lobby official, Washington, DC.

30. Felicity Barringer, "Ruling Undermines Lawsuits Opposition Emissions Controls," *New York Times*, April 3, 2007.

31. Darren Samuelson, "Former EPA Official Details White House Retreat on GHG Regs," *E&E Daily*, July 18, 2008; However, figures such as Connaughton

simultaneously opposed using this authority to actively manage US carbon pollution. Felicity Barringer, "White House Refused to Open Pollutants E-mail," *New York Times*, June 25, 2008.

32. Darren Samuelson and Robin Bravender, "EPA Releases Bush-era Endangerment Document," *New York Times*, October 13, 2009. The Obama administration's Endangerment Finding expanded on the Bush draft by making more expansive claims about greenhouse gas impacts and by treating all greenhouse gases together, rather than each separately.

33. Barringer, "White House Refused to Open Pollutants E-mail."

34. Samuelson and Bravender, "EPA Releases Bush-era Endangerment Finding"; Samuelson, "Former EPA Official Details White House Retreat on GHG Regs."

35. Despite his administration's efforts to stall policy action, Bush took some weak steps. For instance, in May 2007, he ordered executive agencies to cooperate toward reducing greenhouse gas emissions from cars and trucks. White House, "Executive Order: Cooperation among Agencies in Protecting the Environment with Respect to Greenhouse Gas Emissions from Motor Vehicles, Nonroad Vehicles, and Nonroad Engines," May 14, 2007.

36. *This iteration also strengthened its carbon pollution reduction targets.

37. It was cosponsored by Snowe, Collins, and Coleman among Republicans.

38. Interview 87 with senior policy advisor to elected official, Washington, DC. Warner was also alarmed by pine wilt in the Idaho forests where he had once worked as a firefighter.

39. Ibid. When constituents accused him of devastating Virginian lives with the effort, he responded that he was the Senator from Virginia but he needed to take a broader perspective when legislating.

40. Ibid.

41. Ibid.

42. Interview 89 with senior bureaucrat, Washington, DC.

43. *Warner periodically asked what the USCAP position was on a particular issue but the organization lacked specific advice. Interview 87 with senior policy advisor to elected official, Washington, DC; Interview 89 with senior bureaucrat, Washington, DC.

44. Staffers drew analogies to fashioning a highway bill. They conceived of cap and trade as "both an authorization bill and an appropriations bill" in one. Interview 89 with senior bureaucrat, Washington, DC.

45. By contrast, the McCain–Lieberman bills had not negotiated allowance distributions.

46. *It covered all emissions sources that released more than ten thousand tons of CO_2 annually.

47. Interview 87 with senior policy advisor to elected official, Washington, DC; Frank Davies, "Major Global Warming Bill Headed for Senate," *San Jose Mercury News*, December 6, 2007.

48. *Clean Air Act regulatory relief was held back as a bargaining chip for later negotiating leverage. Interview 89 with senior bureaucrat, Washington, DC; Interview 96 with senior policy advisor to elected official, Washington, DC.

49. *Markup debates at subcommittee and committee levels resulted in few design changes.

50. Not all EPW Republicans expressed blanket opposition to emissions trading. Lamar Alexander supported a sectoral approach for the power sector, only opposing Lieberman–Warner because of its economy-wide coverage. Johnny Isakson, who had recently traveled to Greenland to witness the effects of climate change, also adopted a constructive stance despite his ultimate opposition. Interview 87 with senior policy advisor to elected official, Washington, DC.

51. Among Republicans, Senators Stevens (AK) and Murkowski (AK) also cosponsored Bingaman–Specter.

52. See also: Statement of Hon. Jeff Bingaman, US Senator from the State of New Mexico in S. Hrg. 110–12 (110th Congress): "Senator's Perspectives on Global Warming," hearing before the Committee on Environment and Public Works, January 30, 2007; US Senate Committee on Energy and Natural Resources, "Bingaman, Specter on New Climate Change Recommendations," press release, April 19, 2007.

53. *NCEP commissioners included former EPA bureaucrats, politicians, academics, senior officials from Exelon, ConocoPhillips, and Ford, the United Steelworkers of America president, and a representative from NRDC. Thus, carbon polluters were actively involved in NCEP policy design.

54. The safety valve came from NCEP, which had recommended a $10 price. Bingaman–Specter's 2020 targets were identical to final NCEP guidelines, and their 2030 targets were similar. NCEP, *Energy Policy Recommendations to the President and the 110th Congress*, August 2007.

55. Interview 84 with senior union official, Washington, DC. Unlike coal unions, coal companies did not support Lieberman–Warner.

56. Letter from the Cecil Roberts, president of the United Mine Workers of America, May 27, 2008, *Congressional Record* 154, no. 93 (June 6, 2008): S5339.

57. Interview 87 with senior policy advisor to elected official, Washington, DC.

58. The Lieberman–Warner Bill allowed carbon for a portion of each covered entity's compliance obligations. However, Boxer opposed offsets altogether (Pooley 2010).

59. In a complex bargain, senior officials close to both proposals suggested a "strategic allowance reserve." This allowed future allowances to be borrowed from future allocations to moderate short-term carbon prices above pre-agreed thresholds. Letter to Bettina Poirier, David McIntosh, Chelsea Maxwell, Jonathan Black, Tom Dower, and Joellen Darcy from Jason Grumet, NCEP and Tim Profeta, Nicholas Institute for Environmental Policy Solutions, April 7, 2008.

60. *She claimed that legislators could no longer "fiddle while the planet gets ready to burn." US Senate Committee on Environment and Public Works, "Senator Boxer and Environmental Leaders Press Conference on Urgent Need to Address Global Warming," press release, March 12, 2008.

61. Defenders of Wildlife et al., Letter to Members of the Environment and Public Works Committee, November 15,2007.

62. Nathaniel Keohane and Peter Goldmark, *What Will It Cost Us to Protect Ourselves from Global Warming?* Environmental Defense Fund, 2008.

63. Friends of the Earth, "Friends of the Earth Statement on the Lieberman-Warner Substitute," media release, May 21, 2008.

64. Guest essay for grist.com by Carl Pope, executive director of the Sierra Club, "Carl Pope of the Sierra Club Lays Out a Blueprint for an Effective Climate Bill," February 15, 2008.

65. *Letter from Senator Barbara Boxer to members of the US Senate, May 16, 2008.

66. Warner objected to government subsidies more generally.

67. Interview 87 with senior policy advisor to elected official, Washington, DC.

68. Interview 96 with senior policy advisor to elected official, Washington, DC.

69. Interview 89 with senior bureaucrat, Washington, DC.

70. Interview 87 with senior policy advisor to elected official, Washington, DC.

71. Letter to Senators Harry Reid and Barbara Boxer dated June 6, 2008 from Senators Debbie Stabenow, Jay Rockefeller, Carl Levin, Blanche Lincoln, Mark Pryor, Jim Webb, Evan Bayh, Clair McCaskill, Sherrod Brown, and Ben Nelson.

72. *The AFL-CIO demanded a stronger safety valve. Senate Hearing 110–1233 on America's Climate Security Act of 2007, S. 2191, November 8, 13, and 15, 2007.

73. These included subsidies for renewables, coal-based sequestration, advanced vehicles, and biomass fuels. Letter to Senator Barbara Boxer from William Samuel, Director, Department of Legislation, AFL-CIO, November 5, 2007.

74. Letter from Cecil Roberts, president of the United Mine Workers of America, May 27 2008, *Congressional Record* 154, no. 93 (June 6, 2008): S5339.

75. Letter from Alan Reuther, Legislative Director for the United Auto Workers dated June 2, 2008, *Congressional Record* 154, no. 93 (June 6, 2008): S5340.

76. Letter from Leo Gerard, United Steelworkers, and Carl Pope, Sierra Club, under the BlueGreen Alliance banner to US Senators, June 5, 2008.

77. Charles River Associates, *Economic Analysis of the Lieberman-Warner Climate Security Act of 2007*, January 2008; Science Applications International Corporation, *Analysis of the Lieberman-Warner Climate Security Act (S. 2191) using the National Energy Modeling System (NEMS/ACCF/NAM)*, report prepared for American Council for Capital Formation and the National Association of Manufacturers, March 2008.

78. US Chamber of Commerce Letter Opposing the Manager's Amendment to S. 3036, the "Lieberman-Warner Climate Security Act of 2008," June 3, 2008.

79. US Chamber of Commerce, *Wake Up to Climate Legislation* video, https://www .youtube.com/watch?v=XevRKc82soI, accessed February 19, 2015.

80. Letter from the Jay Timmons, executive vice-president of the National Association of Manufacturers, June 3, 2008, *Congressional Record* 154, no. 93 (June 6, 2008): S5339.

81. Letter from Bob Stallman, Farm Bureau, dated May 30, 2008, *Congressional Record* 154, no. 93 (June 6, 2008): S5338.

82. ACCCE was founded in 2007 and spent more $40 million per year to support coal's position in the US energy mix (Pooley 2010).

83. Duke Energy cited insufficient allowances overall and allowance distribution patterns that would give clean energy a windfall (Pooley 2010).

84. Multigroup stakeholder letter to members of the US Senate, June 2, 2008.

85. For example, S. Hrg. 110–12 (110th Congress), "Senator's Perspectives on Global Warming," hearing before the Committee on Environment and Public Works, January 30, 2007.

86. Ibid.

87. *White House, "Statement of Administration Policy: S. 3036—Lieberman-Warner Climate Security Act," June 2, 2008. Bush's first carbon pollution reduction target, announced in April, promised to stabilize US greenhouse gas emissions by 2025; it was partly viewed as defensive posturing during Senate climate debates. Sheryl Gay Stolberg, "Bush Sets Greenhouse Gas Emissions Goals," *New York Times*, April 17, 2008.

88. Technically, a cloture vote on S. 3036, a substitute amendment offered by Boxer to the Lieberman–Warner bill that came out of the EPW committee in December 2007.

89. Biden, Clinton, Coleman, Kennedy, McCain, and Obama.

90. Interview 96 with senior policy advisor to elected official, Washington, DC.

91. *Transcript of Republican presidential debate in Boca Raton, Florida, January 24, 2008.

92. In conversation with the *San Francisco Chronicle* in January 2008, Obama did not mince words in describing his desire to transform the price structure of the US economy: "So if somebody wants to build a coal-powered plant, they can; it's just that it will bankrupt them because they're going to be charged a huge sum for all that greenhouse gas that's being emitted." https://www.youtube.com/watch?time _continue=12&v=4aTf5gjvNvo, accessed June 15, 2019.

93. McCain expressed greater reticence, emphasizing preference for legislative approaches. Jim Efstathiou Jr., "Obama to Declare Carbon Dioxide Dangerous Pollutant," *Bloomberg*, October 16, 2008.

94. Commission on Presidential Debates, "October 15, 2008 Debate Transcript," Hempstead, New York; Commission on Presidential Debates, "October 7, 2008 Debate Transcript," Nashville, Tennessee.

95. For example, on November 18, Obama promised climate policy leadership to the Bipartisan Governors Global Climate Summit, "start[ing] with a federal cap and trade system." The Office of the President-Elect, "President-elect Barack Obama to Deliver Taped Greeting to Bi-partisan Governors Climate Summit," press release, November 18, 2008.

96. President Barack Obama, "Address Before a Joint Session of the Congress," February 24, 2009.

97. Juliet Elperin and Michael Grunwald, "Internal Rifts Cloud Democrats' Opportunity on Warming," *Washington Post*, January 23, 2007.

98. Ibid.

99. Memorandum from John Dingell and Rick Boucher to members of the Energy and Commerce Committee, "Climate Change Discussion Draft Legislation," October 7, 2008.

100. Ibid.

101. The bill also included reserve allowances as a cost-containment mechanism.

102. Patrick O'Connor, "Waxman Dethrones Dingell as Chairman," *Politico*, November 20, 2008.

103. Rather than outlining policy specifics, it empowered the EPA to create an emissions trading scheme to meet these targets.

104. Interview 92 with senior US policy advisor to elected official, Washington, DC.

105. Interview 96 with senior policy advisor to elected official, Washington, DC.

106. Interview 88 with senior policy advisor to elected official, Washington, DC.

107. Unhappy with this compromise, the National Wildlife Federation withdrew from USCAP (Pooley 2010).

108. *As before, the framework also advocated complementary measures, including investments in CCS technology. United States Climate Action Partnership (USCAP), "A Blueprint for Legislative Action: Consensus Recommendations for US Climate Protection Legislation," January 2009.

109. Interview 88 with senior policy advisor to elected official, Washington, DC; Interview 92 with senior US policy advisor to elected official, by phone.

110. Darren Samuelsohn and Katherine Ling, "'Fragile Compromise' of Power Plant CEOs in Doubt as Senate Climate Debate Nears," *New York Times*, August 5, 2009.

111. Ibid.; Edison Electric Institute, "EEI Global Climate Change Points of Agreement," January 14, 2009.

112. *Utilities proposed a "50-50-50" formula for allowance allocation. Merchant coal generators received 50 percent of emissions in free allowances, while all other local distribution companies received the remainder of utility-allocated allowances, 50 percent based on carbon pollution and 50 percent based on overall retail sales.

113. Interview 97 with mid-level US bureaucrat, by phone.

114. *In particular, Sec. 116 of the bill is almost identical to the USCAP proposal, taking its precise thresholds and dates.

115. Obama also cautioned that "if you're giving away carbon permits for free, then basically you're not really pricing the thing and it doesn't work." White House Office of the Press Secretary, "Remarks by the President at the Business Roundtable," March 12, 2009, St. Regis Hotel, Washington, DC. Economists criticized Obama for misunderstanding how cap mechanisms functioned. The administration pushed back by arguing low EU prices resulted from excess allowance allocations. Darren Samuelson, "Obama Erred on Key Cap-and-Trade Features," *New York Times*, March 13, 2009; Ian Talley, "White House Flexibility Signaled on Climate Bill," *Wall Street Journal*, April 9, 2009.

116. This target fell within USCAP bounds—at the maximum range—except for the slightly more ambitious 2050 target of 83 percent rather than 80 percent.

117. *Entities needed to yield 1.25 offset credits for every compliance unit, including international credits as defined in the bill.

118. *It also established a low-carbon fuel standard to reduce life-cycle GHG emissions from transportation fuels 5 percent below 2005 baselines by 2023 and 10 percent by 2030, electric vehicle support, electricity system reforms to support Smart Grids, and support for renewables transmission planning.

119. It created a national strategy for investment in CCS, including efforts to develop oversight of geological sequestration sites, subsidies for early adoption of CCS, support for commercial CCS deployment, and carbon pollution performance standards for coal-fired power plants.

120. *For consumers, it created an energy refund program for up to 150 percent of the poverty line for consumers who qualified for SNAP.

121. Covered sectors were preempted from the Clean Air Act with respect to criteria pollutants, GHG hazardous pollutants, PSD, New Source Review, and Title V permits.

122. Interview 88 with senior policy advisor to elected official, Washington, DC.

123. *Interview 92 with senior US policy advisor to elected official, by phone; Interview 97 with mid-level US bureaucrat, by phone.

124. Interview 95 with elected official, Washington, DC.

125. Interview 88 with senior policy advisor to elected official, Washington, DC.

126. *Waxman agreed to almost all of Boucher's requests, including weakened performance standards for coal-fired power plants (Pooley 2010).

127. C-SPAN, "Representative Rick Boucher on Energy Legislation," May 20, 2009.

128. Interview 84 with senior union official, Washington, DC.

129. American Coalition for Clean Coal Electricity, "ACCCE Statement in Regards to the Passage of the American Clean Energy and Security Act," press release, June 26, 2009; Interview 92 with senior US policy advisor to elected official, by phone.

130. Interview 88 with senior policy advisor to elected official, Washington, DC; Interview 96 with senior policy advisor to elected official, Washington, DC.

131. *Interview 84 with senior union official, Washington, DC.

132. This was the 50-50-50 compromise. EEI believed this was the best deal possible (Pooley 2010).

133. For instance, Exelon estimated first-year revenues of $1.1 billion from the allowance formulas. Exelon, Form 10-Q, Quarterly Report Pursuant to Section 13 or

15(d) of the Securities Exchange Act of 1934, for the Quarterly Period Ended September 30, 2009.

134. Interview 92 with senior US policy advisor to elected official, by phone.

135. *Samuelsohn and Ling, "'Fragile Compromise.'"

136. Darren Samuelson and Ben Geman, "Moderate House Democrats Lay Out Concerns with Draft Climate Bill," *New York Times*, April 22, 2009.

137. Interview 88 with senior policy advisor to elected official, Washington, DC.

138. The entire negotiations therefore had something of an ordering effect, with early sectors being more prioritized with respect to allowance streams. Interview 92 by phone with senior US policy advisor to elected official.

139. Ibid.

140. Samuelson and Geman, "Moderate House Democrats Lay Out Concerns with Draft Climate Bill."

141. Interview 11 with two senior union officials, Washington, DC.

142. Interview 92 with senior US policy advisor to elected official, by phone.

143. *Mike Ross (AR), Jim Matheson (UT), Charlie Melancon (LA), and John Barrow (GA).

144. Interview 92 with senior US policy advisor to elected official, by phone.

145. Stephen Power, "Farm State Wish List Could Hold Key to Waxman-Markey Bill," *Environment Capital Blog, Wall Street Journal*, June 11, 2009.

146. Interview 92 with senior US policy advisor to elected official, by phone.

147. *Strategy memo from Republican Senate EPW staff to House to Senate energy/environment staff, "Re: A Strategy for Climate Change: Consumers vs. Big Business," May 14, 2009.

148. Al Gore's emergence as the foremost voice worldwide on climate change may have also accelerated party polarization on this issue.

149. These Republicans all represented "blue" states.

150. US EPA Office of Atmospheric Programs, "EPA Analysis of the American Clean Energy and Security Act of 2009 H.R. 2454 in the 111th Congress," June 23, 2009.

151. Center for Public Integrity, "Tally of Interests on Climate Bill Tops a Thousand," August 10, 2009.

152. *Center for Public Integrity, "Southern Company Dominates the Climate Lobbying Scene," July 1, 2009.

153. Event-study estimates of the Waxman-Markey bill suggest it would have imposed total costs of between $110 billion to $260 billion on US companies (Meng 2017).

154. But Dow Chemical vigorously supported Waxman–Markey.

155. Samuelsohn and Ling, "'Fragile Compromise' of Power Plant CEOs."

156. Jim Tankersley, "US Chamber of Commerce Seeks Trial on Global Warming," *Los Angeles Times*, August 25, 2009; John Broder, "Storm over the Chamber," *New York Times*, November 18, 2009.

157. Duke Energy had earlier withdrawn from NAM over the issue. Clifford Krauss and Kate Galbraith, "Climate Bill Splits Exelon and US Chamber," *New York Times*, September 28, 2009.

158. Interview 1 with senior business lobby official, Washington, DC.

159. Operating under the name "America's Power Army," these volunteers were tasked with "spontaneously" asking officials about energy policy to pressure Senate Democrats. Anne Mulkern, "'Citizen Army' Carries Coal's Climate Message to Hinterlands," *New York Times*, August 6, 2009.

160. David Fahrenthold, "Congress Discovers Another Forged Advocacy Letter," *Capitol Briefing Blog. Washington Post*, August 18, 2009.

161. Copy of email from API—obtained by Greenpeace—August 2009.

162. API would "coordinate transportation to the venues, if required, for your employees." The first Energy Citizen rally was held in Houston in mid-August 2009, largely attended by oil workers. Clifford Krauss and Jad Mouwad, "Oil Industry Backs Protests of Emissions Bill," *New York Times*, August 18, 2009.

163. The campaign involved TV, radio, and Internet ads. NAM and NFIB, "NFIB Launch Media Campaign Opposing Waxman-Markey Climate Change Bill," press release, August 28, 2009.

164. Pooley 2010. See also "Farm Bureau, "Don't CAP Our Future," About, Facebook page.

165. As Tea Party protestors argued, proponents were "cap and traitors" (Bartosiewicz and Miley 2013). Self-identified Tea Party Republicans reported dramatically lower climate beliefs (Leiserowitz et al. 2011).

166. For instance, AFP organized a Regulation Reality Tour to contest federal climate action. John Broder, "Climate Change Doubt Is a Tea Party Article of Faith," *New York Times*, October 20, 2010.

167. Ronald Brownstein, "GOP's New Senate Class Could Be Conservative Vanguard," *National Journal*, September 25, 2010.

168. Greenpeace USA, "Greenpeace: Waxman–Markey Climate Change Bill Not Strong Enough to Stop Global Warming," May 16, 2009.

169. James Hansen, "G-8 Failure Reflects U.S. Failure on Climate Change," *Huffington Post*, August 9, 2009.

170. Interview 94 with senior policy advisor to elected official, by phone; Interview 96 with senior policy advisor to elected official, Washington, DC.

171. USCAP, "USCAP Statement on Passage of the American Clean Energy Security Act," press release, May 21, 2009.

172. *Interview 92 with senior US policy advisor to elected official, by phone; Interview 96 with senior policy advisor to elected official, Washington, DC. Senate staffers cites insufficient joint planning, and no engagement with House Republicans despite the need for Republican votes to clear the Senate's sixty-vote threshold.

173. *McCain would criticize Waxman–Markey as a "farce" that was insufficiently ambitious because it accommodated carbon polluters through generous allowance allocations. Stephen Moore, "Pulling No Punches," *Wall Street Journal*, August 1, 2009.

174. Interview 99 with senior policy advisor to elected official, Washington, DC.

175. Ibid.

176. An exception was a complex effort in the Energy and Natural Resources Committee that reported out a bill that would have established a renewable energy standard. Senate opponents also worked hard to undermine a potential deal. For instance, Nebraska Republican Mike Johanns successfully amended a 2010 budget resolution in April 2009 to block climate bill consideration through the reconciliation process. The vote was 67–31, with 26 Democrats joining Republicans to ensure climate reforms would require 60 votes.

177. Baucus also wanted strong EPA preemption. Lisa Lerer, "Baucus Opposition Chills Climate Bill," *Politico*, October 29, 2009.

178. Lisa Lerer, "Climate Change on the Back Burner?" *Politico*, November 4, 2009.

179. Alexander Bolton, "Climate Bill Hinges on Ohio's Sen. Brown," *The Hill*, November 1, 2009.

180. Lisa Lerer, "Senate Democrats to WH: Drop Cap and Trade," *Politico*, December 27, 2009.

181. Interview 99 with senior policy advisor to elected official, Washington, DC. The trio made extensive efforts to attract Republicans, including offering Alaska Republican Lisa Murkowski the opportunity to write the bill's oil drilling provisions. Murkowski then demanded opening up the Arctic National Wildlife Refuge for

drilling as a climate bill precondition! Ryan Lizza, "As the World Burns," *New Yorker*, October 11, 2000.

182. *Darren Samuelsohn, "Senate Trio Hopes to Hit Pay Dirt With Carbon 'Fee' on Fuels," *New York Times*, March 3, 2010.

183. Interview 99 with senior policy advisor to elected official, Washington, DC; Lizza, "As the World Burns,"

184. Lizza, "As the World Burns."

185. *Partly because of Graham's reservations, labor was mostly excluded from negotiations, which focused on business community engagement. Interview 99 with senior policy advisor to elected official, Washington, DC.

186. Ibid.

187. Lizza, "As the World Burns."

188. Interview 99 with senior policy advisor to elected official, Washington, DC. Kerry's team once tried to unilaterally sneak labor provisions into the bill draft.

189. Lizza, "As the World Burns."

190. Ibid.

191. Carl Hulse and David Herszenhorn, "Democrats Call Off Climate Bill Effort," *New York Times*, July 22, 2010.

192. These included ConocoPhillips, AIG, BP, Xerox, Marsh, Caterpillar, John Deere, Ford, and GM. USCAP's last news release was dated January 25, 2011.

193. *White House, "Presidential Memorandum—EPA Waiver," Memorandum for the Administrator of the Environmental Protection Agency from President Barack Obama, January 26, 2009.

194. EPA, "United States Files Clean Air Lawsuit against Westar Energy; Complaint Is Part of National Initiative to Stop Illegal Pollution from Coal-fired Power Plants," press release, February 4, 2009.

195. API, "Proposed Endangerment Finding Troubling—API," media release, April 17, 2009.

196. NAM, "On the EPA's Proposed Endangerment Finding, the NAM Comments," June 25, 2009.

197. Manufacturers and Chemical Industry of North Carolina, "Endangerment Finding Raises Stakes for Duke Energy, Industry and NC," December 11, 2009.

198. White House Office of the Press Secretary, "President to Attend Copenhagen Climate Talks," press release, November 25, 2009.

199. Mark Pryor (AR), Evan Bayh (IN), Ben Nelson (NE), Jay Rockefeller (WV), Blanche Lincoln (AR), and Mary Landrieu (LA).

200. Juliet Eilperin, "Peterson and Co. Strike at EPA," *PostCarbon* (blog), *Washington Post*, February 26, 2010.

201. This was cosponsored by Senators Kent Conrad (ND), Byron Dorgan (ND), Tim Johnson (SD), Ben Nelson (NE), Claire McCaskill (MO), and Jim Webb (VA).

202. Juliet Eilperin, "Rockefeller Pushes to Rein in EPA," *PostCarbon* (blog), *Washington Post*, March 4, 2010.

203. Executive Office of the President, *The President's Climate Action Plan*, June 2013.

204. US Environmental Protection Agency, "EPA to Hold Public Listening Sessions on Reducing Carbon Pollution from Existing Power Plants," press release, September 30, 2013.

205. Benjamin Goad, "EPA 'Listening Tour' on Climate Rules Skips States Powered by Coal," *The Hill*, October 9, 2013.

206. Coral Davenport, "EPA Staff Struggling to Create Pollution Rule," *New York Times*, February 4, 2014.

207. The agency's proposal was particularly influenced by NRDC officials. Davenport, "EPA Staff Struggling to Create Pollution Rule."

208. Mark Landler and John Broder, "Obama Outlines Ambitious Plan to Cut Greenhouse Gases," *New York Times*, June 25, 2013.

209. Interview 89 with senior bureaucrat, Washington, DC.

210. Ibid.

211. Interview 1 with senior business lobby official, Washington, DC.

212. Davenport, "EPA Staff Struggling to Create Pollution Rule."

213. American Coalition for Clean Coal Electricity, "ACCCE on Initial Review of 111(d): EPA Carbon Emissions Rule Misses the Mark," press release, June 2, 2014.

214. Mitch McConnell, "States Should Reject Obama Mandate for Clean-Power Regulations," *Lexington Herald-Leader*, op-ed, March 3, 2015.

215. US Chamber of Commerce, *Assessing the Impact of Potential New Carbon Regulations in the United States. Washington, DC: Institute for 21st Century Energy, 2014*; Josh Israel, "Major Companies Distance Themselves from US Chamber Campaign against Obama's Climate Plan," *ThinkProgress*, June 3, 2014.

216. Letter from Business and Innovative Climate & Energy Policy (BICEP) group to President Obama, June 2, 2014.

217. Interview 90 with mid-level environmental official, Washington, DC.

218. Interview with 9 with senior union official, Washington, DC.

219. Ibid.

220. Interview with 84 with senior union official, Washington, DC.

221. Interview 11 with two senior union officials, Washington, DC.

222. Interview 9 with senior union official, Washington, DC; Interview 84 with senior union official, Washington, DC.

223. IBEW, "IBEW to Speak Out against EPA Carbon Emission Rule," press release, July 28, 2014; IBB, "EPA Rules Will Not Effectively Impact World Climate Change"; UMWA, "EPA Existing Source Emissions Rule Puts American Jobs at Risk, Does Nothing to Address Climate Change," press release, June 2, 2014.

224. Interview 84 with senior union official, Washington, DC.

225. The stay was a 5–4 decision just days before the death of Justice Antonin Scalia; had Scalia died even a week later, implementation would have proceeded with a split 4–4 Court.

226. *Steven Greenhouse, "AFL-CIO Backs Keystone Oil Pipeline, if Indirectly," *New York Times*, February 27, 2013.

227. Interview 91 with mid-level environmental official, Washington, DC.

228. Laborer's International Union of North America, "LIUNA leaves BlueGreen Alliance," January 20, 2012.

229. For example, sharp differences were seen in 2008 versus 2012 in Obama's vote share in adjacent Virginia counties with or without coal industries. Nate Cohn, "Why Democrats Have Little to Lose in Taking on the Coal Industry," *Upshot* blog, *New York Times*, June 2, 2014.

6 Climate Policy Conflict in Australia

1. Australian Bureau of Statistics, *History of Coal Mining*, Year Book of Australia 1910.

2. Prior to being supplanted by Indonesia in 2011, Australia was the world's largest coal exporter for twenty-five years. Energy Information Administration (EIA), "Country Profiles: Australia," updated June 21, 2013, http://www.eia.gov/countries/cab.cfm?fips=as, accessed March 14, 2014.

3. Sinclair Davidson and Ashton de Silva, "The Australian Coal Industry—Adding Value to the Australian Economy," April 2013, unpublished report.

4. Interview 61 with energy consultant, Melbourne.

5. EIA, "Country Profiles: Australia."

6. Australian Department of the Environment, "National Greenhouse Gas Inventory—Kyoto Protocol Accounting Framework."

7. Interview 66 with senior business lobby official, Melbourne.

8. The Liberals now govern with National partners even when Liberals alone hold parliamentary majorities. Coalition membership also includes some regional parties, and varies between federal and state levels in complex ways.

9. Interview 43 with senior political advisor to elected official, Sydney.

10. Interview 59 with elected official, Canberra.

11. Interview 43 with senior political advisor to elected official, Sydney.

12. *Interview 63 with mid-level bureaucrat, Melbourne.

13. The Office of National Assessment is Australia's federal intelligence agency.

14. Prominent conflicts included the Franklin hydroelectric dam in the early 1980s and the Wesley Vale pulp and paper mill later that decade. Interview 59 with elected official, Canberra; Interview 71 with senior environmental official, Sydney. Throughout this period, some environmental advocacy groups, for instance the Wilderness Society, avoided climate change altogether. Others, including the Australian Conservation Foundation (ACF), engaged in only abstract-level climate-related lobbying. Interview 59 with elected official, Canberra; Interview 60 with senior environmental official, Canberra; Staples 2009.

15. Interview 60 with senior environmental official, Canberra.

16. *Interview 43 with senior political advisor to elected official, Sydney; Interview 59 with elected official, Canberra. Australia uses a ranked ballot system federally; parties thus compete to be other parties' second choice.

17. The Lavoisier group was founded in 2000 to block Australian climate policy-making and undermine public belief in climate science.

18. In a term endemic to Australia, early actors referred to climate change simply as "greenhouse."

19. Interview 59 with elected official, Canberra. Prominent green agenda critics included Finance Minister Peter Walsh, Resource Minister Peter Cook, and Industry and Technology Minister John Button (Economou 1996).

20. *While the document discussed the climate threat and recommitted Australia to take a leadership role in climate research and international climate negotiations, domestic details focused on energy conservation and transport.

21. *Interview 59 with elected official, Canberra. Australian bureaucrats saw these targets as unrealistic and overly ambitious. Interview 63 with mid-level bureaucrat, Melbourne.

22. Interview 63 with mid-level bureaucrat, Melbourne.

23. *Early Australian climate debates intersected with Hawke's Ecological Sustainable Development (ESD) consultation, initiated in 1989 to diffuse adversarial conflict over environmental policymaking. Interview 59 with elected official, Canberra. The ESD process was a quasi-corporatist effort to bring labor officials, business leaders, and some environmental advocates together to envision future Australian environmental policy. Interview 71 with senior environmental official, Sydney.

24. *As quoted in Staples 2012.

25. Interview 71 with senior environmental official, Sydney. Displaying little patience for environmental policies that threatened established patterns of economic development, Keating was a "dig it up and sell it kind of man." Interview 60 with senior environmental official, Canberra; Interview 63 with mid-level bureaucrat, Melbourne. He opposed Richardson's first Australian carbon pollution reduction target. Peter Walsh, *Confessions of a Failed Finance Minister* (Milsons Point, NSW: Random House Australia, 1995). 212–213. In August 1994, Keating chastised environmentalists for their focus on the "amorphous issue of greenhouse," suggesting they instead celebrate forest conservation victories. ABC national news interview, August 2, 1994. As cited in Taplin 1994.

26. *Both environment and industrial stakeholders were upset by government changes to ESD climate policy recommendations, feeling the government had turned its back on a deliberative consensus.

27. *Interview 71 with senior environmental official, Sydney.

28. Interview 60 with senior environmental official, Canberra. In the final ESD report, the use of taxation and pricing as environmental policy tools was raised, but without specific reference to climate change. Ecologically Sustainable Development Steering Committee, *National Strategy for Ecologically Sustainable Development*, December 1992. Part 3: chapter 20, "Pricing and Taxation."

29. ABC Radio, "Senator Faulkner Receives Support for His First Major Environment Policy Statement," *PM*, June 28, 1994; Interview 60 with senior environmental official, Canberra; Interview 63 with mid-level bureaucrat, Melbourne.

30. Press Release of the Minister for Environment, Sports and Territories, "Financial Review: Greenhouse Story Is Incorrect," January 18, 1995.

31. *The tax was to be included in the May 1995 budget.

32. Contemporaneous economic simulations suggested carbon taxes in excess of $100 AUD per ton of carbon might be necessary for Australia to achieve its 20 percent reduction target (Cornwell and Creedy 1995, 1996). Press Release of the Minister for Environment, Sports and Territories, Senator John Faulkner, "Transcript of Doorstop on Greenhouse," March 1, 1995.

33. ABC Radio, "Business Council of Australia Believes the Government May Introduce a Carbon Tax as a Way to Reduce the Deficit and Win Green Support," *7:30 Report*, January 3, 1995; Interview 60 with senior environmental official, Canberra.

34. *ABC Radio, "Opposition by the Mining Industry to Proposal for a Carbon Tax," *AM*, February 8, 1995. Later in the interview the AMIC's representative David Buckingham asked, rhetorically, "Would Norway be taxing its gas?" Norway was, of course, taxing offshore natural gas in 1995 at the time.

35. For instance, Minister for Primary Industries and Energy Bob Collins publicly rejected the possibility that his government would pass a carbon tax even as the matter was still being debated internally. ABC Radio, "Minister Comments on the Probability of the Drought Being Broken after Heavy Rain, and a Carbon Tax," *7:30 Report*, January 20, 1995.

36. *ABC Radio, "Opposition Rejects the Government's Proposal to Introduce a Carbon Tax," *PM*, January 18, 1995.

37. *There simply wasn't support in his Keating-led Labor Cabinet. Interview 60 with senior environmental official, Canberra; Interview 63 with mid-level bureaucrat, Melbourne.

38. A 1996 government evaluation of the Greenhouse Challenge found that even limited program reductions would mostly have occurred in the policy's absence. George Wilkenfeld and Associated & Economic and Energy Analysis, *Evaluating the Greenhouse Challenge Issues and Options*, Greenhouse Challenge Office, Canberra, 1996.

39. For example, a National Greenhouse Advisory Panel established by Keating in 1994 continued under the Howard government.

40. Keating broke with many world leaders by not attending the Rio Earth summit in 1992.

41. Interview 63 with mid-level bureaucrat, Melbourne. In an unsigned letter to then Environment Minister Faulkner, Keating signaled his concern with Australian abatement targets (Taplin 1994).

42. Interview 63 with mid-level bureaucrat, Melbourne.

43. For instance Resources and Energy Minister Warwick Parer had major investments in coal mining and had been a senior coal executive before his election (Hamilton 2007).

44. Interview 59 with elected official, Canberra.

45. These are known as the party's "wet" factions. Interview 71 with senior environmental official, Sydney.

46. Interview 50 with senior bureaucrat, Sydney; Interview 63 with mid-level bureaucrat, Melbourne.

47. Australian Government Department of Prime Minister and Cabinet, "Statement by the Hon John Howard MP: Safeguarding the Future—Australia's Response to Climate Change." November 20, 1997.

48. Interview 63 with mid-level bureaucrat, Melbourne.

49. Ministers from Transport and Finance also sat in on AGO council meetings. Interview 50 with senior bureaucrat, Sydney.

50. Interview 57 with senior Australian bureaucrat, by Skype.

51. Ibid. These relationships, for instance with WWF, created rifts in the environmental community, since Hill's actions directed funding away from environmental campaigns and activism support.

52. Interview 50 with senior bureaucrat, Sydney; Interview 63 with mid-level bureaucrat, Melbourne.

53. *The debates overlaid factional fights between "wets" and "drys," most clearly symbolized when Hill was minister of the environment and Alexander Downer the minister of industry. Interview 63 with mid-level bureaucrat, Melbourne.

54. *Interview 52 with mid-level bureaucrat, Canberra.

55. Australian civil servants would later discover these documents had been used by EU bureaucrats to design their own emissions trading scheme. Interview 50 with senior bureaucrat, Sydney.

56. Early emissions trading debates mobilized Australian climate skeptics, including the Lavoisier group's creation in 1999 by former Labor Finance Minister Peter Walsh and Liberal Party insider Hugh Morgan. In 2000, Morgan compared the AGO emissions trading briefs to Nazi propaganda (Hamilton 2007).

57. Interview 57 with senior Australian bureaucrat, by Skype.

58. *Interview 50 with senior bureaucrat, Sydney; Interview 56 with senior bureaucrat, Canberra.

59. Interview 50 with senior bureaucrat, Sydney; Interview 57 with senior Australian bureaucrat, by Skype.

60. Interview 50 with senior bureaucrat, Sydney.

61. *Australian National Audit Office, *The Administration of Major Programs: Australian Greenhouse Office*, Audit Report No. 34 2003–04. Performance audit.

62. *Australian National Audit Office, *Administration of Climate Change Programs: Department of Environment, Water, Heritage and the Arts, Department of Climate Change and Energy Efficiency, Department of Resources, Energy and Tourism* Audit Report No. 26 2009–10. Performance audit.

63. Interview 57 with senior Australian bureaucrat, by Skype.

64. *This mobilization was led by the resource and mining sector, particularly AIGN (Hamilton 2007).

65. Interview 60 with senior environmental official, Canberra.

66. Interview 63 with mid-level bureaucrat, Melbourne.

67. Ibid.

68. During the Kyoto run-up, only one person focused on climate across the entire department, working on modeling international negotiations' macroeconomic impacts. Interview 56 with senior bureaucrat, Canberra. Treasury officials were only marginally involved in the AGO, seen as a "spending" agency with little economic impact. Interview 50 with senior bureaucrat, Sydney.

69. *Interview 51 with senior bureaucrat, Canberra.

70. Interview 48 with Australian elected official, by phone. Interview 51 with senior bureaucrat, Canberra. Interview 63 with mid-level bureaucrat, Melbourne.

71. *Interview 63 with mid-level bureaucrat, Melbourne; Interview 69 with senior bureaucrat, Sydney.

72. Interview 63 with mid-level bureaucrat, Melbourne.

73. *GGAS enactment has been linked NSW Premier Carr's personal commitment to climate policymaking. Interview 42 with Australian energy consultant, by Skype; Interview 45 with senior policy advisor to political party, Canberra.

74. *Interview 61 with energy consultant, Melbourne.

75. Initially called the Interjurisdictional Working Group on Emissions Trading, the effort was rebranded "national" to annoy federal Liberals and "task force" to assume more legitimacy. Interview 65 with senior bureaucrat, Melbourne.

76. Interview 61 with energy consultant, Melbourne; Interview 65 with senior bureaucrat, Melbourne. Officials also struggled to agree on policy timing and allowance allocation decisions.

77. *Interview 61 with energy consultant, Melbourne. The NETT process succeeded in its efforts to propel emissions trading back onto the Australian agenda. Yet,

its report had limited impact. Labor's 2007 election win preempted the need for pressure from states. NETT's final report was also weakened by state-level officials concerned about policy costs.

78. The bureaucracy championed the policy more aggressively than political leaders, anticipating future policy need. Interview 51 with senior bureaucrat, Canberra.

79. *Interview 67 with senior business lobby official, Sydney.

80. Interview 65 with senior bureaucrat, Melbourne; Interview 69 with senior bureaucrat, Sydney.

81. Interview 55 with two senior bureaucrats, Canberra; Interview 75 with senior policy advocate, Melbourne.

82. Interview 70 with senior environmental official, Sydney; Interview 76 with senior environmental official, by Skype.

83. Interview 74 with senior policy advocate, Sydney.

84. Interview 55 with two senior bureaucrats, Canberra; Interview 69 with senior bureaucrat, Sydney.

85. Interview 82 with Australian elected official, by phone.

86. Interview 55 with two senior bureaucrats, Canberra; Interview 59 with elected official, Canberra.

87. The group was led by Peter Shergold, then head of the Department of Prime Minister and Cabinet. It also included the secretaries of the environment, treasury, and industry departments.

88. *Interview 69 with senior bureaucrat, Sydney. Press Release of Office of the Prime Minister, "Prime Ministerial Task Group on Emissions Trading," December 10, 2006.

89. Ibid.

90. Interview 69 with senior bureaucrat, Sydney, Interview 55 with two senior bureaucrats, Canberra.

91. Ibid.

92. *Interview 55 with two senior bureaucrats, Canberra; Interview 56 with senior bureaucrat, Canberra.

93. Interview 61 with energy consultant, Melbourne.

94. Ibid.

95. *Submission of the Australian Coal Association to the Prime Ministerial Task Group on Emissions Trading, March 2007.

96. Of almost 150 submissions received from industry groups and some environmental groups, not a single submission was received from any of Australia's major trade unions.

97. Australian Government, *Report of the Task Group on Emissions Trading*, Department of Prime Minister and Cabinet, May 31, 2007, 6.

98. Only agriculture and land use were excluded until technical considerations could be worked through.

99. Interview 56 with senior bureaucrat, Canberra.

100. This may partly have been the result of the report's tone, which communicated different messages to different constituencies without resolving tensions. For instance, it emphasized both the need to preserve Australian industrial competitiveness and to act aggressively to counter climate risks. Interview 56 with senior bureaucrat, Canberra.

101. Interview 69 with senior bureaucrat, Sydney.

102. Ibid.; Interview 82 with Australian elected official, by phone.

103. Interview 69 with senior bureaucrat, Sydney.

104. This was particularly true of Western Australian industrial actors. Interview 43 with senior political advisor to elected official, Sydney.

105. Ibid.

106. Interview 56 with senior bureaucrat, Canberra.

107. *Liberal Party of Australia, *A Climate Resilient Future: A Clean Energy Plan for Australia*, November 20, 2007. Election Policy Document.

108. Interview 42 with Australian energy consultant, by Skype.

109. Michelle Grattan, "Howard Takes Heat over Kyoto Revelation," *The Age*, October 28, 2007.

110. Interview 98, interview with elected official, Cambridge, MA.

111. *Environmental groups that had fought Howard over Kyoto were skeptical about the sincerity of Howard's ETS commitment. Interview 78 with senior environmental official, Sydney.

112. This included the right-aligned Australian Workers Union.

113. Transcript of "Opening Remarks to the National Climate Change Summit," Parliament House, Canberra, March 31, 2007, by Leader of the Labor Party Kevin Rudd.

114. *In doing so, Rudd chose not to begin deliberations from either the NETT or PMTGET starting point, instead reinitiating policy design discussions.

115. The new secretary of the department, Martin Parkinson, had been the deputy secretary of the Treasury and the head of the PMTGET's bureaucratic secretariat. Other senior Treasury officials joined Martin in leadership ranks.

116. Interview 46 with elected official, Canberra. This intellectual hierarchy created some resentment among environment officials. Interview 57 with senior Australian bureaucrat, by Skype.

117. Interview 51 with senior bureaucrat, Canberra; Interview 80 with senior bureaucrat, by Skype.

118. *Initial policy design involved more than twenty separate Cabinet submissions, each corresponding to a single chapter or topic in the eventual Green Paper.

119. *Department of Climate Change, *Carbon Pollution Reduction Scheme* Green Paper, July 2008.

120. About a thousand in Australia at the time, covering around 75 percent of Australia's total emissions.

121. Even the formula used to classify emission-intensive trade-exposed industries (EITE) remained largely unaltered relative to the PMTGET report. Interview 45 with senior policy advisor to political party, Canberra.

122. Interview 82 with Australian elected official, by phone. Howard is reported to have exclaimed that Rudd simply copied his ETS proposal. Interview 55 with two senior bureaucrats, Canberra; Interview 56 with senior bureaucrat, Canberra.

123. Interview 54 with senior business lobby official, Melbourne.

124. *Interview 56 with senior bureaucrat, Canberra.

125. The introduction of the CPRS was linked to a one-cent reduction in fuel taxes to offset any effects of the CPRS on gasoline prices, as well as to assistance measures for low- and middle-income households.

126. *Interview 66 with senior business lobby official, Melbourne; Interview 75 with senior policy advocate, Melbourne. In another major difference, direct emissions from agriculture were brought into the scheme in a more concrete way.

127. Interview 56 with senior bureaucrat, Canberra.

128. For instance, a representative from a major business in Australia was surprised at one government consultation that carbon pricing would increase the price of gasoline. Interview 56 with senior bureaucrat, Canberra. Even the most impacted industries sometimes had a limited policy understanding.

129. Interview 54 with senior business lobby official, Melbourne; Interview 56 with senior bureaucrat, Canberra.

130. Interview 67 with senior business lobby official, Sydney.

131. Interview 98 with elected official, Cambridge, MA.

132. A partial exception was the CFMEU, whose coal mining constituents were clear climate policy losers, and who had been engaged in climate debates since the Rio summit. Other industrial unions only began to perceive a policy threat during Green Paper debates.

133. Interview 72 with senior union official, Sydney.

134. *AWU's opposition to the CPRS was also a function of strategic positioning within Australian politics.

135. Interview 72 with senior union official, Sydney.

136. At the time ACTU president Sharan Burrow was firmly in the pro-climate camp, having previously been a central figure in the Australian Education Union.

137. Interview 72 with senior union official, Sydney.

138. Interview 73 with senior union official, Sydney. This maintained the strong voice of industrial unions at the heart of policy bargaining.

139. The congress agreed to ensure that "the just transition process protects jobs in carbon intensive industries." ACTU, *A Fair Society: Environment and Climate Change Policy*, ACTU Congress, June 2–4, 2009, 2.

140. *Interview 58 with senior union official, Melbourne; Interview 70 with senior environmental official, Sydney. The Climate Institute is a policy-oriented think tank in Australia.

141. Interview 76 with senior environmental official, by Skype.

142. For instance, differences about emissions standards for power generators were managed this way. Interview 58 with senior union official, Melbourne.

143. Interview 58 with senior union official, Melbourne; Interview 70 with senior environmental official, Sydney.

144. Interview 68 with senior policy advocate, Canberra; Interview 74 with senior policy advocate, Sydney.

145. Department of Climate Change, *Carbon Pollution Reduction Scheme: Australia's Low Pollution Future* White Paper, December 15, 2008.

146. Interview 56 with senior bureaucrat, Canberra; Interview 65 with senior bureaucrat, Melbourne.

147. Interview 55 with two senior bureaucrats, Canberra; Interview 56 with senior bureaucrat, Canberra.

148. *For example, the threshold at which EITE's were eligible for free permit allocation was lowered and made more flexible.

149. Pensioners and other allowance recipients would see a 2.5 percent payment increase. For households, the Low Income Tax Offset was increased to $390. Changes were also made to the Family Tax Benefit system. The government estimated that 89 percent of low-income households would receive compensation at a rate of 120 percent of their estimated cost of living increases. Department of Climate Change, *CPRS*, 15.

150. Interview 98 with elected official, Cambridge, MA.

151. While all the SCCC partners were present during negotiations, SCCC members participated as a single actor, jointly representing the perspectives they could agree on. Interview 70 with senior environmental official, Sydney. Despite the representation of carbon polluting sectors on both industry and labor sides, there was almost no advance work between labor and capital to jointly represent sectoral interests. Interview 66 with senior business lobby official, Melbourne.

152. Interview 56 with senior bureaucrat, Canberra.

153. Interview 74 with senior policy advocate, Sydney.

154. Agreeing to this less ambitious target was controversial within ACF, even after leadership endorsed the amended CPRS package. Interview 70 with senior environmental official, Sydney.

155. Interview 75 with senior policy advocate, Melbourne.

156. Interview 56 with senior bureaucrat, Canberra.

157. Interview 66 with senior business lobby official, Melbourne.

158. Ibid.

159. The ACCI remained bitterly opposed, because of its single-minded focus on keeping down energy costs as well as idiosyncratic organizational issues. Interview 74 with senior policy advocate, Sydney. The bargain prompted the Minerals Council to intensify its public campaign. Though some member companies would privately concede the deal was reasonable, they kept up their public pressure to shift the distributive costs even further away from major fossil fuel interests.

160. Interview 74 with senior policy advocate, Sydney.

161. Interview 55 with two senior bureaucrats, Canberra.

162. *For example, Liberal Cory Bernardi promoted the prominent skeptical book *Thank God for Carbon*.

163. Joyce also promoted public tours by climate skeptics, presenting anti-climate messages to rural audiences across the country. ABC, program transcript of Sarah Ferguson's report "Malcolm and the Malcontents," *Four Corners*, first broadcast November 9, 2009.

164. *Negotiations initially were led by Andrew Robb, who stepped away for health reasons and was replaced by Ian MacFarlane.

165. *He proposed an intensity-based baseline credit scheme for the electricity sector. Interview 66 with senior business lobby official, Melbourne.

166. *Interview 66 with senior business lobby official, Melbourne.

167. Ibid.; Interview 74 with senior policy advocate, Sydney.

168. *Interview 66 with senior business lobby official, Melbourne; Interview 70 with senior environmental official, Sydney.

169. *Sandra O'Malley, "Turnbull Threatens to Quit over ETS." *The Age*, October 1, 2009.

170. ABC, "Malcolm and the Malcontents," *Four Corners*.

171. Interview 83 with senior policy advisor to elected official, by phone.

172. Interview 74 with senior policy advocate, Sydney.

173. EITEs moved from 90 percent free permits and 60 percent free permits to 94.5 percent free permits and 66 percent free permits.

174. *Interview 66 with senior business lobby official, Melbourne.

175. For example, $1.23 billion AUD was directed through the Coal Sector Adjustment Scheme to provide permits for coal mines with fugitive emissions, reducing mine liability from $20 AUD to $5 AUD per unit of saleable coal. An additional $20 million AUD to $270 million AUD went to a Coal Sector Abatement Fund. "Details of Proposed CPRS Changes," November 24, 2009.

176. TCAP constituted a 75 percent increase in assistance to the electricity sector. The funds were to be used over two years to compensate for marginal increases in electricity costs to manufacturing and mining associated with CPRS introduction.

177. Interview 70 with senior environmental official, Sydney.

178. Interview 76 with senior environmental official, by Skype.

179. Interview 74 with senior policy advocate, Sydney.

180. Interview 55 with two senior bureaucrats, Canberra. This allowance workaround was firmly entrenched in the bureaucracy's design strategy as early as the PMTGET. Interview 69 with senior bureaucrat, Sydney.

181. Interview 75 with senior policy advocate, Melbourne.

182. Interview 43 with senior political advisor to elected official, Sydney; Interview 46 with elected official, Canberra.

183. Interview 73 with senior union official, Sydney.

184. Interview 56 with senior bureaucrat, Canberra.

185. Interview 66 with senior business lobby official, Melbourne.

186. Interview 46 with elected official, Canberra.

187. Interview 43 with senior political advisor to elected official, Sydney; Interview 74 with senior policy advocate, Sydney. Labor felt it was stealing the Green's electoral turf, and the greens were taking far-left positions as a result of political ambitions, not because they were good policy. Interview 98 with elected official.

188. Negotiating with the Liberals over the ETS was seen as preferable to being "blackmailed" by the Greens. Interview 43 with senior political advisor to elected official, Sydney.

189. Interview 76 with senior environmental official, by Skype.

190. Interview 45 with senior policy advisor to political party, Canberra.

191. Interview 53 with elected official, Canberra.

192. Interview 45 with senior policy advisor to political party, Canberra. Generally, the Greens pushed a 40 percent reduction below 1990-level target by 2020. The Green Party had its own divided beliefs about climate policy instruments. Generally, there was only superficial buy-in for emissions trading as a policy instrument. A major faction would have preferred carbon taxes, perhaps at a rate near to $25 AUD per ton CO_2.

193. Ibid.

194. Ibid.

195. Ibid.

196. Interview 53 with elected official, Canberra.

197. *By establishing a short-term fixed price period, the Greens believed the emissions trading architecture could be set up immediately while Australian parties continued to debate emissions reduction trajectories. Australian Green Party, "Greens Propose Garnaut's Interim Solution to Break CPRS Deadlock," press release, January 21, 2010.

198. Interview 46 with elected official, Canberra' Interview 47 with elected official, Canberra.

199. Interview 43 with senior political advisor to elected official, Sydney.

200. *Interview 57 with senior Australian bureaucrat, by Skype. Lenore Taylor, "ETS off the Agenda until Late Next Term," *Sydney Morning Herald*, April 27, 2010. A number of interacting forces contributed to this weakening: the massive drought had broken suddenly in 2009, power prices rose as a result of transmission infrastructure investments, and the global financial crisis continued to pose economic challenges. Interview 47 with elected official, Canberra; Interview 49 with senior agency official, Sydney.

201. Interview 72 with senior union official, Sydney.

202. Interview 66 with senior business lobby official, Melbourne.

203. Interview 53 with elected official, Canberra; Interview 60 with senior environmental official, Canberra; Interview 67 with senior business lobby official, Sydney. Gillard's senior staff were disengaged and did not show "a lot of strong conviction on the issue."

204. Though she is reported to have argued for a more conditional commitment on global action (Chalmers 2013); Interview 75 with senior policy advocate, Melbourne.

205. Interview 45 with senior policy advisor to political party, Canberra.

206. Tim Leslie, "Climate Panel a 'Cynical Delaying Tactic,'" *ABC News*, July 24, 2010.

207. Turnbull continued to criticize Abbott's climate leadership, even in the midst of the campaign. "Coalition Climate Policy Not Ideal: Turnbull," *ABC News*. July 22, 2010.

208. Leader of the Opposition Tony Abbott, "Direct Action on the Environment and Climate Change," press release, February 2, 2010.

209. Interview 66 with senior business lobby official, Melbourne.

210. Ibid.

211. Interview 74 with senior policy advocate, Sydney; Interview 75 with senior policy advocate, Melbourne.

212. Interview 53 with elected official, Canberra.

213. Interview 45 with senior policy advisor to political party, Canberra.

214. Interview 44 with government advisor, Canberra.

215. *Their participation was proposed by the Greens, who thought experts would help increase the ambition of the committee by anchoring its deliberations in climate science and economics.

216. In leaked copies of the "Blue Book"—the bureaucratic transition document prepared in case the opposition (Liberals) won the 2010 election—the bureaucracy remained committed to emissions trading, questioning the cost and efficiency of Direct Action-oriented policies. Government of Australia, Blue Book, 2010.

217. Labor leaders did engage in consultative roundtables as MPCCC discussions unfolded; however, social partner comments did not substantially change committee debates or policy decisions. Interview 45 with senior policy advisor to political party, Canberra.

218. Interview 75 with senior policy advocate, Melbourne.

219. *New Minister for Energy and Climate Change Greg Combet had been the senior industrial officer and then president of ACTU. He was seen by union leaders, particularly industrial unions, as a sympathetic figure. Interview 43 with senior political advisor to elected official, Sydney; Interview 54 with senior business lobby official, Melbourne.

220. *Pro-climate Burrow had left as head of the ACTU in 2010, removing a major pro-climate figure from a labor position of influence. Interview 72 with senior union official, Sydney.

221. Interview 58 with senior union official, Melbourne.

222. *"Union Threatens Carbon Tax Revolt," *ABC News*, April 14, 2011.

223. Interview 73 with senior union official, Sydney.

224. Interview 67 with senior business lobby official, Sydney. Industry is deeply skeptical of Greens because they have a platform outside of environmental issues that is seen as left of Labor and therefore anti-business, rooted in their background of some factions in the Communist party. Interview 77 with senior business leader, Melbourne.

225. Interview 67 with senior business lobby official, Sydney.

226. *Interview 70 with senior environmental official, Sydney.

227. *Interview 74 with senior policy advocate, Sydney.

228. Interview 45 with senior policy advisor to political party, Canberra.

229. Interview 44 with government advisor, Canberra.

230. There were subtle differences. For instance, EITE permit allocations changed a bit. However, it is difficult to parse whether the changes shaped policy ambition, because of different accounting rules. Interview 56 with senior bureaucrat, Canberra.

231. Interview 44 with government advisor, Canberra.

232. Interview 48 with Australian elected official, by phone.

233. This suggestion was not taken seriously by other stakeholders and was never included in Treasury modeling. Interview 44 with government advisor, Canberra; Interview 53 with elected official, Canberra.

234. The 5 percent goal was anticipated to be a 30 percent reduction compared to a business-as-usual scenario. Interview 49 with senior agency official, Sydney.

235. *Interview 53 with elected official, Canberra.

236. Interview 48 with Australian elected official, by phone; Interview 67 with senior business lobby official, Sydney.

237. Interview 46 with elected official, Canberra.

238. Interview 58 with senior union official, Melbourne; Interview 66 with senior business lobby official, Melbourne.

239. The steel industry was facing economic pressures as a result of currency volatility. Interview 73 with senior union official, Sydney; *ABC News*, "AWU—Steel Must Be Exempt from Carbon Tax," April 15, 2011.

240. Interview 72 with senior union official, Sydney. Proponents brag that it ended up 102 percent compensation for the sector.

241. Interview 53 with elected official, Canberra. They were pushed to do this, in part, by the ACF.

242. *Interview 67 with senior business lobby official, Sydney.

243. *Ibid.

244. Interview 65 with senior bureaucrat, Melbourne.

245. Interview 45 with senior policy advisor to political party, Canberra.

246. Some individual components were legislated separately over the following year. For instance, legislation creating the Clean Energy Finance Corporation was passed on June 25, 2012 and legislation creating the Climate Change Authority was passed on July 1, 2012.

247. Coalition talking points from September 2011 emphasized Gillard's dishonesty and inflated estimates of the economic harm that the policy threatened: Coalition in Confidence, "Carbon Tax Legislation—Talking Points," September 2011.

248. Gillard allegedly made the admission in an effort to avoid word games. Interview 59 with elected official, Canberra.

249. ABC, "Turnbull Discusses Broadband and Climate Policy," *Lateline*, May 18, 2011.

250. Carbon polluters saw the Minerals Council had been successful at vocal public campaigns, and felt more comfortable publicly opposing reform efforts. Interview 77 with senior business leader, Melbourne.

251. *The consulting company itself became worried when senior Liberals reacted antagonistically to early report drafts that criticized the Direct Action plan. The final report recommendations were so broad they were politically meaningless. Interview 66 with senior business lobby official, Melbourne.

252. Pro-climate forces in the business community had few incentives to speak up. Interview 54 with senior business lobby official, Melbourne.

253. Interview 45 with senior policy advisor to political party, Canberra.

254. Interview 80 with senior bureaucrat, by Skype.

255. Interview 53 with elected official, Canberra.

256. While implementing the CPM legislation, bureaucrats discovered that they could introduce the legislated price because of arcane technicalities that had not been addressed in the initial legislation. However, Oakeshott refused to support legislation correcting these issues, having always objected to the price floor during the MPCCC. Interview 45 with senior policy advisor to political party, Canberra; Interview 53 with elected official, Canberra. The result was that another key demand of the Green Party had to be dropped from the package.

257. *The idea was to allow European and Australian companies to exchange pollution permits across the two systems. Europe supported the idea as a way to manage EU-permit oversupply problems, since all internal models showed that Australia would be a net permit importer. Interview 56 with senior bureaucrat, Canberra.

258. Interview 66 with senior business lobby official, Melbourne.

259. Ibid.

260. Again, the move was supported privately by some carbon-intensive businesses, but did not see benefit in expressing this publicly.

261. The commission fundraised to continue as the privately funded Climate Council, becoming the most successful crowdsourced campaign in Australian history. Interview 44 with government advisor, Canberra.

262. Abbott would later appoint a panel to review Australia's renewable energy targets in February 2014, led by climate change skeptic Dick Warburton.

263. Due to a quirk of the Australian system, senators elected during September only take their seats the following July, creating a nine-month lame-duck session.

264. Climate Change Authority, *Targets and Progress Review Final Report*, February 2014.

265. However, the conditions proved more flexible than anyone had anticipated. Interview 65 with senior bureaucrat, Canberra. One of the critical components of this shift was the inclusion of carbon budget considerations in the CEFP deal, which became a powerful justification for early action that opponents had not foreseen. Interview 75 with senior policy advocate, Melbourne.

266. The chamber's unique proportional representation and transferable ballot electoral system often provide voice to small parties.

267. Lenore Taylor, "PUP Senators Will Vote to Repeal Carbon Price but Back Emissions Trading," *Guardian*, June 25, 2014.

268. Interview 51 with senior bureaucrat, Canberra; Interview 55 with two senior bureaucrats, Canberra.

269. Interview 69 with senior bureaucrat, Sydney.

270. Lenore Taylor. "'Grand Bargain' May Secure Enough Support for Direct Action to Pass Senate," *Guardian*, August 22, 2014.

271. Ibid.

272. In practice, this shares structural similarities with a baseline credit scheme.

273. For instance, internally, anti-climate members of the Liberal Cabinet pushed for the ERF to establish three-year contracts that would be unusable; advocates worried that even the five-year contracts offered under the current ERF are unlikely to be attractive to many facilities. Interview 61 with energy consultant, Melbourne.

274. Lenore Taylor, "Emissions Trading Will Be Back if Direct Action Proves Ineffective," *Guardian*, October 29, 2014.

275. Lenore Taylor, "Direct Action: Coalition Secures $2.5bn Plan amid Fears over Emissions Target," *Guardian*, October 29, 2014.

276. Palmer United Party, "Palmer Saves Emissions Trading Scheme," press release, October 30, 2014.

277. *Climate Institute, *A Review of Subsidy and Carbon Price Approaches to Greenhouse Gas Emission Reduction*, SH43458 Emission Reduction Fund Policy Review, August 14, 2013; Climate Institute, *Coalition Climate Policy and the National Climate Interest* Policy Brief, August 2013.

278. Interview 61 with energy consultant, Melbourne; Interview 64 with farm lobby official, Melbourne; Interview 66 with senior business lobby official, Melbourne.

279. Interview 75 with senior policy advocate, Melbourne; Interview 77 with senior business leader, Melbourne.

280. Interview 70 with senior environmental official, Sydney; Interview 62 with mid-level environmental official, Melbourne.

281. Greg Hunt and Rufus Black, "A Tax to Make Polluters Pay: The Application of Pollution Taxes within the Australian Legal System," May 20, 1990.

282. Lenore Taylor, "Turnbull Government Signals New Approach to Climate Policy," *Guardian*, September 21, 2015.

7 Exploring the Theory's Generalizability

1. See footnote 13 in Hatch 1995.

2. As in Norway, the United States, and Australia, green groups prioritized other environmental issues and were not major drivers of climate policymaking during this period (Hatch 1995).

3. These exemptions were later reduced to 40 percent in the mid-2000s.

4. German feed-in-tariffs had first been introduced in 1991 under Chancellor Kohl.

5. The same consumer-focused emphasis shaped the 2002 Combined Heat and Power Law (*Kraft-Wärme-Kopplungs Modernisierungsgesetz*) that promoted co-generation facilities (Hughes and Urpelainen 2015).

6. It did maintain Green Party support.

7. EU Commission, C(2004) 2515/2 final, July 7, 2004.

8. Shaping Germany's Future: Coalition treaty between CDU/CSU and SPD, February 2014.

9. Arne Delfs, "Merkel Embraces Coal as Rookie Lawmaker Makes Mark on Policy," *Bloomberg Business*, December 12, 2013.

10. Swantje Küchler and Rupert Wronski, *Was Strom wirklich kostet*, January 2015. Forum Ökologish-Soziale Marktwirtschaft and Greenpeace Energy.

11. Erika Körner, "Germany Denies Plans to Close Old Coal Plants in Sprint to 2020 Targets," *Euractiv*, November 11, 2014.

12. Michael Bauchmäseller, "Gabriel läutet Ausstieg aus der Kohlekraft ein," *Süddeutsche Zeitung*, March 19, 2015.

13. Arthur Neslen, "German Backlash Grows against Coal Power Clampdown," *The Guardian*, April 14, 2015.

14. Caroline Copley and Markus Wacket, "Germany to Review Jobs Impact of Coal Levy Before Decision," *Reuters*, April 24, 2015.

15. Schoppa (1991) describes a zoku as an "unofficial clique of Diet members."

16. This relative autonomy has declined over time, but Japan still stands out relative to other countries in this respect.

17. Before 2001, METI was known as the Ministry of International Trade and Industry (MITI).

18. This partly stems from postwar rules limiting civil service tenure that shifted increasing numbers of then MITI personnel into industry (Wallace 1995).

19. Japan sought to host COP 3 in part to position itself for a Security Council seat. This decision did not reflect national climate policymaking prioritization (Oshitani 2006).

20. However, industrial coal used in process industries (iron, steel, coke, and cement) was exempted.

21. Speech on the environment by Prime Minister Taro Aso, June 10, 2009.

22. The government subsidized one-third of emissions-reduction project costs for JVETS participants.

23. Local-level regional emissions trading schemes were established in Tokyo (2010) and neighboring Saitama (2011).

24. However, this shadow case focuses only on federal policymaking.

25. Chris Morris, "Liberal Green Shift Is 'Green Shaft,' Says Harper," *Toronto Star*, August 14, 2008.

26. Joanna Smith, "Carbon Tax Would Hurt Poor, NDP Says," *Toronto Star*, May 23, 2008.

27. Jason Dion, David Sawyer, and Philip Gass, "A Climate Gift or a Lump of Coal? The Emission Impacts of Canadian and U.S. Greenhouse Gas Regulations in the Electricity Sector," International Institute for Sustainable Development, September 18, 2014.

28. CBC Radio, "Preston Manning Argues Conservatives Should Support Carbon Pricing," *The Current*, December 3, 2014.

29. James McCarthy, "Now Down to Just 100 Active Members—the Decline of the Once-Mighty NUM," *Wales Online*, December 26, 2015.

30. For instance, see Thatcher's General Assembly address on November 8, 1989. Environmental taxes more broadly had emerged on the agenda into the late 1980s after media attention to a Department of Energy report on the topic; the topic was then described in the Conservative government's 1992 environment White Paper (Jordan et al. 2003).

31. Consistent with theoretical predictions about a policy's distributive burden in pluralist countries, even these modest tax reforms were coupled to consumer compensation, here in the form of revenue for homeowner insulation subsidies.

32. Initially, a function of pkg or pkW—starting at 0.43p/kWh on electricity, 0.15p/kWh on natural gas, 1.17p/kg coal, and 0.96 liquid petroleum gas (heating source).

33. If private sector actors failed to meet these negotiated benchmarks, they would be forced back into the CCL system for at least two years.

34. Business Strategies. *Climate Change Levy: Impact on the UK economy.* July 1999.

35. Labor actors also felt they had received insufficient backing from green groups during the CCL debate (Carter 2008).

36. Until that time, even green groups had not positioned climate as a clear lobbying priority (Carter 2014).

37. By virtue of increasing levels of "euroscepticism," leaders also opposed EU-level environmental reforms in the UK (Carter 2009).

38. This included suggestions that Cameron himself moved away from green issues in internal policy planning processes. Rowena Mason, "David Cameron at Centre of 'Get Rid of all the Green Crap' Storm," *The Guardian*, November 21, 2013.

8 Disrupting the Logic of Double Representation

1. IPCC, "Summary for Policymakers," in *Global Warming of 1.5°C.* An IPCC Special Report on the impacts of global warming of 1.5°C above pre-industrial levels and related global greenhouse gas emission pathways, in the context of strengthening the global response to the threat of climate change, sustainable development, and efforts to eradicate poverty. Ed. V. Masson-Delmotte et al. (Geneva: GenevaWorld Meteorological Organization, 2018).

2. Darren Samuelson and Ben Geman, "Moderate House Democrats Lay Out Concerns with Draft Climate Bill," *New York Times*, April 22, 2009.

3. Interview 59, elected official, Canberra.

4. As a Statoil executive recently declared: "The oil that should be produced from now to the end should, to a very large extent, come from Norway, because it's good oil. It's safe and secure and delivered from us with low carbon emissions." Henry Fountain, "Norwegians Turn Ambivalent on Statoil, Their Economic Bedrock," *New York Times*, December 30, 2014.

5. For a variant of this perspective, see also Victor 2011.

6. White House Office of the Press Secretary, "U.S.-China Joint Announcement on Climate Change," November 11, 2014.

7. Hyunjoo Jin, "Hyundai Union Head Fears GM-like Crisis; Says Electric Cars Destroy Jobs," *Reuters*, March 26, 2018.

8. Letter from Sean McGarvey, president of North America's Building Trades Unions, to Richard Trumka, president of AFL-CIO, September 14, 2016.

References

Abramowitz, A., and K. Saunders. 2008. "Is Polarization a Myth?" *Journal of Politics* 70 (2): 542–555.

Adams, S. 2006. "Promotion, Competition, Captivity: The Political Economy of Coal." *Journal of Policy History* 18 (1): 74–95.

Aidt, T., and B. Dallal. 2008. "Female Voting Power: The Contribution of Women's Suffrage to the Growth of Social Spending in Western Europe (1869–1960)." *Public Choice* 134 (3–4): 391–417.

Aklin, M. 2016. "Re-exploring the Trade and Environment Nexus through the Diffusion of Pollution." *Environmental and Resource Economics* 64 (4): 663–682.

Aklin, M., and J. Urpelainen. 2013. "Political Competition, Path Dependence, and the Strategy of Sustainable Energy Transitions." *American Journal of Political Science* 57 (3): 643–658.

Allern, E. 2010. *Political Parties and Interest Groups in Norway*. Colchester, UK: ECPR Press.

Andersen, M., and D. Liefferink. 1999. *European Environmental Policy*. Manchester, UK: Manchester University Press.

Bailey, I., and S. Rupp. 2005. "Geography and Climate Policy: A Comparative Assessment of New Environmental Policy Instruments in the UK and Germany." *Geoforum* 36 (3): 387–401.

Bailey, I., I. MacGill, R. Passey, and H. Compston. 2012. "The Fall (and Rise) of Carbon Pricing in Australia: A Political Strategy Analysis of the Carbon Pollution Reduction Scheme." *Environmental Politics* 21 (5): 691–711.

Balco, G. 1999. "Hot Air and Congress." *Eos* 80 (29): 320–320.

Barrett, S. 2003. *Environment and Statecraft*. Oxford, UK: Oxford University Press.

Barrett, S. 2006. "Climate Treaties and 'Breakthrough' Technologies." *American Economic Review* 96 (2): 22–25.

Barrett, S., and R. Stavins. 2003. "Increasing Participation and Compliance in International Climate Change Agreements." *International Environmental Agreements* 3 (4): 349–376.

Bartosiewicz, P., and M. Miley. 2013. "The Too Polite Revolution: Why the Recent Campaign to Pass Comprehensive Climate Legislation Failed." Paper prepared for Symposium on the Politics of Americas Fight Against Global Warming, February 4, 2013, Harvard University, Cambridge, MA.

Bättig, M., and T. Bernauer. 2009. "National Institutions and Global Public Goods: Are Democracies More Cooperative in Climate Change Policy?" *International Organization* 63 (2): 281–308.

Baumgartner, F., J. Berry, M. Hojnacki, D Kimball, and B. Leech. 2009. *Lobbying and Policy Change: Who Wins, Who Loses, and Why*. Chicago: University of Chicago Press.

Bernstein, S. 2001. *The Compromise of Liberal Environmentalism*. New York: Columbia University Press.

Beuermann, C., and T. Santarius. 2006. "Ecological Tax Reform in Germany." *Energy Policy* 34 (8): 917–929.

Bowen, A., and J. Rydge. 2011. *Climate-Change Policy in the United Kingdom*. OECD Economic Department Working Paper 886.

Boykoff, M. T. 2011. *Who Speaks for the Climate?* Cambridge, UK: Cambridge University Press.

Breetz, H. 2013. "Fueled by Crisis: US Alternative Fuel Policy, 1975–2007." PhD thesis. Massachusetts Institute of Technology.

Breetz, H., M. Mildenberger, and L. Stokes. 2018. "The Political Logics of Clean Energy Transitions." *Business and Politics* 20 (4): 492–522.

Broadbent, J. 1999. *Environmental Politics in Japan*. Cambridge, UK: Cambridge University Press.

Brulle, R. J. 2014. "Institutionalizing Delay: Foundation Funding and the Creation of US Climate Change Counter-Movement Organizations." *Climatic Change* 122 (4): 681–694.

Brulle, R., J. Carmichael, and J. Jenkins. 2012. "Shifting Public Opinion on Climate Change: An Empirical Assessment of Factors Influencing Concern over Climate Change in the US, 2002–2010." *Climatic Change* 114 (2): 169–188.

Bruvoll, A., and B. Larsen. 2002. "Greenhouse Gas Emissions in Norway: Do Carbon Taxes Work?" Statistics Norway, Research Department Discussion Paper 337.

Bulkeley, H. 2000. "The Formation of Australian Climate Change Policy: 1985–1995." In *Climate Change in the South Pacific*, ed. A. Gillespie and W. Burns, 33–50. New York: Springer.

Bulkeley, H. 2001. "No Regrets?: Economy and Environment in Australia's Domestic Climate Change Policy Process." *Global Environmental Change* 11 (2): 155–169.

Burgmann, V. 2013. "From 'Jobs versus Environment' to 'Green-Collar Jobs.'" In *Trade Unions in the Green Economy*, ed. N. Räthzel and D. Uzzell, 131–145. Oxon, UK: Routledge.

Carter, N. 2008. "Combating Climate Change in the UK." *Political Quarterly* 79 (2): 194–205.

Carter, N. 2009. "Vote Blue, Go Green? Cameron's Conservatives and the Environment." *Political Quarterly* 80 (2): 233–242.

Carter, N. 2014. "The Politics of Climate Change in the UK." *Wiley Interdisciplinary Reviews: Climate Change* 5 (3): 423–433.

Carter, N. 2015. "The Greens in the UK General Election of 7 May 2015." *Environmental Politics* 24 (6): 1055–1060.

Carter, N., and B. Clements. 2015. "From 'Greenest Government Ever' to 'Get Rid of all the Green Crap': David Cameron, the Conservatives and the Environment." *British Politics* 10 (2): 204–225.

Cass, L. R. 2006. *The Failures of American and European Climate Policy*. Albany: SUNY Press.

Chalmers, J. 2013. *Glory Daze: How a World-Bearing Nation Got so Down on Itself*. Melbourne, Australia: Melbourne University Press.

Cheon, A., and J. Urpelainen. 2013. "How Do Competing Interest Groups Influence Environmental Policy?" *Political Studies* 61 (4): 874–897.

Christoff, P., and R. Eckersley. 2011. "Comparing State Responses." In *Oxford Handbook of Climate Change and Society*, ed. J. Dryzek, R. Norgaard, and D. Schlosberg, 431–448. Oxford, UK: Oxford University Press.

Clements, B. 2012. "Exploring Public Opinion on the Issue of Climate Change in Britain." *British Politics* 7 (2): 183–202.

Clements, B. 2014. "Political Party Supporters' Attitudes towards and Involvement with Green Issues in Britain." *Politics* 34 (4): 362–377.

Conconi, P., and C. Perroni. 2002. "Issue Linkage and Issue Tie-in in Multilateral Negotiations." *Journal of International Economics* 57 (2): 423–447.

Cornwell, A., and J. Creedy. 1995. "A Carbon Tax for Australia." *Economic Papers* 14 (4): 16–28.

Cornwell, A., and J. Creedy. 1996. "Carbon Taxation, Prices and Inequality in Australia." *Fiscal Studies* 17 (3): 21–38.

Crowther, T., K. Todd-Brown, C. Rowe, W. Wieder, J. Carey, M. Machmuller, B. Snoek et al. 2016. "Quantifying Global Soil Carbon Losses in Response to Warming." *Nature* 540 (7631): 104–108.

Darkin, B. 2006. "Pledges, Politics and Performance: An Assessment of UK Climate Policy." *Climate Policy* 6 (3): 257–274.

Daugbjerg, C., and G. T. Svendsen. 2001. *Green Taxation in Question*. New York: Palgrave Macmillan.

Dryzek, J. S. 2013. *The Politics of the Earth*. Oxford, UK: Oxford University Press.

Duit, A. 2016. "The Four Faces of the Environmental State." *Environmental Politics* 25 (1): 69–91.

Dupuis, J., and R. Biesbroek. 2013. "Comparing Apples and Oranges: The Dependent Variable Problem in Comparing and Evaluating Climate Change Adaptation Policies." *Global Environmental Change* 23 (6): 1476–1487.

Economou, N. 1996. "Australian Environmental Policy Making in Transition: The Rise and Fall of the Resource Assessment Commission." *Australian Journal of Public Administration* 55 (1): 12–22.

Eidlin, B. 2016. "Why Is There No Labor Party in the United States?" *American Sociological Review* 81 (3): 488–516.

Enloe, C. 1975. *The Politics of Pollution in a Comparative Perspective*. New York: McKay.

Erlandson, D. 1994. "The BTU Tax Experience: What Happened and Why It Happened." *Pace Environmental Law Review* 12:173–184.

Esping-Andersen, G. 1993. *The Three Worlds of Welfare Capitalism*. New York: John Wiley & Sons.

Feygina, I., J. Jost, and R. Goldsmith. 2010. "System Justification, the Denial of Global Warming, and the Possibility of System-sanctioned Change." *Personality and Social Psychology Bulletin* 36 (3): 326–338.

Fitzpatrick, T., and M. Cahill. 2002. *Environment and Welfare*. New York: Palgrave Macmillan.

Geddes, B. 1990. "How the Cases You Choose Affect the Answers You Get: Selection Bias in Comparative Politics." *Political Analysis* 2 (1): 131–150.

Gelbspan, R. 1997. *The Heat Is On: The High Stakes Battle over Earth's Threatened Climate.* Boston: Addison Wesley.

Giddens, A. 2009. *The Politics of Climate Change.* Cambridge, UK: Cambridge University Press.

Gifford, R. 2011. "The Dragons of Inaction: Psychological Barriers That Limit Climate Change Mitigation and Adaptation." *American Psychologist* 66 (4): 290–302.

Godal, O., and B. Holtsmark. 2001. "Greenhouse Gas Taxation and the Distribution of Costs and Benefits: The Case of Norway." *Energy Policy* 29 (8): 653–662.

Gore, A. 1993. *Earth in Balance.* Boston: Houghton Mifflin.

Guber, D. 2012. "A Cooling Climate for Change? Party Polarization and the Politics of Global Warming." *American Behavioral Scientist* 24 (1): 93–115.

Gulbrandsen, L., A. Underdal, D. Victor, and J. Wettestad. 2017. "Theory and Method." In *The Evolution of Carbon Markets*, ed. J. Wettestad and L. Gulbrandsen 13–29. Oxon, UK: Routledge.

Hacker, J., and P. Pierson. 2002. "Business Power and Social Policy: Employers and the Formation of the American Welfare State." *Politics & Society* 30 (2): 277–325.

Hajer, M. 1995. *The Politics of Environmental Discourse.* Oxford, UK: Clarendon Press.

Hall, P. A., and D. Soskice. 2001. *Varieties of Capitalism: The Institutional Foundations of Comparative Advantage.* Oxford, UK: Oxford University Press.

Hamilton, C. 2007. *Scorcher: The Dirty Politics of Climate Change.* Melbourne, Australia: Black Inc.

Han, H., and C. Barnett-Loro. 2018. "To Support a Stronger Climate Movement, Focus Research on Building Collective Power." *Frontiers in Communication* 3 (55), https://doi.org/10.3389/fcomm.2018.00055.

Hansen, J., M. Sato, P. Kharencha, D. Beerling, R. Berner, V. Masson-Delmotte, M. Pagani et al. 2008. "Target Atmospheric CO_2: Where Should Humanity Aim?" *Open Atmospheric Science Journal* 2:217–231.

Harring, N., and J. Sohlberg. 2017. "The Varying Effects of Left–Right Ideology on Support for the Environment: Evidence from a Swedish Survey Experiment." *Environmental Politics* 26 (2): 278–300.

Harrison, K. 2010a. "The Comparative Politics of Carbon Taxation." *Annual Review of Law and Social Science* 6:507–529.

Harrison, K. 2010b. "The Struggle of Ideas and Self-interest in Canadian Climate Policy." In *Global Commons, Domestic Decisions*, ed. K. Harrison and L. Sundstrom, 168–200. Cambridge, MA: MIT Press.

Harrison, K. 2012. "A Tale of Two Taxes: The Fate of Environmental Tax Reform in Canada." *Review of Policy Research* 29 (3): 383–407.

Harrison, K., and L. Sundstrom. 2010a. "Conclusion: The Comparative Politics of Climate Change." In *Global Commons, Domestic Decisions*, ed. K. Harrison and L. Sundstrom, 261–290. Cambridge, MA: MIT Press.

Harrison, K., and L. M. Sundstrom. 2010b. "Introduction: Global Commons, Domestic Decisions." In *Global Commons, Domestic Decisions*, ed. K. Harrison and L. Sundstrom, 1–22. Cambridge, MA: MIT Press.

Hatch, M. 1995. "The Politics of Global Warming in Germany." *Environmental Politics* 4 (3): 415–440.

Hawke, B. 1994. *The Hawke Memoirs*. London: William Heinemann.

Heal, G. 2000. *Nature and the Marketplace: Capturing the Value of Ecosystem Services.* Washington, DC: Island Press.

Heitzig, J., K. Lessmann, and Y. Zou. 2011. "Self-enforcing Strategies to Deter Free-riding in the Climate Change Mitigation Game and Other Repeated Public Good Games." *Proceedings of the National Academy of Sciences* 108 (38): 15739–15744.

Hoberg, G., and K. Harrison. 1994. "It's Not Easy Being Green: The Politics of Canada's Green Plan." *Canadian Public Policy* 20 (2): 119–137.

Hodne, F. 1975. *An Economic History of Norway, 1815–1970*. Trondheim, Norway: Tapir Press.

Hoel, M., and A. De Zeeuw. 2010. "Can a Focus on Breakthrough Technologies Improve the Performance of International Environmental Agreements?" *Environmental and Resource Economics* 47 (3): 395–406.

Hoffmann, M. 2011. *Climate Governance at the Crossroads*. Oxford, UK: Oxford University Press.

Hovden, E., and G. Lindseth. 2002. "Norwegian Climate Policy 1989–2002." In *Realizing Rio in Norway*, ed. W. Lafferty, M. Nordskog, and H. Aakre, 143–168. Oslo, Norway: ProSus.

Hu, J., W. Crijns-Graus, L. Lam, and A. Gilbert. 2015. "Ex-ante Evaluation of EU ETS during 2013–2030: EU-Internal Abatement." *Energy Policy* 77:152–163.

Hughes, L. 2012. "Climate Converts: Institutional Redeployment, Industrial Policy, and Public Investment in Energy in Japan." *Journal of East Asian Studies* 12 (1): 89–117.

Hughes, L., and J. Urpelainen. 2015. "Interests, Institutions, and Climate Policy: Explaining the Choice of Policy Instruments for the Energy Sector." *Environmental Science & Policy* 54:52–63.

Hulme, M. 2010. *Why We Disagree about Climate Change*. Cambridge, UK: Cambridge University Press.

Inglehart, R. 2015. *The Silent Revolution: Changing Values and Political Styles among Western Publics*. Princeton, NJ: Princeton University Press.

Iversen, T., and F. Rosenbluth. 2010. *Women, Work, and Politics: The Political Economy of Gender Inequality*. New Haven, CT: Yale University Press.

Iversen, T., and D. Soskice. 2006. "Electoral Institutions and the Politics of Coalitions: Why Some Democracies Redistribute More Than Others." *American Political Science Review* 100 (2): 165–181.

Jacobs, A. 2008. "The Politics of When: Redistribution, Investment and Policy Making for the Long Term." *British Journal of Political Science* 38 (2): 193–220.

Jacobs, A. 2011. *Governing for the Long Term*. Cambridge, UK: Cambridge University Press.

Jacobs, A. 2016. "Policy Making for the Long Term in Advanced Democracies." *Annual Review of Political Science* 19:433–454.

Jahn, D. 2016. *The Politics of Environmental Performance*. Cambridge, UK: Cambridge University Press.

Javeline, D. 2014. "The Most Important Topic Political Scientists Are Not Studying: Adapting to Climate Change." *Perspectives on Politics* 12 (2): 420–434.

Jenkins, J. 2014. "Political Economy Constraints on Carbon Pricing Policies: What Are the Implications for Economic Efficiency, Environmental Efficiency, and Climate Policy Design?" *Energy Policy* 69:467–477.

Jones, C., and D. Levy. 2007. "North American Business Strategies Towards Climate Change." *European Management Journal* 25 (6): 428–440.

Jordan, A., R. Wurzel, R. Zito, and L. Brückner. 2003. "Policy Innovation or 'Muddling Through'? 'New' Environmental Policy Instruments in the United Kingdom." *Environmental Politics* 12 (1): 179–200.

Kahan, D. 2015. "Climate-Science Communication and the Measurement Problem." *Political Psychology* 36 (S1): 1–43.

Kamieniecki, S. 2006. *Corporate America and Environmental Policy*. Palo Alto, CA: Stanford University Press.

Karapin, R. 2016. *Political Opportunities for Climate Policy*. Cambridge, UK: Cambridge University Press.

Kasa, S. 2000. "Policy Networks as Barriers to Green Tax Reform: The Case of CO_2-Taxes in Norway." *Environmental Politics* 9 (4): 104–122.

Kasa, S., and A. Underthun. 2010. "Navigation in New Terrain with Familiar Maps." *Environment and Planning A* 42 (6): 1328–1345.

Kelemen, R., and D. Vogel. 2010. "Trading Places: The Role of the United States and the European Union in International Environmental Politics." *Comparative Political Studies* 43 (4): 427–456.

Kemfert, C. 2004. "Climate Coalitions and International Trade." *Energy Policy* 32 (4): 455–465.

Keohane, R. 2015. "The Global Politics of Climate Change: Challenge for Political Science." *PS: Political Science & Politics* 48 (1): 19–26.

Keohane, R. O., and D. G. Victor. 2011. "The Regime Complex for Climate Change." *Perspectives on Politics* 9 (1): 7–23.

King, R., and A. Borchardt. 1994. "Red and Green: Air Pollution Levels and Left Party Power in OECD Countries." *Environment and Planning C* 12 (2): 225–241.

Knight, J. 1992. *Institutions and Social Conflict*. Cambridge, UK: Cambridge University Press.

Kroll, S., and J. Shogren. 2008. "Domestic Politics and Climate Change: International Public Goods in Two-level Games." *Cambridge Review of International Affairs* 21 (4): 563–583.

Kuramochi, T. 2015. "Review of Energy and Climate Policy Developments in Japan Before and After Fukushima." *Renewable and Sustainable Energy Reviews* 43: 1320–1332.

Layzer, J. 2007. "Deep Freeze: How Business Has Shaped the Global Warming Debate in Congress." In *Business and Environmental Policy*, ed. M. Kraft and S. Kamieniecki, 93–125. Cambridge, MA: MIT Press.

Layzer, J. A. 2012. *Open for Business: Conservatives' Opposition to Environmental Regulation*. Cambridge, MA: MIT Press.

Leggett, J. 2001. *The Carbon War*. New York: Routledge.

Leigh, A. 2000. "Factions and Fractions: A Case Study of Power Politics in the Australian Labor Party." *Australian Journal of Political Science* 35 (3): 427–448.

Leiserowitz, A. 2006. "Climate Change Risk Perception and Policy Preferences: The Role of Affect, Imagery, and Values." *Climatic Change* 77 (1–2): 45–72.

Leiserowitz, A., E. Maibach, C. Roser-Renouf, and J. Hmielowski. 2011. "Politics & Global Warming: Democrats, Republicans, Independents, and the Tea Party." Yale University and George Mason University. New Haven, CT: Yale Project on Climate Change Communication.

Lesbirel, S. 1990. "Implementing Nuclear Energy Policy in Japan." *Energy Policy* 18 (3): 267–282.

Lijphart, A. 1999. *Patterns of Democracy*. New Haven, CT: Yale University Press.

Lindblom, C. E. 1982. "The Market as Prison." *Journal of Politics* 44 (2): 324–336.

Lipscy, P. 2019. "The Institutional Politics of Energy and Climate Change." Book manuscript, Stanford University.

Lockwood, M. 2013. "The Political Sustainability of Climate Policy: The Case of the UK Climate Change Act." *Global Environmental Change* 23 (5): 1339–1348.

Lund, B. 2009. *Verdivalg i Krig og Fred: 12 Essays om Ledelse og Overlevelse*. Oslo, Norway: Conflux.

Lundqvist, L. 1980. *The Tortoise and the Hare: Clean Air Policies in the United States and Sweden*. Ann Arbor: University of Michigan Press.

Madden, N. 2014. "Green Means Stop: Veto Players and Their Impact on Climate-Change Policy Outputs." *Environmental Politics* 23 (4): 570–589.

Martin, C., and D. Swank. 2012. *The Political Construction of Business Interests*. Cambridge, UK: Cambridge University Press.

Matthews, M. 2001. "Cleaning up Their Acts: Shifts of Environment and Energy Policies in Pluralist and Corporatist States." *Policy Studies Journal* 29 (3): 478–498.

McAdam, D. 2017. "Social Movement Theory and the Prospects for Climate Change Activism in the United States." *Annual Review of Political Science* 20:189–208.

McCright, A., and R. Dunlap. 2011. "Cool Dudes: The Denial of Climate Change among Conservative White Males in the United States." *Global Environmental Change* 21 (4): 1163–1172.

McCright, A., R. Dunlap, and S. Marquart-Pyatt. 2016. "Political Ideology and Views about Climate Change in the European Union." *Environmental Politics* 25 (2): 338–358.

McGlade, C., and P. Ekins. 2015. "The Geographical Distribution of Fossil Fuels Unused When Limiting Global Warming to 2°C." *Nature* 517:187–193.

McKean, M. A. 1981. *Environmental Protest and Citizen Politics in Japan*. Berkeley: University of California Press.

McLean, I. 2008. "Climate Change and UK Politics: From Brynle Williams to Sir Nicholas Stern." *Political Quarterly* 79 (2): 184–193.

Meadowcroft, J. 2005. "From Welfare State to Ecostate." In *The State and the Global Ecological Crisis*, ed. J. Barry and R. Eckersley, 3–23. Cambridge, MA: MIT Press.

Meadowcroft, J. 2012. "Greening the State?" In *Comparative Environmental Politics*, ed. P. Steinberg and S. VanDeveer, 63–87. Cambridge, MA: MIT Press.

Meckling, J. 2011. *Carbon Coalitions: Business, Climate Politics, and the Rise of Emissions Trading*. Cambridge, MA: MIT Press.

Meckling, J., N. Kelsey, E. Biber, and J. Zysman. 2015. "Winning Coalitions for Climate Policy." *Science* 349 (6253): 1170–1171.

Meckling, J., T. Sterner, and G. Wagner. 2017. "Policy Sequencing Toward Decarbonization." *Nature Energy* 2 (12): 918–922.

Meng, K. C. 2017. "Using a Free Permit Rule to Forecast the Marginal Abatement Cost of Proposed Climate Policy." *American Economic Review* 107 (3): 748–784.

Mettler, S. 2011. *The Submerged State: How Invisible Government Policies Undermine American Democracy*. Chicago: University of Chicago Press

Midttun, A. 1988. "The Negotiated Political Economy of a Heavy Industrial Sector: The Norwegian Hydropower Complex in the 1970s and 1980s." *Scandinavian Political Studies* 11 (2): 115–144.

Midttun, A., and O. Hagen. 1997. "Environmental Policy as Democratic Proclamation and Corporatist Implementation." *Scandinavian Political Studies* 20 (3): 285–310.

Mildenberger, M., J. Marlon, P. Howe, and A. Leiserowitz. 2017. "The Spatial Distribution of Republican and Democratic Climate Opinions at State and Local Scales" *Climatic Change* 145 (3–4): 539–548.

Milkoreit, M. 2017. *Mindmade Politics: The Cognitive Roots of International Governance*. Cambridge, MA: MIT Press.

Milne, J. 2008. "Carbon Taxes in the United States: The Context for the Future." *Vermont Journal of Environmental Law* 10:1–30.

Moe, T. 2010. "Norwegian Climate Policies 1990–2010." Center for International Climate and Environment Research (CICERO). *Policy Note* (2010): 3.

Moen, S. E. 2009. "Innovation and Production in the Norwegian Aluminum Industry." In *Innovation, Path Dependency, and Policy*, ed. J. Fagerberg, D. Mowery, and B. Verspagen, 149–178. Oxford, UK: Oxford University Press.

Montgomery, W. 1972. "Markets in Licenses and Efficient Pollution Control Programs." *Journal of Economic Theory* 5 (3): 395–418.

Mooney, C. 2007. "An Inconvenient Assessment." *Bulletin of the Atomic Scientists* 63 (6): 40–47.

Nawrotzki, R. 2012. "The Politics of Environmental Concern: A Cross-National Analysis." *Organization & Environment* 25 (3): 286–307.

Neumayer, E. 2003. "Are Left-Wing Party Strength and Corporatism Good for the Environment?" *Ecological Economics* 45 (2): 203–220.

Newell, P. 2008. "Civil Society, Corporate Accountability and the Politics of Climate Change." *Global Environmental Politics* 8 (3): 122–153.

Newell, P., and M. Paterson. 2010. *Climate Capitalism*. Cambridge, UK: Cambridge University Press.

Norgaard, K. 2011. *Living in Denial: Climate Change, Emotions, and Everyday Life*. Cambridge, MA: MIT Press.

Obach, B. 2004a. *Labor and the Environmental Movement*. Cambridge, MA: MIT Press.

Obach, B. 2004b. "New Labor Slowing the Treadmill of Production?" *Organization & Environment* 17 (3): 337–354.

Olson, M. 1965. *The Logic of Collective Action*. Cambridge, MA: Harvard University Press.

O'Neill, K. 2000. *Waste Trading among Rich Nations: Building a New Theory of Environmental Regulation*. Cambridge, MA: MIT Press.

Oreskes, N., and E. M. Conway. 2010. *Merchants of Doubt*. New York: Bloomsbury Publishing USA.

Oshitani, S. 2006. *Global Warming Policy in Japan and Britain*. Manchester, UK: Manchester University Press.

Paterson, M. 2001. "Risky Business: Insurance Companies in Global Warming Politics." *Global Environmental Politics* 1 (4): 18–42.

Pearse, G. 2007. *High and Dry: John Howard, Climate Change and the Selling of Australia's Future*. London: Penguin UK.

Pempel, T., and K. Tsunekawa. 1979. "Corporatism without Labor? The Japanese Anomaly." In *Trends Toward Corporatist Intermediation*, ed. P. Schmitter and G. Lehmbruch, 231–270. Beverly Hills, CA: Sage.

Peters, G., G. Marland, C. Le Quéré, T. Boden, J. Canadell, and M. Raupach. 2012. "Rapid Growth in CO_2 Emissions after the 2008–2009 Global Financial Crisis." *Nature Climate Change* 2 (1): 2–4.

Pierson, P. 1994. *Dismantling the Welfare State?: Reagan, Thatcher and the Politics of Retrenchment*. Cambridge, UK: Cambridge University Press.

Pierson, P. 2007. "The Costs of Marginalization of Qualitative Methods in the Study of American Politics." *Comparative Political Studies* 40 (2): 146–169.

Piltz, R. 2005. "Toward a Second US National Climate Change Assessment." *Eos* 86 (52): 550–551.

Pooley, E. 2010. *The Climate War.* New York: HarperCollins.

Prasad, M. 2013. *The Land of Too Much: American Abundance and the Paradox of Poverty.* Cambridge, MA: Harvard University Press.

Rabe, B. G. 2004. *Statehouse and Greenhouse: The Emerging Politics of American Climate Change Policy.* Washington, DC: Brookings Institution Press.

Rabe, B. G. 2018. *Can We Price Carbon?* Cambridge, MA: MIT Press.

Raymond, L. 2016. *Reclaiming the Atmospheric Commons: The Regional Greenhouse Gas Initiative and a New Model of Emissions Trading.* Cambridge, MA: MIT Press.

Reitan, M. 1998. "Ecological Modernisation and Realpolitik: Ideas, Interests and Institutions." *Environmental Politics* 7 (2): 1–26.

Riksrevisjonen. 2010. *The Office of the Auditor General's Investigation into Target Achievement in Climate Policy.* Riksrevisjonen Document 3:5 (2009–2010). April 15.

Royden, A. 2002. "US Climate Change Policy under President Clinton: A Look Back." *Golden Gate University Law Review* 32:415–478.

Sandler, T. 2004. *Global Collective Action.* Cambridge, UK: Cambridge University Press.

Schattschneider, E. E. 1960. *The Semi-Sovereign People.* New York: Holt, Rinehart and Winston.

Scheve, K., and D. Stasavage. 2006. "Religion and Preferences for Social Insurance." *Quarterly Journal of Political Science* 1 (3): 255–286.

Schleich, J., et al. 2001. "Greenhouse Gas Reductions in Germany—Lucky Strike or Hard Work?" *Climate Policy* 1 (3): 363–380.

Schmitter, P. 1974. "Still the Century of Corporatism?" *Review of Politics* 36 (1): 85–131.

Schnaiberg, A. 1980. *The Environment: From Surplus to Scarcity.* Oxford, UK: Oxford University Press.

Schoppa, L. 1991. "Zoku Power and LDP Power: A Case Study of the Zoku Role in Education Policy." *Journal of Japanese Studies* 17 (1): 79–106.

Schreurs, M. 1997. "Japan's Changing Approach to Environmental Issues." *Environmental Politics* 6 (2): 150–156.

Schreurs, M. 2002. *Environmental Politics in Japan, Germany, and the United States.* Cambridge, UK: Cambridge University Press.

Scruggs, L. 2003. *Sustaining Abundance: Environmental Performance in Industrial Democracies.* Cambridge, UK: Cambridge University Press.

Scruggs, L., and S. Benegal. 2012. "Declining Public Concern about Climate Change: Can We Blame the Great Recession?" *Global Environmental Change* 22 (2): 505–515.

Shipan, C., and W. Lowry. 2001. "Environmental Policy and Party Divergence in Congress." *Political Research Quarterly* 54 (2): 245–263.

Shulman, S. 2008. *Undermining Science: Suppression and Distortion in the Bush Administration*. Berkeley and Los Angeles: University of California Press.

Silbey, J. 2010. "American Political Parties: History, Voters, Critical Elections and Party Systems." In *The Oxford Handbook of American Political Parties and Interest Groups*, ed. L. Maisel and J. Berry, 97–120. Oxford, UK: Oxford University Press.

Sinha, M. 2016. *The Slave's Cause: A History of Abolition*. New Haven, CT: Yale University Press.

Skjærseth, J., and J. Wettestad. 2008. *EU Emissions Trading: Initiation, Decisionmaking and Implementation*. Farnham, UK: Ashgate Publishing Ltd.

Skjærseth, J., and J. Wettestad. 2010. "Fixing the EU Emissions Trading System? Understanding the Post-2012 Changes." *Global Environmental Politics* 10 (4): 101–123.

Skocpol, T. 2013. "Naming the Problem: What It Will Take to Counter Extremism and Engage Americans in the Fight against Global Warming." Unpublished manuscript.

Skocpol, T., and A. Hertel-Fernandez. 2016. "The Koch Network and Republican Party Extremism." *Perspectives on Politics* 14 (3): 681–699.

Smith, M. 1993. *Pressure, Power and Policy: State Autonomy and Policy Networks in Britain and the United States*. Cambridge, UK: Cambridge University Press.

Smith, M. 2000. *American Business and Political Power*. Chicago: University of Chicago Press.

Staples, J. 2009. "Australian Government Action in the 1980s." In *Climate Change: On for Young and Old*, ed. H. Sykes, 112–118. Albert Park, Australia: Future Leaders.

Staples, J. 2012. "Environmental Policy, Environmental NGOs and the Keating Government." Paper presented to Australian Political Studies Conference, Hobart, Tasmania.

Stavins, R. 2003. "Experience with Market-Based Environmental Policy Instruments." *Handbook of Environmental Economics* 1:355–435.

Steinberg, P., and S. VanDeveer. 2012. *Comparative Environmental Politics: Theory, Practice, and Prospects*. Cambridge, MA: MIT Press.

Sterman, J. 2008. "Risk Communication on Climate: Mental Models and Mass Balance." *Science* 322 (5901): 532–533.

Stern, N. 2007. *Stern Review: The Economics of Climate Change*. Cambridge, UK: Cambridge University Press.

Stevis, D. 2013. "US Labour Unions Confront Climate Change." In *Trade Unions in the Green Economy*, ed. N. Räthzel and D. Uzzell, 179–195. Oxon, UK: Routledge.

Stimson, J. 2015. *Tides of Consent: How Public Opinion Shapes American Politics*. Cambridge, UK: Cambridge University Press.

Stokes, L. 2020. *Short Circuiting Policy: Organized Interests in the American States and the Erosion of Clean Energy Laws*. Oxford, UK: Oxford University Press.

Svendsen, G., C. Daugbjerg, L. Hjøllund, and A. Pedersen. 2001. "Consumers, Industrialists and the Political Economy of Green Taxation: CO_2 Taxation in OECD." *Energy Policy* 29 (6): 489–497.

Sweeney, S. 2013. "US Trade Unions and the Challenge of 'Extreme Energy.'" In *Trade Unions in the Green Economy*, ed. N. Räthzel and D. Uzzell, 196–213. Oxon, UK: Routledge.

Takao, Y. 2012. "The Transformation of Japan's Environmental Policy." *Environmental Politics* 21 (5): 772–790.

Taplin, R. 1994. "Greenhouse: An Overview of Australian Policy and Practice." *Australian Journal of Environmental Management* 1 (3): 142–155.

Thatcher, M. 2002. *Statecraft*. New York: Harper-Collins.

Thelen, K. 2014. *Varieties of Liberalization and the New Politics of Social Solidarity*. Cambridge, UK: Cambridge University Press.

Tiberghien, Y., and M. Schreurs. 2007. "High Noon in Japan: Embedded Symbolism and Post-2001 Kyoto Protocol Politics." *Global Environmental Politics* 7 (4): 70–91.

Tjernshaugen, A. 2011. "The Growth of Political Support for CO_2 Capture and Storage in Norway." *Environmental Politics* 20 (2): 227–245.

Tjernshaugen, A., and O. Langhelle. 2009. "Technology as Political Glue: CCS in Norway." In *Caching the Carbon*, ed. J. Meadowcroft and O. Langhelle, 98–124. Cheltenham, UK: Edward Elgar.

Toyne, P., and S. Balderstone. 2003. "The Environment." In *The Hawke Legacy: A Critical Retrospective*, ed. S. Ryan and T. Bramston, 170–183. North Melbourne, Australia: Pluto Press.

Tranter, B. 2011. "Political Divisions over Climate Change and Environmental Issues in Australia." *Environmental Politics* 20 (1): 78–96.

Tranter, B., and K. Booth. 2015. "Scepticism in a Changing Climate: A Cross-national Study." *Global Environmental Change* 33:154–164.

Trumbull, G. 2012. *Strength in Numbers: The Political Power of Weak Interests*. Cambridge, MA: Harvard University Press.

Tsebelis, G. 2002. *Veto Players*. Princeton, NJ: Princeton University Press.

Unruh, G. 2000. "Understanding Carbon Lock-in." *Energy Policy* 28 (12): 817–830.

Urpelainen, J. 2012. "The Strategic Design of Technology Funds for Climate Cooperation." *Environmental Science & Policy* 15 (1): 92–105.

Van Asselt, H., N. Kanie, and M. Iguchi. 2009. "Japan's Position in International Climate Policy: Navigating between Kyoto and the APP." *International Environmental Agreements* 9 (3): 319–336.

Victor, D. 2011. *Global Warming Gridlock*. Cambridge, UK: Cambridge University Press.

Vogel, D. 1986. *National Styles of Regulation: Environmental Policy in Great Britain and the United States*. Ithaca, NY: Cornell University Press.

Walker, K. 2002. "Environmental Policy in Australia." In *Environmental Politics and Policy in Industrialized Countries*, ed. U. Desai, 251–293. Cambridge, MA: MIT Press.

Wallace, D. 1995. *Environmental Policy and Industrial Innovation*. London, UK: Earthscan.

Weart, S. 2008. *The Discovery of Global Warming*. Cambridge, MA: Harvard University Press.

Weber, E., and P. Stern 2011. "Public Understanding of Climate Change in the United States." *American Psychologist* 66 (4): 315–328.

Whitman, C. 2005. *It's My Party Too: The Battle for the Heart of the GOP and the Future of America*. New York: Penguin.

Wicken, O. 2009. "The Layers of National Innovation Systems: The Historical Evolution of a National Innovation System in Norway." In *Innovation, Path Dependency, and Policy: The Norwegian Case*, ed. J. Fagerberg, D. Mowery, and B. Verspagen, 38–56. Oxford, UK: Oxford University Press.

Wilensky, H. 2002. *Rich Democracies: Political Economy, Public Policy, and Performance*. Berkeley: University of California Press.

Yandle, B., and S. Buck. 2002. "Bootleggers, Baptists, and the Global Warming Battle." *Harvard Environmental Law Review* 26:177–244.

Young, O. 2002. *The Institutional Dimensions of Environmental Change: Fit, Interplay and Scale*. Cambridge, MA: MIT Press.

Index

American and Comparative Environmental Policy

Sheldon Kamieniecki and Michael E. Kraft, series editors

Russell J. Dalton, Paula Garb, Nicholas P. Lovrich, John C. Pierce, and John M. Whiteley, *Critical Masses: Citizens, Nuclear Weapons Production, and Environmental Destruction in the United States and Russia*

Daniel A. Mazmanian and Michael E. Kraft, editors, *Toward Sustainable Communities: Transition and Transformations in Environmental Policy*

Elizabeth R. DeSombre, *Domestic Sources of International Environmental Policy: Industry, Environmentalists, and U.S. Power*

Kate O'Neill, *Waste Trading among Rich Nations: Building a New Theory of Environmental Regulation*

Joachim Blatter and Helen Ingram, editors, *Reflections on Water: New Approaches to Transboundary Conflicts and Cooperation*

Paul F. Steinberg, *Environmental Leadership in Developing Countries: Transnational Relations and Biodiversity Policy in Costa Rica and Bolivia*

Uday Desai, editor, *Environmental Politics and Policy in Industrialized Countries*

Kent Portney, *Taking Sustainable Cities Seriously: Economic Development, the Environment, and Quality of Life in American Cities*

Edward P. Weber, *Bringing Society Back In: Grassroots Ecosystem Management, Accountability, and Sustainable Communities*

Norman J. Vig and Michael G. Faure, editors, *Green Giants? Environmental Policies of the United States and the European Union*

Robert F. Durant, Daniel J. Fiorino, and Rosemary O'Leary, editors, *Environmental Governance Reconsidered: Challenges, Choices, and Opportunities*

Paul A. Sabatier, Will Focht, Mark Lubell, Zev Trachtenberg, Arnold Vedlitz, and Marty Matlock, editors, *Swimming Upstream: Collaborative Approaches to Watershed Management*

Sally K. Fairfax, Lauren Gwin, Mary Ann King, Leigh S. Raymond, and Laura Watt, *Buying Nature: The Limits of Land Acquisition as a Conservation Strategy, 1780–2004*

Steven Cohen, Sheldon Kamieniecki, and Matthew A. Cahn, *Strategic Planning in Environmental Regulation: A Policy Approach That Works*

Michael E. Kraft and Sheldon Kamieniecki, editors, *Business and Environmental Policy: Corporate Interests in the American Political System*

Joseph F. C. DiMento and Pamela Doughman, editors, *Climate Change: What It Means for Us, Our Children, and Our Grandchildren*

Christopher McGrory Klyza and David J. Sousa, *American Environmental Policy, 1990–2006: Beyond Gridlock*

John M. Whiteley, Helen Ingram, and Richard Perry, editors, *Water, Place, and Equity*

Judith A. Layzer, *Natural Experiments: Ecosystem-Based Management and the Environment*

Daniel A. Mazmanian and Michael E. Kraft, editors, *Toward Sustainable Communities: Transition and Transformations in Environmental Policy*, second edition

Henrik Selin and Stacy D. VanDeveer, editors, *Changing Climates in North American Politics: Institutions, Policymaking, and Multilevel Governance*

Megan Mullin, *Governing the Tap: Special District Governance and the New Local Politics of Water*

David M. Driesen, editor, *Economic Thought and U.S. Climate Change Policy*

Kathryn Harrison and Lisa McIntosh Sundstrom, editors, *Global Commons, Domestic Decisions: The Comparative Politics of Climate Change*

William Ascher, Toddi Steelman, and Robert Healy, *Knowledge in the Environmental Policy Process: Re-Imagining the Boundaries of Science and Politics*

Michael E. Kraft, Mark Stephan, and Troy D. Abel, *Coming Clean: Information Disclosure and Environmental Performance*

Paul F. Steinberg and Stacy D. VanDeveer, editors, *Comparative Environmental Politics: Theory, Practice, and Prospects*

Judith A. Layzer, *Open for Business: Conservatives' Opposition to Environmental Regulation*

Kent Portney, *Taking Sustainable Cities Seriously: Economic Development, the Environment, and Quality of Life in American Cities*, second edition

Raul Lejano, Mrill Ingram, and Helen Ingram, *The Power of Narrative in Environmental Networks*

Christopher McGrory Klyza and David J. Sousa, *American Environmental Policy: Beyond Gridlock*, updated and expanded edition

Andreas Duit, editor, *State and Environment: The Comparative Study of Environmental Governance*

Joseph F. C. DiMento and Pamela Doughman, editors, *Climate Change: What It Means for Us, Our Children, and Our Grandchildren*, second edition

David M. Konisky, editor, *Failed Promises: Evaluating the Federal Government's Response to Environmental Justice*

Leigh Raymond, *Reclaiming the Atmospheric Commons: Explaining Norm-Driven Policy Change*

Robert F. Durant, Daniel J. Fiorino, and Rosemary O'Leary, editors, *Environmental Governance Reconsidered: Challenges, Choices, and Opportunities*, second edition

James Meadowcroft and Daniel J. Fiorino, editors, *Conceptual Innovation in Environmental Policy*

Barry G. Rabe, *Can We Price Carbon?*

Kelly Sims Gallagher and Xiaowei Xuan, *Titans of the Climate: Explaining Policy Process in the United States and China*

Matto Mildenberger, *Carbon Captured: How Business and Labor Control Climate Politics*